HANDBOOK OF U.S. COLORANTS FOR FOODS, DRUGS, AND COSMETICS

HANDBOOK OF U.S. COLORANTS FOR FOODS, DRUGS, AND COSMETICS

DANIEL M. MARMION

A WILEY-INTERSCIENCE
PUBLICATION

JOHN WILEY & SONS
New York • Chichester
Brisbane • Toronto

Library of Congress Cataloging in Publication Data:

Marmion, Daniel M   1935-
    Handbook of U.S. Colorants for Foods, Drugs, and Cosmetics.

    "A Wiley-Interscience publication."
    Includes index.
    1. Coloring matter in food.   2. Coloring mattter.
I.   Title.

TP456.C65M37   664'.06   78-10949
ISBN 0-471-04684-1

Printed in the United States of America

10 9 8 7 6 5 4 3 2 1

PREFACE

Because of their widespread use and economic importance and the frequent controversies centered around them, much has been written about the colorants used in foods, drugs, and cosmetics. Unfortunately, what has been written is widely distributed throughout the literature. What follows is an attempt to gather together as much of this information as possible. Hopefully, this collection will serve as a manual for those who manufacture colorants, regulate their use, incorporate them into their products, study their effects, or consume the myriad of articles in which they are found. No such manual exists now.

The colorants considered here are, for the most part, only those now in use in the United States. A small number of recently delisted colorants are discussed, either because they were delisted after this work was published or because it was felt that they might still exist in products on the market and could still be of some interest. A few others not used in the United States are considered in certain analytical discussions, because their similarity to U.S. colorants might make the procedures adaptable to American products.

This handbook is divided into three parts. *Part A* provides a general background of color additives and includes information on their history and regulation, lists of currently permitted colorants, their description, properties, areas of use, specifications, and other items of interest. *Part B* deals with colorant analysis. The treatment is extensive, because the purity requirements imposed on color additives have generated a vast number of procedures. Most are given in detail; however, a few of the less important ones are summarized in the bibliographies following the various sections. Topics covered include identification, strength, moisture, metals, insolubles, inorganic salts, and colored as well as colorless impurities. *Part C,* including the resolution of mixtures and the analysis of commercial products, is somewhat of a potpourri designed to give the reader enough of a background to be able to deal with the nearly infinite number of possible situations with which he might be confronted.

Throughout this work the nomenclature is what is commonly employed in connection with color additives. Although many of the terms may appear unorthodox, they are from the jargon of the industry and will be familiar to people working in the field.

DANIEL M. MARMION

Buffalo, New York,
January, 1979

CONTENTS

HANDBOOK OF U.S. COLORANTS FOR FOODS, DRUGS, AND COSMETICS

PART **A** HISTORY REGULATION, DESCRIPTION AND USE

Chapter 1 History; Colorants in Use Today

Color is as common in our environment as the air we breathe. In fact, it is so prevalent that we are not always aware just how much we depend on it. Color is important to man as a means of identification, as a method of judging quality, and for its basic esthetic value. Consequently, it is no wonder that for centuries color has played a prominent role in three of the things most important to man—his food, medicine, and physical appearance.

History is replete with accounts of the widespread application of color additives. Paintings in Egyptian tombs dating as far back as 1500 B.C. depict the making of colored candy. Pliny the Elder tells us that wine was artifically colored four centuries before the birth of Christ, whereas the coloring of spices and condiments is known to have been practiced at least 500 years ago.

The use of colorants in drugs undoubtedly has as long a history since color has been associated with disease and its treatment since antiquity. Many such practices are documented in Egyptian papyri. The use of colorants in cosmetics was probably more widespread and certainly better documented than their application in either foods or drugs. Archeologists have evidence that Egyptians used green ore of copper as an eye shadow as early as 5000 B.C. Egyptian women are also known to have used henna to dye their hair, carmine to redden their lips, and kohl, an antimony compound, to blacken eyebrows, lids, and lashes. Thousands of years ago it was common practice in India to tint faces yellow with saffron and to dye feet red with henna. In similar times Chinese women used vegetable extracts to dye their feet, cheeks, and the tips of their tongues, whereas the men and women of Asia Minor painted their faces with litmus and marshmallow. Romans used white lead and chalk on their faces and blue and gold dyes on their hair and beards.

Until the middle of the nineteenth century the colorants used in foods, drugs, and cosmetics were materials easily obtainable from natural sources, that is, animals, vegetables, and minerals. In 1856 Sir William Henry Perkin discovered the first synthetic organic dyestuff, mauve, and soon a host of new and different colorants was added to the artist's palette.

The use of some of these in foods began in Europe almost immediately and was soon extended to drugs and cosmetics. French wines, for example, were colored with fuchsine, a triphenylmethane dye, as early as 1860. The

3

United States first legalized the use of synthetic organic dyes in foods by an act of Congress that authorized the addition of coloring matter to butter (August 2, 1886). The second such recognition came some 10 years later when on June 6, 1896 Congress recognized coloring matter as a legitimate constituent of cheese. By 1900 Americans were eating a wide variety of artificially colored products including ketchup, jellies, cordials, butter, cheese, ice cream, candy, sausage, noodles, and wine. The use of colorants in drug and cosmetic products was also on the increase.

This proliferation in the use of color additives was soon recognized as a threat to the nation's health. Of particular concern was the fact that poisonous substances such as chrome yellow, Martius Yellow, and quicksilver vermillion were sometimes incorporated into foods and that dyes were frequently used to hide poor quality and to add weight or bulk to certain items. Of equal concern was the fact that often little or no control was exercised over the purity of the colorants used in foods, and dyes found unsatisfactory for textiles were sometimes deliberately channeled into food products. Public awareness that such materials as arsenic acid and mercury were employed in the manufacture of various colorants soon created a fear of coal-tar dyes that lingers even today. Because of increasing public concern some measures were taken by the food manufacturers to police their own industry. An example was the list published in 1899 by the National Confectioners Association of coloring matters that they considered unfit for coloring foods. However, the effect of such actions by industry was marginal, and it was soon obvious that governmental control was necessary.

The first effective step taken by the government to check such practices was when, under the Appropriations Act of 1900 for the Department of Agriculture, the Bureau of Chemistry was given funds to investigate the relationship of coloring matters to health and to establish principles that should be followed to govern their use. Results came quickly with the issuance by the Secretary of Agriculture of a series of Food Inspection Decisions (FID). One (FID 4[3c], issued August 6, 1904) declared a food as adulterated "if it be colored, powdered or polished with intent to deceive or to make the aritcle appear of better quality than it really is". Another exempted fabricated confections from this adulteration proviso, except in those cases where the candy contained a colorant that might lead the consumer to believe that a naturally colored ingredient was present when in fact is was not. This regulation made it necessary to declare on the label the presence of such substances as imitation chocolate (FID 29, issued September 27, 1905). A third decision (FID 39, issued May 1, 1906) contained the first direct statement by the department concerning a coal-tar dye considered unsafe in foods. In effect, it stopped importation of macaroni colored with Martius Yellow.

At about the same time a thorough study was undertaken by the Department of Agriculture to determine which dyes, if any, were safe for use in foods and what restrictions should be placed on their use. This task was monumental, to say the least, and eventually included a study of the chemistry and physiology of the then nearly 700 extant coal-tar dyes as well as the laws of various countries and states regarding their use in food products. Most of this investigation was done under the guidance of Dr. Bernard C. Hesse, whose findings were reflected in the Food and Drugs Act of 1906. This act, plus FID No. 76 (July 13, 1907) put an end to the indiscriminate use of dangerous and impure coloring matters in foods. Among other things, this new legislation required that only colors of known composition, examined physiologically and showing no unfavorable results, could be used in foods. Seven dyes were subsequently listed for use, including:

Original Name	Current Name
Amaranth	—
Ponceau 3R	—
Orange I	—
Erythrosine	FD&C Red No. 3
Naphthol Yellow S	Ext. D&C Yellow No. 7
Light Green SF Yellowish	—
Indigo Disulfo Acid, Sodium Salt	FD&C Blue No. 2

The new regulations also establish a system for certification of synthetic organic food colors by the Department of Agriculture. Certification was not mandatory, but dye manufactures soon found it to their benefit to have their products certified; the first certification took place on April 1, 1908.

Because of the increased needs of industry, the next three decades witnessed a continual growth in the use and number of color additives. The list of colors certifiable for use in foods was expanded to include the following:

Original Name	Current Name	Year Added
Tartrazine	FD&C Yellow No. 5	1916
Sudan I	—	1918
Butter Yellow	—	1918
Yellow AB	—	1918
Yellow OB	—	1918
Guinea Green B	—	1922
Fast Green FCF	FD&C Green No. 3	1927
Ponceau SX	FD&C Red No. 4	1929
Sunset Yellow FCF	FD&C Yellow No. 6	1929
Brilliant Blue FCF	FD&C Blue No. 1	1929

In 1938 a new law came into being, the Federal Food, Drug, and Cosmetic Act of 1938, instituting several new and important practices. First, it clearly stated that, henceforth, the use of any uncertified coal-tar color in any food, drug, or cosmetic shipped in interstate commerce was strictly forbidden. This restriction applied regardless of the inherent toxicity of the colorant. In effect, the colorants that could be used were limited, certification became mandatory, and governmental control was extended to the coloring of drugs and cosmetics. Next, it created three categories* of coal-tar colors:

FD&C colors—those certifiable for use in coloring foods, drugs, and cosmetics.

D&C colors—dyes and pigments considered safe in drugs and cosmetics when in contact with mucous membranes or when ingested.

Ext. D&C colors—those colorants that, because of their oral toxicity, were not certifiable for use in products intended for ingestion, but were considered safe for use in products externally applied.

Passage of the 1938 Act launched a new series of scientific investigations and public hearings regarding the safety of the colorants then on the market. These efforts culminated in the publication in September 1940 of Service and Regulatory Announcement, Food, Drug, and Cosmetics No. 3, which listed specific colorants that could be used along with specifications and regulations relating to their manufacture, labeling, certification, and sale.

In the early 1950s, just when it appeared that the situation with regard to color additives was finally under control, new difficulties developed. The problems were precipitated by two events: a new round of pharmacological testing of food colors by the Food and Drug Administration (FDA)[†] and a number of cases of sickness in children who had reportedly eaten candy and popcorn colored with excessive amounts of dye. The new animal-feeding studies undertaken by the FDA were conducted at higher levels and for longer test periods than any experiments previously conducted and resulted in unfavorable findings for FD&C Orange No. 1, FD&C Orange No. 2 and FD&C Red No. 32.

The disputes that followed centered around the FDA interpretation of the 1938 act, which states that "The Secretary shall promulgate regulations providing for the listing of coal-tar colors which are harmless and suitable for use in food...." The FDA felt that "harmless" here meant that a colorant must be safe regardless of the amount used, that is, harmless per se. On the basis of this argument the FDA delisted the colorants in question. Meanwhile the food-color manufacturers argued that the FDA interpretation of the law was too strict, that a color additive need only be harmless when properly used, and that the FDA should establish safe limits. They also contended that the conditions used for the new animal feeding tests were too severe.

After a series of legal battles in the lower courts the problem was finally

---

*In surveying the colorants in use at the time it was discovered that several manufacturers were selling the same dyes under different names. To clearly differentiate between a textile-grade colorant and a certified colorant with the same chemical structure but having a different level of purity, and to prevent giving one manufacturer an advantage over his competitors by selecting his trade name as the official designation of a colorant to be allowed under the 1938 law, the terms FD&C, D&C, and Ext. D&C were invented.

†The FDA, which enforces the law governing color additives, was created by the Agricultural Appropriations Act of 1931.

taken to the Supreme Court, which ruled that under the 1938 law, the FDA did not have the authority to establish limits of use for colorants and that they were obligated to decertify or delist a color if any quantity of it caused harm even though lesser amounts were perfectly safe. The FDA's hands were tied. A review of the remaining colors was started, and soon several more were delisted, including FD&C Yellow Nos. 1-4. It was immediately and painfully obvious that the existing law on certifiable colors was unworkable and that the entire house of cards was about to collapse.

Through the efforts of the Certified Color Industry Committee* and the FDA a new law was formulated, the Color Additives Amendments of 1960 (Public Law 86-618). Basically, the amendments provided a much needed breathing spell. For one thing, they allowed for the continued use of existing color additives pending the completion of investigations needed to ascertain their suitability for listing as "permanent" colorants. Equally as important, they authorized the Secretary of Health, Education, and Welfare to establish limits of use, thus eliminating the controversial "harmless per se" interpretation formerly employed. Other features eliminated any distinction under the law between "coal-tar" colors and other color additives and empowered the Secretary to decide which colors must be certified and which could be exempted from certification based on their relationship to public health.

Under provisions of the new law the producers and consumers of the color additives were obliged to provide the necessary scientific data to obtain "permanent" listing of a color additive. Because of the expense involved, testing was started on only those colors that were of economic importance and, consequently, many previously certifiable colors were eventually delisted by default. The deadline or closing date for providing this data has been extended several times by the secretary, using powers granted to him by the amendments.

To date many colorants not requiring certification and a few certified colors have been "permanently" listed. The remainder of the colorants continue to be listed provisionally. Those currently in use and their status are shown in Table 1-3, and a chronological history of synthetic certifiable food colors is given in Table 4. These lists are accurate as of January 1, 1979 but are subject to change by both addition and deletion. Such changes as well as any changes in the regulations discussed in Chapter 3 are routinely published in the Federal Register.† Additional information as to what colorants can be used and the regulations pertaining to them can be obtained from the FDA, Division of Colors and Cosmetics, 200 C St., S. W. Washington, D. C. 20204.

*An informal, unincorporated association comprised of most of the food-color manufactures in the United States. The committee was formed to deal with regulatory and legislative problems affecting the entire industry and involving the FDA.
†The Federal Register is published by the office of the Federal Register, National Archives and Records Service, General Services Administration, Washington, D. C. 20408. It is distributed only through the Superintendent of Documents, U.S. Government Printing Office, Washington, D. C. 20402.

TABLE 1 COLORANTS PERMITTED IN FOODS

Food and Drug Administration Official Name	Color Index Number	Limitations[a]	Current Status
SUBJECT TO CERTIFICATION			
FD&C Blue No. 1	42090		Listed
FD&C Blue No. 2	73015		Provisional
FD&C Green No. 3	42053		Provisional
FD&C Red No. 3	45430		Listed
FD&C No. 40	16035		Listed
FD&C Yellow No. 5	19140		Listed
FD&C Yellow No. 6	15985		Provisional
Citrus Red No. 2	12156	Orange skins Only; 2.0 ppm max., based on the weight of the whole fruit	Listed
Orange B[b]	19235	Sausage and frankfurter casings or surfaces only; 150 ppm max., based on the weight of the finished product	Listed
EXEMPT FROM CERTIFICATION			
Annatto Extract	75120		Listed
β-Apo-8'-Carotenal	40820	Maximum—15 mg/lb of solid or semisolid food, or pint of liquid food	Listed
Canthaxanthin	40850	Maximum—30 mg/lb of solid or semisolid food, or pint of liquid food	Listed
Caramel			Listed
β-Carotene	75130		Listed
Carrot Oil			Listed
Cochineal Extract and Carmine	75470		Listed
Corn Endosperm Oil		Chicken feed only	Listed
Dehydrated Beets (Beet Powder)			Listed
Dried Algae Meal		Chicken feed only	Listed
Ferrous Gluconate		Ripe olives only	Listed
Fruit Juice			Listed
Grape Skin Extract		Beverages only	Listed
Paprika			Listed
Paprika Oleoresin			Listed
Riboflavin			Listed
Saffron	75100		Listed
Synthetic Iron Oxide	77491 77492 77499	Dog and cat food only; 0.25% max.	Listed
Tagetes Meal and Extract	75125	Chicken feed only	Listed
Titanium Dioxide	77891	1% Maximum in finished food	Listed

8

TABLE 1 Continued

Food and Drug Administration Official Name	Color Index Number	Limitations[a]	Current Status
Toasted Partially Defatted Cooked Cottonseed Flour			Listed
Turmeric	75300		Listed
Turmeric Oleoresin	75300		Listed
Ultramarine Blue	77007	Salt for animal feed only; 0.5% max.	Listed
Vegetable Juice			Listed

[a] No color additive or product containing one can be used in the area of the eye, in surgical sutures or injections unless so stated.
[b] A proposal was made in October, 1978 to delist Orange B for use in foods on the grounds that it may contain traces of 2-naphthylamine, a material considered by many to be a carcinogen. A final ruling on this proposal will probably be made while this book is in press.

TABLE 2 COLORANTS PERMITTED IN DRUGS

Food and Drug Administration Official Name	Color Index Number	Limitations[a]	Current Status
SUBJECT TO CERTIFICATION			
FD&C Blue No. 1	42090	Ingested drugs	Listed
		Other uses	Provisional
FD&C Blue No. 2	73015	Nylon sutures only; 1% max.	Listed
		Ingested drugs	Provisional
		Other uses	Provisional
FD&C Green No. 3	42053		Provisional
FD&C Red No. 3	45430	Ingested drugs	Listed
		Other uses	Provisional
FD&C Red No. 4	14700	Externally applied drugs only	Listed
FD&C Red. No. 40	16035		Listed
FD&C Yellow No. 5	19140	Ingested Drugs	Listed
		Other uses	Provisional
FD&C Yellow No. 6	15985		Provisional
D&C Blue No. 4	42090	Externally applied drugs only	Listed
D&C Blue No. 6	73000	Sutures only; polyethylene terephthalate sutures for general surgical use, 0.2% max.; plain or chromic collagen absorbable sutures for general surgical use, 0.25% max.; plain or chromic	Listed

TABLE 2 Continued

Food and Drug Administration Official Name	Color Index Number	Limitations[a]	Current Status
		collagen absorbable sutures for opthalmic surgical use, 0.5% max.; polypropylene surgical sutures for general surgical use, 0.5% max.	
D&C Blue No. 9	69825	Cotton and silk sutures only; 2.5% max.	Listed
D&C Green No. 5	61570	Nylon 66 and Nylon 6 sutures only; 0.6% max.	Listed
		Other uses	Provisional
D&C Green No. 6	61565	Polyethylene terephthalate sutures, 0.75% max, and polyglycolic acid sutures, 0.1% max.	Listed
		Other uses	Provisional
D&C Green No. 8	59040	Externally applied drugs only; 0.01% max.	Listed
D&C Orange No. 4	15510	Externally applied drugs only	Listed
D&C Orange No. 5	45370:1	Ingested and/or internally used products, 0.75 mg max. as pure dye per daily dosage or use	Provisional
D&C Orange No. 10	45425:1		Provisional
D&C Orange No. 11	45425		Provisional
D&C Orange No. 17	12075		Provisional
D&C Red No. 6	15850		Provisional
D&C Red No. 7	15850:1		Provisional
D&C Red No. 8	15585	Ingested and/or internally used products, 0.75 mg max. as pure dye per daily dosage or use	Provisional
D&C Red No. 9	15585:1		Provisional
D&C Red No. 17	26100	Externally applied drugs only	Listed
D&C Red No. 19	45170	Ingested and/or internally used products, 0.75 mg max. as pure dye per daily dosage or use	Provisional
D&C Red No. 21	45380:2		Provisional
D&C Red No. 22	45380		Provisional
D&C Red No. 27	45410:1		Provisional
D&C Red No. 28	45410		Provisional
D&C Red No. 30	73360		Provisional
D&C Red No. 31	15800:1	Externally applied drugs only	Listed
D&C Red No. 33	17200	Ingested and/or internally used products, 0.75 mg max. as pure dye per daily dosage or use	Provisional
D&C Red No. 34	15880:1	Externally applied drugs only	Listed
D&C Red No. 36	12085	Ingested and/or internally used products, 1.7 mg max. as pure dye per daily dosage or use	Provisional

TABLE 2 Continued

Food and Drug Administration Official Name	Color Index Number	Limitations[a]	Current Status
D&C Red No. 37	45170:1	Ingested and/or internally used products, 0.75 mg max. as pure dye per daily dosage or use	Provisional
D&C Red No. 39	13058	Externally applied quarternary ammonium germicides only; 0.1% max.	Listed
D&C Violet No. 2	60725	Externally applied drugs only	Listed
D&C Yellow No. 7	45350:1	Externally applied drugs only	Listed
D&C Yellow No. 8	45350	Externally applied drugs only	Listed
D&C Yellow No. 10	47005		Provisional
D&C Yellow No. 11	47000	Externally applied drugs only	Listed
Ext. D&C Yellow No. 7	10316	Externally applied drugs only	Listed
[Phthalocyaninato (2−)] copper	74160	Polypropylene sutures only; 0.5% max.	Listed

EXEMPT FROM CERTIFICATION

Alumina	77002		Listed
Aluminum Powder	77000	Externally applied drugs only (1)	Listed
Annatto Extract	75120	(1)	Listed
Bismuth Oxychloride	77163	Externally applied drugs only (1)	Listed
Bronze Powder	77440	Externally applied drugs only (1)	Listed
Calcium Carbonate	77220		Listed
Canthaxanthin	40850		Listed
Caramel			Listed
β-Carotene	75130	(1)	Listed
Chromium-Cobalt- Aluminum Oxide		Polyethylene sutures only; 2% max.	Listed
Chromium Hydroxide Green	77289	Externally applied drugs only (1)	Listed
Chromium Oxide Greens	77288	Externally applied drugs only (1)	Listed
Cochineal Extract and Carmine	75470		Listed
Copper Powder	77400	Externally applied drugs only (1)	Listed
Dihydroxyacetone		Externally applied drugs only	Listed
Ferric Ammonium Citrate		With pyrogallol in plain and chromic catgut sutures only; 3% max. (of combination)	Listed
Ferric Ammonium Ferrocyanide		Externally applied drugs only (1)	Listed
Guanine	75170	Externally applied drugs only (1)	Listed
Logwood Extract	75290	Nylon 66, Nylon 6, and silk sutures only; 1.0% max.	Listed
Mica	77019	Externally applied drugs only (1)	Listed
Potassium Sodium Copper Chlorophyllin	75810	Dentifrices only; 0.1% max.	Listed
Pyrogallol	76515	With ferric ammonium citrate in plain and chromic catgut sutures; 3% max. (of combination)	Listed

TABLE 2 Continued

Food and Drug Administration Official Name	Color Index Number	Limitations[a]	Current Status
Prophyllite		Externally applied drugs only	Listed
Synthetic Iron Oxide	77491	5 mg/day (as Fe) in drugs that	Listed
	77492	are ingested	
	77499		
Talc	77019		Listed
Titanium Dioxide	77891	(1)	Listed
Zinc Oxide	77947	Externally applied drugs only (1)	Listed

[a]No color additive or product containing one can be used in the area of the eye, in surgical sutures or injections unless so stated. (1) May also be used in those drugs intended for use in the area of the eye.

TABLE 3 COLORANTS PERMITTED IN COSMETICS

Food and Drug Administration Official Name	Color Index Number	Limitations[a]	Current Status
SUBJECT TO CERTIFICATION			
FD&C Blue No. 1	42090		Provisional
FD&C Green No. 3	42053		Provisional
FD&C Red No. 3	45430		Provisional
FD&C Red No. 4	14700	External use only	Listed
FD&C Red No. 40	16035		Listed
FD&C Yellow No. 5	19140		Provisional
FD&C Yellow No. 6	15985		Provisional
D&C Blue No. 4	42090	External use only	Listed
D&C Brown No. 1	20170	External use only	Listed
D&C Green No. 5	61570		Provisional
D&C Green No. 6	61565		Provisional
D&C Green No. 8	59040	External use only; 0.01% max.	Listed
D&C Orange No. 4	15510	External use only	Listed
D&C Orange No. 5	45370:1	6% Maximum in lipstick; external use only	Provisional
D&C Orange No. 10	45425:1		Provisional
D&C Orange No. 11	45425		Provisional
D&C Orange No. 17	12075	6% Maximum in lipstick; external use only	Provisional
D&C Red No. 6	15850		Provisional
D&C Red No. 7	15850:1		Provisional
D&C Red No. 8	15585	6% Maximum in lipstick; external use only	Provisional
D&C Red No. 9	15585:1	6% Maximum in lipstick; external use only	Provisional
D&C Red No. 17	26100	External use only	Listed
D&C Red No. 19	45170	6% Maximum in lipstick; external use only	Provisional

TABLE 3 Continued

Food and Drug Administration Official Name	Color Index Number	Limitations[a]	Current Status
D&C Red No. 21	45380:2		Provisional
D&C Red No. 22	45380		Provisional
D&C Red No. 27	45410:1		Provisional
D&C Red No. 28	45410		Provisional
D&C Red No. 30	73360		Provisional
D&C Red No. 31	15800:1	External use only	Listed
D&C Red No. 33	17200	6% Maximum in lipstick; external use only	Provisional
D&C Red No. 34	15880:1	External use only	Listed
D&C Red No. 36	12085	3% Maximum in lipstick; external use only	Provisional
D&C Red No. 37	45170:1	External use only	Provisional
D&C Violet No. 2	60725	External use only	Listed
D&C Yellow No. 7	45350:1	External use only	Listed
D&C Yellow No. 8	45350	External use only	Listed
D&C Yellow No. 10	47005		Provisional
D&C Yellow No. 11	47000	External use only	Listed
Ext. D&C Violet No. 2	60730	External use only	Listed
Ext. D&C Yellow No. 7	10316	External use only	Listed

EXEMPT FROM CERTIFICATION

Food and Drug Administration Official Name	Color Index Number	Limitations[a]	Current Status
Aluminum Powder	77000	External use only (1)	Listed
Annatto	75120	(1)	Listed
Bismuth Oxychloride	77163	(1)	Listed
Bronze Powder	77440	(1)	Listed
Caramel			Provisional
Carmine	75470		Provisional
β-Carotene	75130	(1)	Listed
Chromium Hydroxide Green	77289	External use only (1)	Listed
Chromium Oxide Greens	77288	External use only (1)	Listed
Copper Powder	77400	(1)	Listed
Dihydroxyacetone		External use only	Listed
Disodium EDTA-Copper		Shampoos only	Listed
Ferric Ammonium Ferrocyanide	77510 77520	External use only (1)	Listed
Ferric Ferrocyanide (Iron Blue)		External use only (1)	Provisional
Guaiazulene		External use only	Listed
Guanine	75170	(1)	Listed
Henna	75480	Hair dyes only, not near eye	Listed
Lead Acetate		Hair dyes only	Provisional
Manganese Violet	77742	(1)	Listed
Mica	77019	(1)	Listed
Potassium Sodium Copper Chlorophyllin	75810	Dentifrices only; 0.1% max. (2)	Listed
Pyrophyllite		External use only	Listed
Synthetic Iron Oxides	77491 77492 77499	(1)	Listed

TABLE 3 Continued

Food and Drug Administration Official Name	Color Index Number	Limitations[a]	Current Status
Titanium Dioxide	77891	(1)	Listed
Ultramarine Blue	77007	External use only (1)	Listed
Ultramarine Green	77013	External use only (1)	Listed
Ultramarine Pink	77007	External use only (1)	Listed
Ultramarine Red	77007	External use only (1)	Listed
Ultramarine Violet	77007	External use only (1)	Listed
Zinc Oxide	77947	(1)	Listed

[a]No color additive or product containing one can be used in the area of the eye, in surgical sutures or injections unless so stated. (1) May also be used in cosmetics intended for use in the area of the eye. (2) Can only be used in combination with certain substances.

TABLE 4   CHRONOLOGICAL HISTORY OF SYNTHETIC FOOD COLORS IN THE UNITED STATES

Year Listed for Food Use	Common Name	FDA Name	Color Index Number	Year Delisted	Currently Permitted in Food
1907	Ponceau 3R	FD&C Red No. 1	16155	1961	No
1907	Amaranth	FD&C Red No. 2	16185	1976	No
1907	Erythrosine	FD&C Red No. 3	45430	—	Yes
1907	Orange I	FD&C Orange No. 1	14600	1956	No
1907	Naphthol Yellow S	FD&C Yellow No. 1	10316	1959	No
1907	Light Green SF Yellowish	FD&C Green No. 2	42095	1966	No
1907	Indigotine	FD&C Blue No. 2	73015	—	Yes
1916	Tartrazine	FD&C Yellow No. 5	19140	—	Yes
1918	Sudan I	—	12055	1918	No
1918	Butter Yellow	—		1918	No
1918	Yellow AB	FD&C Yellow No. 3	11380	1959	No
1918	Yellow OB	FD&C Yellow No. 4	11390	1959	No
1922	Guinea Green B	FD&C Green No. 1	42085	1966	No
1927	Fast Green FCF	FD&C Green No. 3	42053	—	Yes
1929	Ponceau SX	FD&C Red No. 4	14700	1976	No
1929	Sunset Yellow FCF	FD&C Yellow No. 6	15985	—	Yes
1929	Brilliant Blue FCF	FD&C Blue No. 1	42090	—	Yes
1939	Naphthol Yellow S potassium salt	FD&C Yellow No. 2	10316	1959	No
1939	Orange SS	FD&C Orange No. 2	12100	1956	No
1939	Oil Red XO	FD&C Red No. 32	12140	1956	No
1950	Benzyl Violet 4B	FD&C Violet No. 1	42640	1973	No
1959	Citrus Red No. 2	Citrus Red No. 2	12156	—	Yes
1966	Orange B	Orange B	19235	—	Yes
1971	Allura [a] Red AC	FD&C Red No. 40	16035	—	Yes

[a]Registered trademark of Buffalo Color Corporation.

Chapter 2 Areas of Use

COLORING FOOD

Basic Reasons

One might ask—with all the problems associated with the manufacture and sale of color additives—"why bother?" The answer to this question is, of course, not a simple one and is not the same in all cases.

Color is added to food either because it has no natural color of its own, because its natural color is lost or drastically altered as a result of processing or storage, or because the color of the foodstuff varies greatly with the season of the year or the geographic origin of the product. Whatever the reason is in any particular case, the overall objective of coloring food is to make it recognizable and appealing to the consumer so that he will buy it.

Color has always been associated with the quality of foods. Green bananas and grapefruit, for example, are recognized immediately as immature, whereas anything but bright red beef is usually considered suspect. Wise shoppers automatically shun excessively brown or spotted produce and, in general, select the brightest, most uniformly colored products available. In fact, the color of foods is so important to us and so well fixed in our minds that some serve as reference standards when speaking of certain hues—lemon yellow, eggshell white, cherry red, and pea green, to mention only a few. Consequently, to the average consumer off-color foods have come to mean off-grade foods.

Regional and Seasonal Problems

The problems of the dairy and citrus fruit industries are typical of those encountered with products produced in different areas of the country or at different times of the year.

Consider the growing of oranges. In many parts of the United States the soil and weather conditions are such that chlorophyll continuously forms in the fruit as well as in the leaves of the trees resulting in mature oranges that are substantially greener than the same variety of orange produced in regions of the country with different growing conditions. Florida Valencia oranges, for example, mature in the latter part of March when the weather is favorable to the development of chlorophyll, which is produced in such quantities in the fruit peel that it eventually turns pale and green. In fact, most varieties of

Florida oranges tend to be green, suggesting immaturity, even though they contain the proper ratio of solids to acid for fully nutritious, mature fruit.

The necessity of coloring these oranges to make them comparable in appearance and thus as commercially acceptable as naturally orange-colored fruit from other areas of the country was recognized years ago and began on a commercial scale about 1934. Today the peels of those (and only those) oranges not intended for processing continue to be dyed where necessary. The percentage of the total crop colored varies from year to year and depends largely on the weather.

The problems of the diary farmer are no less complicated. Approximately 90% of the yellow color in milk is due to the presence of β-carotene, a fat-soluble carotenoid extracted from feed by cows. As is well known, summer milk is more yellow than winter milk. This is largely due, of course, to seasonal feeding practices in which cows grazing on lush green pastures in the spring and summer months consume much highler levels of carotenoids than do cows barn fed on hay and grain in the fall and winter. The problem is further complicated since various breeds of cows and even individual animals differ in the efficiency in which they extract β-carotene from feed and in the degree to which they convert it into colorless vitamin A. The differences in the color of milk are more obvious in products made from milk fat, since here the yellow color is concentrated. Thus unless standardized through the addition of colorant, products like butter and cheese show a wide variation in shade and in many cases appear unsatisfactory to the consumer. In addition to standardizing the color of butter and certain yellow cheeses by the addition of yellow colorants, it is frequently necessary to use various amounts of blue or green colorants when making Gorgonzola, Nuworld, Provolone, Blue, and various other cheeses in order to neutralize the yellow of the curd used to prepare them. Other products whose natural color varies enough to make standardization of their color desirable include the shells of certain kinds of nuts and the skins of red and sweet potatoes and ripe olives.

Process and Storage Difficulties

Often the process used to prepare a food leads to the formation of a color in the product, the depth of which depends largely on the time, temperature, pH, air exposure, and other parameters experienced during processing. Here again it is deemed necessary to supplement the color of the product to ensure its uniformity from batch to batch. Items that fall into this category include certain beers, blended whiskies, brown sugars, table syrups, toasted cereals, and baked goods.

The storage of food can also be a problem since natural pigments often deteriorate with time due to exposure to light, heat, air and moisture or because of interaction of the components of the product with each other or with the packaging material. The color of maraschino cherries, for example, fares so poorly with storage that they are routinely artificially colored.

Colorless Foods

The major use of color additives is in products containing little or no color of their own. These includes many liquid and powdered beverages, gelatin desserts, candies, ice creams, sherberts, icings, jams, jellies, and snack foods. Without the addition of color to some of these—gelatin desserts and soft drinks, for example—all flavors of the particular product would be colorless, unidentifiable, and probably unappealing to the consumer.

Miscellaneous Uses

Inks used by inspectors to stamp the grade or quality on meat must, by law, be made from food-grade colors, as must any dyes used in wrapping or packaging materials that come in direct contact with a food. Pet foods, too, if colored, must contain only those colorants recognized by the FDA as suitable for the purpose.

COLORING DRUGS

Compared to the food and cosmetic industries, pharmaceuticals are a minor though important consumer of colorants. Originally, dyes were used in drugs to make them more appealing to the consumer by adding color to otherwise colorless products, by masking unsatisfactory natural colors, and by standardizing the appearance of drugs whose color varied from batch to batch as a consequence of the manufacturing process, a difference in the color of the raw materials used, or both. Some drugs, of course, contained added color for cosmetic purposes, as in the case of the skin-tone dyes added to certain creams and ointments used to treat disorders such as acne.

Although colors are still added to drugs for these purposes, the major use of colorants in pharmaceuticals currently is to provide the manufacturer with a simple means of identifying his products so that they are not inadvertently mixed during manufacture and shipment. Since no industry-wide standards exist for coloring drugs, each manufacturer has been free to develop and use the inhouse scheme that best fits his needs. Many such codes have been devised, and so today the same product frequently appears on the market under several color forms.

COLORING COSMETICS

The reasons for using color additives in cosmetics are perhaps more obvious than the reasons for their use in either foods or drugs. Products such as aftershave lotions, hair tonics, and soaps contain additives purely for esthetic reasons. In many cases, though, the colorant is a major, functional part of a cosmetic, often comprising half of its total weight. Some cosmetics, including eyebrow pencils, nail polishes, and rouges, are really little more than colorants mixed with one or more materials that serve as binders, vehicles, or diluents to give the product desirable application properties but that have little inherent cosmetic value.

Compared to foods and drugs, cosmetics usually contain much higher amounts of colorants. Although foods and drugs seldom contain more than a few to several hundred parts per million (ppm) of color, cosmetics often contain several percent of colorants.

BIBLIOGRAPHY

AMBANELLI, G. Ind. conserve (Parma) *32*, 75-80 (1957). Use of Artificial Colors in Foodstuffs. The color additives acceptable in Italy.

ANSTEAD, D. F. Cosmetic Colours. In *Handbook of Cosmetic Science.* Pergamon Press, New York, 1963, pp. 101-118. A brief description of colors used in cosmetics.

ANSTEAD, D. F. J. Soc. Cosmet. Chemists *10*, 1-20 (1959). Pigments, Lakes and Dyes in Cosmetics. A general review, including regulations in the United States and Great Britain.

BAINBRIDGE, W. C. Ind. Eng. Chem. *18*, 1329-1331 (1926). Development of the Food Color Industry in the United States. Interesting historically.

CALVERY, H. O. Am. J. Pharm. *114*, 324-349 (1942). Coal-Tar Colors, Their Use in Foods, Drugs and Cosmetics. Outdated but interesting historically.

CALZOLARI, C., COASSINI, L., LOKAR, L. Quaderni Macerol. *1*, 89-131 (1962). Synthetic Food Colors. Reviews the regulation of food colors in various countries, the toxicity of the intermediates used to prepare them, and the toxicity of the degradation products of colorants.

CLARK, G. R. Proc. Sci. Sect. Toilet Goods Assoc. *35*, 24-25 (1961). Some Technical Problems in the Cosmetic Color-Additive Field. Outdated but interesting historically.

Color Additives Guide. The Pharmaceutical Manufacturers Association, 1155 15th St., N. W., Washington, D. C. 20005. A listing of the dyes and pigments permitted in 44 countries and the European Economic Community.

DAMON, G. E., JANSSEN, W. F. FDA Consumer, July-August 1973, pp. 15-21. Additives for Eye Appeal. A little of the history and regulation of food colors.

Food Colors. National Academy of Science, Washington, D. C., 1971. A general treatment of food colors, including their history, use, regulation, safety, and properties.

GOTO, R. Yuki Gosei Kagaku Kyokai Shi *24*, 493-500 (1966). Food Colors. A review of the kinds, properties, and applications of food colors.

HESSE, B. C. Coal-Tar Colors Used in Food Products. Bureau of Chemistry,

Bulletin No. 147, February 10, 1912. Results of the Hesse study made at the turn of the century.

HOLTZMAN, H. Am. Perfumer Cosmet. *78*, 27-31 (1963). The Current Color Palette. A somewhat outdated review of the permitted certified and non-certified color additives.

KASPRZAK, F., GLEBKO, B. Chemik *19*, 267-273 (1966). Dyes for Foods, Pharmaceuticals and Cosmetics. Natural and synthetic dyes produced in Poland and other countries are described and compared.

KOCH, L. Am. Perfumer Cosmet. *82*, 35-40 (1967). Some Legal, Chemical and Physical Aspects of Permitted Color additives. A brief review.

LIEBER, H. *The Use of Coal-Tar Colors in Food Products*. H. Lieber & Company New York, 1904. Interesting historically.

NOONAN, J. Color Additives in Foods. In *Handbook of Food Additives*, The Chemical Rubber Company, Cleveland, Ohio, 1968, pp. 25-49. Food colors—their description, properties, regulation, and use.

Public Law No. 717. 75th U.S. Congress, 3rd Session, S. 5 (1938), Section 402. The Federal Food, Drug, and Cosmetic Act of 1938.

REYNOLDS, H., EIDUSON, H., WEATHERWAX, J., DECHERT, D. Anal. Chem. *44*, 22A-24A, 26A, 28A, 31A-34A (1972). FDA Chemistry for Consumers. A review of the history, current structure, and function of the Food and Drug Administration.

SAGARIN, E., Ed. *Cosmetics—Science and Technology*. Interscience, New York, 1957. A good history of the development and use of cosmetics. Includes some treatment of the colorants used.

SOLODUKHIN, A. I. Proizv. Isol'z Vitaminov, Antibiotikov Biol. Aktivn. Veshchestv 145-181 (1965). Production and Use of Food Dyes. A review of the synthetic and natural food dyes used in the Soviet Union.

SOUCI, S. W. Z. Lebensm. Forsch. *108*, 189-195 (1958). The Color Committee of the Deutsche Forschungsgemeinschaft. List of Pigments and Dyes for Cosmetics. Toxicological data on dyes and their suitability for food in various countries.

Specifications for Identity and Purity of Food Additives, Vol. II, Food Colors. Food and Agriculture Organization of the United Nations, Rome, 1963. Some background into the international control of food colors and a description of some color additives and methods for their analysis.

SWARTZ, C. J., COOPER, J. J. Pharm. Sci. *51*, 89-99 (1962). Colorants for Pharmaceuticals. A general review of the colorants and their properties and uses.

U.S. Supreme Court. 358 U.S. 153, December 15, 1958. The court ruling that established the "harmless per se" principle—a color additive had to be harmless regardless of the quantity used.

VODOZ, C. A. Food Technol. *24*, 42-53 (1970). Intentional Food Additives in Europe.

WHITE, H. J., Jr., Ed. *Proceedings of the Perkin Centennial*, September 10, 1956, New York, Sponsored by the American Association of Textile Chemists and Colorists. Includes chapters on the use, properties, and reasons for using color additives in various foods, drugs, and cosmetics.

ZUCKERMAN, S. Colors for Foods, Drugs, and Cosmetics, in *Encyclopedia of Chemical Technology*, 2nd ed., Vol. 5, Wiley, New York, 1964, pp. 857-884. A review of certified colors from the standpoint of regulation, use, specifications, and properties.

Chapter **3** Regulations Governing Use

The regulations governing color additives are too complex and transitory to attempt an authoritative, in-depth review of them here. Nevertheless, a discussion of some of the generalities involved is useful for gaining insight into the kinds of problems associated with their use.

LISTED AND PROVISIONALLY LISTED COLORANTS

From a legal standpoint colorants can be divided into two groups—those listed for use and those provisionally listed. Listed additives are colors that have been sufficiently evaluated to convince FDA of their safety for the application intended. Provisionally listed colorants, on the other hand, are dyes and pigments that are not considered unsafe but that nevertheless have not undergone all the tests required by the Color Additives Amendments of 1960 to establish their eligibility for "permanent" listing. Currently, these colors can only be used in those applications in which they were used prior to enactment of the 1960 amendments. Their status is reviewed about once each year and, if sufficient reason exists and if the manufacturers or consumers of these colors request it, their provisional listing status is extended pending completion of the required scientific investigations.

To develop and evaluate a colorant and obtain "permanent" listing status for it may take from 5-7 years, depending on its intended use. In the case of an FD&C color, for example, a complete evaluation ususally includes long-term internal and external toxicological testing, chemical and shelf-stability studies, and trial runs in typical commercial products. Toxicological testing might include 2-year feeding studies in dogs and rats, repeated dermal application tests on rabbits and mice, and two-generation reproduction studies with rats. In each case the test animals are compared with control groups with respect to survival, appearance, behavior, appetite, elimination, organ weights and ratios, tissue structure, skeletal structure, and other variables, depending on the test involved. Where reproduction studies are concerned, the offspring are similarly evaluated. Chemical and shelf-stability studies include the determination of the effect of light, heat, time, acids, alkalis, moisture, and so on on the as-is dye, whereas application studies involve

REQUEST FOR CERTIFICATION OF
A BATCH OF STRAIGHT COLOR ADDITIVE

Date. . . . .

Division of Color Technology
HFF-430, Bureau of Foods
Food and Drug Administration
200 C St., S. W.
Washington, D. C. 20204

In accordance with the regulations promulgated under the Federal Food, Drug, and Cosmetic Act, we hereby make application for the certification of a batch of straight color additive.

Name of color. . . . .
                    (As listed in 21 CFR Part 74)

Batch number . . . .
                    Manufacturer's number)

Batch weigh . . . . pounds

Batch manufactured by . . . .

. . . . at. . . . .
                (Name and address of actual manufacturer)

How stored pending certification . . . .
                (State conditions of storage, with kind
                and size of containers, location, etc.)

Certification requested of this color for use in . . . .

. . . .
                    (State proposed uses)

Required fee, $ . . . . . (drawn to the order of Food and Drug Administration.

The accompanying sample was taken after the batch was mixed in accordance with 21 CFR80.22 and is accurately representative thereof.

(Signed) . . . .

by . . . .

. . . .
                    (Title)

determining the stability and efficacy of the colorant in the kinds of products and types of containers in which it is intended for use. These data, as well as information relating to the manufacture, analysis, control, and packaging of the color, and the proposed specifications and anticipated levels of use of it are incorporated into a petition to the FDA for their review. Public notice of the filing of the petition and the FDA ruling on it is given in the Federal Register. If the petition is found to be complete and convincing, the color is listed for use in the kinds of products tested and petitioned for—foods, drugs, cosmetics, or all three.

CERTIFIED COLORANTS AND COLORANTS EXEMPT FROM CERTIFICATION

A further distinction between color additives is made relative to whether there is requirement for FDA certification. In general only synthetic organic colorants are now subject to certification, whereas natural organic and inorganic colorants such as turmeric and titanium dioxide are not. This exemption from certification for a particular colorant holds whether the colorants is truly obtained from natural sources or is synthetically produced, as in the case of natural and synthetic β-carotene.

If a color requires certification prior to its sale, an appropriate size representative sample of each batch, along with a request for certification must be submitted to the FDA, Color Certification Branch, to see if it conforms to the specifications and other conditions established for it. An example of such a request form is shown on page 22. In the case of straight colors, including lakes, a fee is charged for this certification service at the rate of $0.15 per pound of the batch, with a minimum charge of $100. If the batch is found satisfactory, a lot number is assigned to it and a certificate of certification is issued. (see the sample at the top of page 24). This certificate is valid so long as the regulations pertaining to the storage, packaging, labeling, distribution, and use of the lot are strictly adhered to.

SPECIFICATIONS

All colorants in use today have specifications that must be met before they can be sold. In the case of the provisionally listed colors, these specifications are only temporary in that they will undoubtedly be revised if and when the colorants are removed from the provisional lists. Specifications for a synthetic aromatic organic dye, a synthetically produced natural colorant and an inorganic pigment, are given as examples in the text that follows.

FD&C Red No. 40

Specifications: FD&C Red No. 40 shall conform to the following specifica-

DEPARTMENT OF HEALTH, EDUCATION, AND WELFARE
PUBLIC HEALTH SERVICE
FOOD AND DRUG ADMINISTRATION
WASHINGTON, D.C. 20204

LOT NO. **AA-** 3003

DATE March 21, 1977

Mr. John Doe
Divisional Vice President
The XYZ Chemical Company
22 Industrial Drive
Cincinnati, OH 45237

TOXICOLOGY SAMPLE

COLOR ADDITIVE CERTIFICATE

The batch of Color Additive described below is hereby certified to you. The use of this color is subject to the terms, conditions, and restrictions set forth in the Federal Food, Drug and Cosmetic Act and the regulations thereunder.

NAME OF COLOR	MFR BATCH NO.	QUANTITY IN LBS	CERT % PURE COLOR	CERTIFIED FOR USE IN
FD&C Yellow #6	3800-J	1212	91	Foods, drugs & cosmetics

Keith S. Heine

FOR THE
COMMISSIONER OF FOOD AND DRUGS.

cc: CIN-DO

FORM FDH 3000 (1/75)   PREVIOUS EDITION IS OBSOLETE.

tions and shall be free from impurities other than those named to the extend that such other impurities may be avoided by good manufacturing practice:

Sum of volatile matter (at $135°C$) and chloride and sulfates (calculated as sodium salts)—Not more than 14.0%.

Water-insoluble matter—Not more than 0.2%.

Higher sulfonated subsidiary colors (as sodium salts)—Not more than 1.0%.

Lower sulfonated subsidiary colors (as sodium salts)—Not more than 1.0%.

Disodium salt of 6-hydroxy-5-[(2-methoxy-5-methyl-4-sulfophenyl)azo]-8-(2-methoxy-5-methyl-4-sulfophenoxy)-2-naphthalenesulfonic acid—Not more than 1.0%.

Sodium salt of 6-hydroxy-2-naphthalenesulfonic acid (Scaheffer's salt)—Not more than 0.3%.

4-Amino-5-methoxy-o-toluenesulfonic acid—Not more than 0.2%.

Disodium salt of 6,6'-oxybis(2-naphthalenesulfonic acid)—Not more than 1.0%.

Lead (as Pb)—Not more than 10 ppm.

Arsenic (as As)—Not more than 3 ppm.

Total color—Not less than 85.0%.

$β$-Apo-8'-Carotenal

Specifications: $β$-Apo-8'-carotenal shall conform to the following specifications:

Physical State—Solid.

One percent solution in chloroform—Clear.

Melting point (decomposition)—136-140°C (corrected).

Loss of weight on drying—Not more than 0.2%.

Residue on ignition—Not more than 0.2%.

Lead (as Pb)—Not more than 10 ppm.

Arsenic (as As)—Not more than 1 ppm.

Assay (spectrophotometric)—96-101%.

Titanium Dioxide

Specifications: Titanium dioxide shall conform to the following specifications:

Lead (as Pb)—Not more than 10 ppm.

Arsenic (as As)—Not more than 1 ppm.

Antimony (as Sb)—Not more than 2 ppm.

Mercury (as Hg)—Not more than 1 ppm.

Loss on ignition at 800°C (after drying for 3 hr at 105°C)—Not more than 0.5%.

Water-soluble substances—Not more than 0.3%.

Acid soluble substances—Not more than 0.5%.

Titanium dioxide—Not less than 99.0% after drying for 3 hr at 105°C.

Lead, arsenic, and antimony—Determined in the solution obtained by boiling 10 g of the titanium dioxide for 15 min in 50 ml of 0.5N hydrochloric acid.

In addition to individual specifications, general specifications have been written for provisionally listed certifiable colors:

GENERAL SPECIFICATIONS

	Maximum %
FD&C Colors	
Lead	0.001
Arsenic (as As_2O_3)	0.00014
Heavy metals (except lead and arsenic)	
(precipitated as sulfides)	Trace
Mercury	0.0001
D&C and Ext. D&C Colors	
Lead	0.002
Arsenic (as As_2O_3)	0.0002
Heavy metals (except lead and arsenic)	
(precipitated as sulfides)	0.003
Colors that are barium salts—	
Soluble barium in dilute HCl (as $BaCl_2$)	0.05

The limit of 1 ppm mercury placed on colors intended for use in foods was established by a letter from the Acting Director of the Division of Colors and Cosmetics to the certified color manufacturers on February 9, 1970. This action was the first step taken to replace the somewhat nebulous "heavy metals" specifications with concrete limits for specific metals.

USE RESTRICTIONS

There are numerous restrictions on the use of color additives. They cannot, for example, be employed to deceive the public by adding weight or bulk to a product or by hiding quality. In addition, colorants or products containing them can't be used in the area of the eyes, in injections, or in surgical sutures without special permission. Exceptions to these regulations include FD&C Blue No. 2, D&C Green No. 6 and [Phthalocyaninato (2−)] copper, which are listed for use in certain surgical sutures, and various provisionally and permanently listed colorants exempt from certification that were allowed in eye makeup prior to 1960 and thus can still be used for that purpose.

Other restrictions pertaining to the areas of use and the quantities of colorants allowed in products are specified in regulations for particular additives. Citrus Red No. 2, for example, can only be used to color the skins of oranges not intended for processing, whereas pyrophyllite can be used only to color drugs that are to be externally applied. A special case of the restricted use of a colorant is that of FD&C Red No. 4. Although it is designated as an FD&C colorant (implying that it can be generally used in foods), its use is now limited to coloring externally applied drugs and cosmetics only. FD&C Red No. 4 can no longer be used to color foods and ingested drugs at all. So many limitations have crept into the system that the designations FD&C, D&C, and Ext. D&C no longer have the meaning they once had.

The amount of a color additive allowed in a product depends on both the colorant and the article being colored. For example, TiO_2 when used to color foods cannot exceed 1% by weight of the food product. On the other hand, there is no numerical limit set on its use in the coloring of ingested or externally applied drugs. Similarly, ultramarine blue may be used to color salt intended for animal feed, but not in amounts exceeding 0.5% by weight of the salt. When numerical limits for the use of colorants are not specified, the amount allowed is controlled by "good manufacturing practice"—an ill-defined term that in effect says that you can not use more of a colorant in a product than the level necessary to achieve the desired effect. Today, the excessive use of colorants is rarely a problem since manufacturers are not likely to waste costly additives and, at the same time, run the risk of making their products appear unnatural.

INTERNATIONAL CONTROL

Attempts have been made to regulate color additives on an international basis. Some years ago the European Economic Communtiy prepared a list of approved colorants in the hope that this would pave the way for a greater universality of colors, but to date their attempts have met with little success.

More important is the work of the Joint FAO/WHO Expert Committee on Food Additives. In 1955 at a conference of the Food and Agriculture Organization of the United Nations and the World Health Organization the two groups recommended that they collect and disseminate information on food additives. A joint committee was formed; at its fourth session held in Rome in 1959 and at sessions held in 1964, 1966, and 1969 this committee considered specifications for the identity and purity of food additives, including food colors. This action culminated in the publication of monographs entitled *Specifications for Identity and Purity of Food Additives*, Vol. II, *Food Colors*, and *Specifications for Identity and Purity and Toxicological Evaluation of Food Colors*.

Chapter 4 Certified Colors

Presently, all certified colors are factory-prepared materials belonging to one of several different chemical classes. Although a few such as D&C Blue No. 6 (indigo) are known in nature, certified colors owe their commercial importance to man's ability to synthetically produce them.

Because of the starting materials used in their manufacture in the past, certified colors have also been known as *coal-tar dyes*. Today, since most of the raw materials used in their preparation are obtained from petroleum, this term no longer applies.

Compared to noncertified color additives, certified colors are a cheaper, brighter, more uniform, and better characterized group of dyestuffs with higher tinctorial strengths and a wider range of hues. They are available as is ("primary colors") and in admixture with other certified colors ("secondary mixes"). Most are sold in various forms, including powder, granuals, aluminum lakes, solutions, and dispersions, depending on the colorant and its intended use.

CHEMICAL CLASSIFICATIONS

Azo colors comprise the largest group of certified colorants. They are characterized by the presence of one or more azo bonds (-N=N-) and are synthesized by the coupling of a diazotized primary aromatic amine to a component capable of coupling, usually a naphthol. Certifiable azo colors can be subdivided into four groups: insoluble unsulfonated pigments, soluble unsulfonated dyes, insoluble sulfonated pigments, and soluble sulfonated dyes.

Unsulfonated pigments such as D&C Orange No. 17 and D&C Red No. 36 are insoluble directly on coupling and contain no groups capable of salt formation. Each contains a chlorine or nitro group in the *ortho* position relative to the azo group, resulting in a sterically hindered molecule with low solubility and excellent light stability.

The unsulfonated dyes are insoluble in water but soluble in aromatic solvents. This group includes Citrus Red No. 2 and D&C Red No. 17.

Insoluble sulfonated pigments are made from colorants that contain a sulfonic acid group that is easily converted into an insoluble metal salt. In most cases the sulfonic acid group is *ortho* to the diazo further reducing the solubilizing characteristics of the sulfonic grouping. The shade of these products is affected by the metal incorporated into the molecule and the physical characteristics of the colorants. D&C Red Nos. 7, 9, and 34 are insoluble sulfonated pigments.

The soluble azo dyes contain one or more sulfonic acid groups. Their degree of water solubility is determined by the number of sulfonic groups present and their position in the molecule. FD&C Red No. 40 and D&C Orange No. 4 belong in this class.

Anthraquinone colorants all contain the following structure:

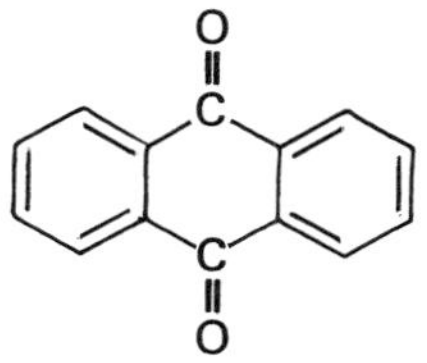

Included in this grouping are D&C Green No. 5, a water-soluble sulfonate, D&C Green No. 6, an unsulfonated water-insoluble compound, and D&C Violet No. 2, a water-insoluble hydroxyanthraquinone. Anthraquinone color additives, in general, are light stable and have good physical and chemical properties for use in cosmetics.

There are three color additives of the indigoid type, including D&C Blue No. 6 (indigo, an insoluble pigment), FD&C Blue No. 2 (the water-soluble disodium sulfonate derivative of indigo), and D&C Red No. 30 (an insoluble thioindigoid). All are related to the basic indigo structure. D&C Blue No. 6 has the following structure:

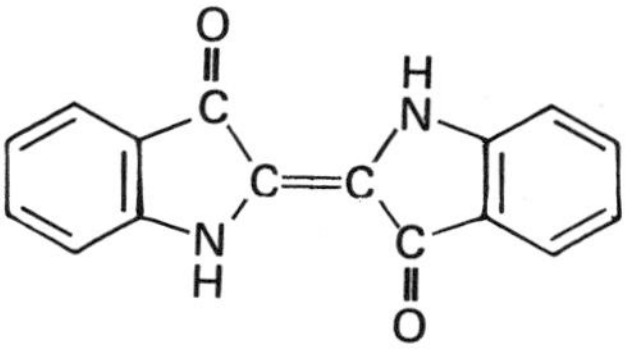

Three dyes are triaryl—or triphenylmethanes. Each, like FD&C Blue No. 1, consists of three aromatic rings attached to a central carbon atom. All are water-soluble, anionic, sulfonated systems. FD&C Blue No. 1 has the following structure:

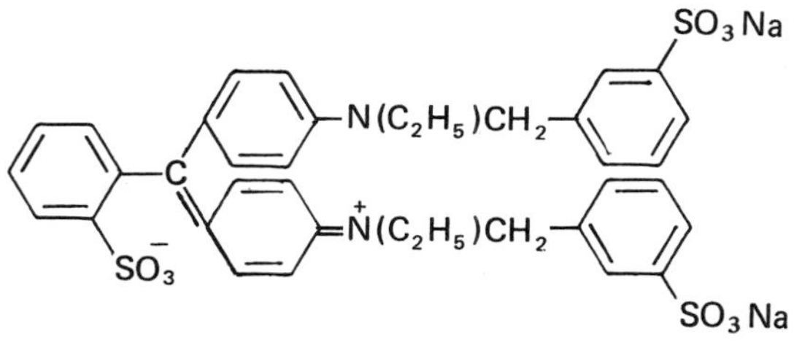

The second largest group of color additives are the xanthenes, which are characterized by the following structure:

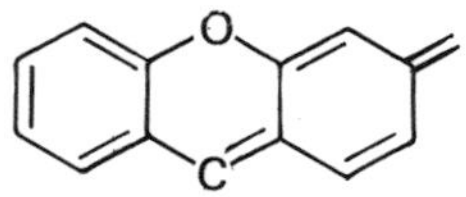

Xanthene colors can be either acidic or basic. Acid xanthenes are known to exist in two tautomeric forms. The phenolic type, or "fluorans", are free-acid structures such as D&C Orange No. 10 and D&C Red No. 21. Most have poor water solubility. In contrast to these, the quinoids or xanthenes are usually the highly water soluble sodium salt counterparts of the fluorans such as D&C Orange No. 11 and D&C Red No. 22.

D&C Red No. 19 is the only certifiable basic xanthene colorant.

Two of the remaining colorants on the list of certifiables are quinolines, the solvent-soluble D&C Yellow No. 11, and its water-soluble sulfonated derivative, D&C Yellow No. 10. Both are derived from quinaldine by condensation with phthalic anhydride. D&C Yellow No. 11 has the following structure:

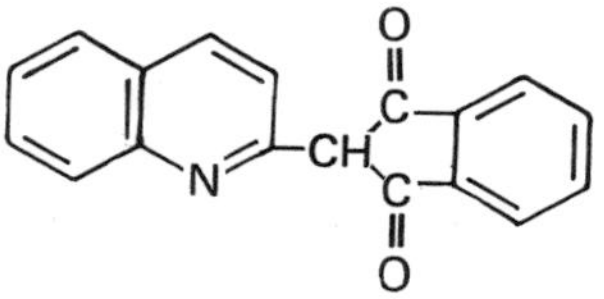

Two others—FD&C Yellow No. 5 and Orange B—are pyrazolones that contain the following common group:

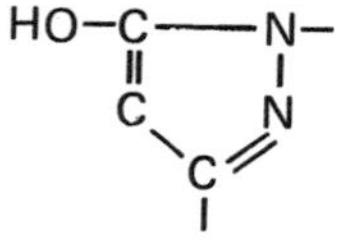

The pyrazolones may also be classified as azo dyes since each contains an -N=N- group.

One nitro dye (Ext. D&C Yellow No. 7), one pyrene colorant (D&C Green No. 8) and one phthalocyanine dye [[Phthalocyaninato (2-)] copper] complete the list of certifiable colors.

The certifiable colors in use today are shown on pp. 53-73.

LAKES

A special group of color additives are the lakes. These are pigments formed by the precipitation and extension of a soluble dye onto an approved insoluble base or substratum. In the case of D&C and Ext. D&C colors, this substratum may be alumina, blanc fixe, gloss white, clay, titanium dioxide, zinc oxide, talc, rosin, aluminum benzoate, calcium carbonate, or any combination of two or more of these materials. Currently, alumina is the only approved substratum for preparing FD&C lakes. Lakes are marketed as is and in admixture with other lakes and are usually available with various pure dye contents and moisture levels.

Unlike dyes that color objects through their adsorption or attachment from solution to the material being colored, lakes, like other pigments, impart color by dispersing them in the medium to be colored. As a consequence of this pigment-like character both the shade and the tinctorial strength of lakes are highly dependent on the conditions used in their manufacture as well as their physical properties including their particle size and crystal structure.

Since there are no solvent-soluble FD&C colors, the FD&C lakes have proven particularly valuable for coloring water-repelling foods such as fats, gums and oils, and for coloring food-packaging materials including lacquers, plastic films, and inks from which soluble dyes would be quickly leached. Similarly useful applications have been found for D&C and Ext. D&C lakes in their respective areas of application.

Other properties of lakes that enhance their usefulness include their opacity, ability to be incorporated into products in the dry state, and stability toward heat and light. Such properties have made possible the more effective and more efficient preparation of candy and pill coatings and have eliminated the need to remove the excess moisture frequently required when coloring dry products with soluble dyes. They have also made possible the coloring of certain products that because of their nature, method of preparation, or method of storage, cannot be colored with ordinary color additives.

PROPERTIES

To fully appreciate the properties of the color additives in use today it is helpful to first outline the requirements of a good colorant. In doing this, though, one must realize that since the potential areas and conditions of use for most additives are so numerous, it is next to impossible to define the perfect colorant and even more difficult to produce such a product. Nevertheless, it is generally recognized that at least the following criteria must be met if a colorant is to be useful.

1. It must be safe at the levels used and under the conditions used.
2. It must not impart any offensive property to a product.
3. It must be stable.
4. It must be non-reactive with the products and containers in which it is used.

5. It must be easy to apply to products.
6. It should be cheap.
7. It should have a high tinctorial strength.

The degree of safety required of a color additive is obviously dependent on the areas and frequency of use intended. Realistically, the toxicity of an Ext. D&C color used in hand soaps, rouges, and other products applied to the surface of the body ought not cause as much concern as the toxicity of a D&C colorant used to color drugs that are to be ingested. On the other hand, no stone can be left unturned in proving the safety of an FD&C colorant intended for use in our food supply.

For the sake of this discussion we presume that all colors permitted in the United States are nontoxic when used as the law allows. It is important to note, though, that not all colors considered "safe" in this country are considered as such in other parts of the world, and vice versa. The reasons for this vary but are frequently related to the ground rules employed in testing them. Here in the United States, for example, it is believed that when studying a colorant's toxicity it should be tested in a manner analogous to the conditions under which it will be used. Consequently, since it is most important that food colors be safe when ingested, animal-feeding studies play the key role in their evaluation. By contrast, scientists in other parts of the world often place a great deal of emphasis on the effects of injecting a solution of the proposed colorant under the skin of test subjects. Understandably, since the mechanisms involved in these tests are so different, the conclusions drawn from them have often also been different and have resulted in the establishment of lists of permitted colors more or less on national or regional bases. Although it is not always clear which school of thought is right when decisions are reached regarding the toxicity of a colorant, it is certain that the failure to be aware of and to understand the reasoning used to make these decisions has more than once caused undue public concern over the safety of color additives in use in the United States. The publicity often given to unscientific and inconclusive independent studies of the toxicity of colors has simply added to the confusion.

Offensive properties that can conceivably be transferred to a product by a colorant include taste and odor, whether it is the taste or odor of the colorant itself or of trace impurities in the colorant that have extremely low taste or odor thresholds. In the case of foods and drugs this is not likely to be much of a problem, since the amount of colorant used in such products is usually low. However, in the case of highly colored cosmetics including lipsticks, face powders, rouges, and other substances used in the area of the mouth and nose, the problem is at least potentially more serious.

An even more serious problem can result from the instability of a colorant, whether it is inherent instability or instability caused by reaction of the dyestuff with a product or a product's container. Generally, color additives have shown excellent stability when stored in the dry state. For example, most certified food colors show no degradation after storage periods of 15 years or more. Unfortunately, the stability of a colorant stored neat is no guarantee of its stability in a product. Consequently, use tests must still be performed and on an individual product/colorant basis.

Many factors can and indeed do contribute to the instability of colorants. Trace metals, for example, including zinc, tin, aluminum, iron, and copper are known to cause fading of some additives. Azo dyes in particular are troublesome in this regard in that they often react with food cans and at a rate proportional to their concentration, causing corrosion of the container and a corresponding loss in the food's dye content. Some colors lack stability in retorted protein foods, whereas others are attacked by reducing and oxidizing agents, including certain invert sugars, aldehydes, and peroxides as well as ascorbic acid, which is a flavor antioxidant. Acid dyes are frequently incompatable with the quaternary salts used in various cosmetics.

Light, of course, is the enemy of all coloring agents, and color additives are no exception. As in the case of the general or overall stability of color additives, the stability of a colorant toward light, either neat or in solution, is not necessarily the same as its stability toward light in a product. Various ingredients, including aldehydic flavors, reducing sugars, and perfume oils, are known to enhance the effects of light on some colorants, whereas, ironically, others prove to be more light stable in a product than alone. Several methods are used to minimize the effects of light on colorants in products, including packaging in light-proof containers, the incorporation of ultraviolet (UV) absorbers into the products, the use of color lakes, and the careful selection of the other ingredients used in the product. No precaution, of course, is better than choosing the correct color for the job in the first place. In general, the resistance to light of dyestuffs now in use as color additives decreases in the order: quinoline-anthraquinone-triphenylmethane-azo-fluoran and pyrene.

The pH value must also be considered when choosing a colorant, since not all of them can be used at all pH values. FD&C Red No. 3, for example, precipitates from acid solution whereas FD&C Green No. 3 turns blue under alkaline conditions. Lakes often show amphoteric properties, with both acids and alkalis tending to solubilize the inorganic substrate releasing free colorant. Other colors exhibit less drastic yet important pH-related changes in their properties, including shifts in shade, variations in shelf life, changes in solubility, and loss of tinctorial strength.

The ease with which a colorant can be applied to a product or, for that matter, the ability to use a colorant for a particular application at all is a function of both the colorant's structure and the product's matrix. Unfortunately, there are no universally useful colorants, and a compromise must almost always be made in their design and selection. The water-soluble FD&C colors, for example, which are so very useful in water-based foodstuffs—including soft drinks and gelatin desserts—are of only limited value in fatty foods (except as the lakes), since the same functional groups that render these dyes water soluble also limit their fat and solvent solubility. Analogous problems exist in the use of D&C and Ext. D&C colorants, whether they are pigments, dyes, or lakes. Thus it is important in choosing a colorant or, for that matter, in developing a new colorant to seriously consider the application properties desired based on the uses it will be put to.

The other properties most desirable in a color additive—low cost and high tinctorial strength—are, for the most part, closely related. The tinctorial strength or coloring power of a dyestuff determines the amount and thus the cost of the colorant that must be added to a product to achieve a particular effect. A colorant's tinctorial strength is an inherent property of its chemical structure and cannot be changed, although maximum use can be made of it by selecting the physical form, vehicle, and conditions under which it is used.

The cost per pound of the dyestuff is determined like that for any other product by the cost of the raw materials, equipment, and labor needed to produce it, as well as the supply and demand of the colorant. To these expenses must be added the additional cost needed to ensure the ultrahigh purity required of such colorants as well as the cost of certification. All these factors combine to make certified color additives far more expensive than typical technical dyestuffs. The saving feature, of course, is that in most cases relatively little colorant is needed to achieve the desired depth of shade in a product, and thus the cost of the colorant adds relatively little to the cost of the finished product. With this background the data presented in Tables 5-15 should be somewhat more meaningful.

Production and Use

The pounds of each colorant certified by the FDA over the past few years can be found in Table 16. The primary FD&C colors obviously dominate this picture, since they alone account for 80% or more of the total number of pounds of colorant certified during any one year.

The major areas in which certified colors are used are shown in Table 17 and 18. The picture Table 17 presents is not complete in that it neither accounts for the pounds of color exported and sold to jobbers nor reflects the usage of the relatively new RD&C Red No. 40; nevertheless, it provides a good indication of current practice. Based on the maximum color concnetrations shown in these tables and the total annual production of food in each food category, the total certified color that might be ingested per person per day is estimated to be 53.5 mg.

COLORANTS

The identities shown for the following colorants are those assigned by the Food and Drug Administration, as they appear in the Code of Federal Regulations (21CFR 1.1). Often, the name given is not the best, reflecting certain inconsistencies in the nomenclature system used to arrive at them, but

TABLE 5 WATER SOLUBILITY OF FD&C COLORS

Federal Name	Common Name	2°C		25°C		60°C	
		g/100 ml	oz/gal	g/100 ml	oz/gal	g/100 ml	oz/gal
FD&C Blue No. 1	Brill. Blue FCF	20.0	26.0	20.0	26.0	20.0	26.0
FD&C Blue No. 2	Sod. Indigo Dis.	0.8	1.0	1.6	2.1	2.2	2.9
FD&C Green No. 3	Fast Green FCF	20.0	26.0	20.0	26.0	20.0	26.0
FD&C Red No. 3	Erythrosine	9.0	11.7	9.0	11.7	17.0	22.1
FD&C Red No. 4	Ponceau SX	4.7	6.1	11.0	14.3	11.0	14.3
FD&C Red No. 40	Allura [a] Red AC	18.0	23.4	22.0	28.6	26.0	33.8
FD&C Yellow No. 5	Tartrazine	3.8	4.9	20.0	26.0	20.0	26.0
FD&C Yellow No. 6	Sunset Yellow FCF	19.0	24.7	19.0	24.7	20.0	26.0

[a]Registered trademark of Buffalo Color Corporation

TABLE 6 ALCOHOL SOLUBILITY OF FD&C COLORS

Federal Name	100% Alcohol				75% Alcohol				50% Alcohol				25% Alcohol			
	25°C		60°C		25°C		60°C		25°C		60°C		25°C		60°C	
	g/100 ml	oz/gal	g/100 ml	oz/gal	g/100 ml	oz/gal	g/100 ml	oz/gal	g/100 ml	oz/gal	g/100 ml	oz/gal	g/100 ml	oz/gal	g/100 ml	oz/gal
FD&C Blue No. 1	0.15	0.20	0.15	0.20	20.0	26.0	20.0	26.0	20.0	26.0	20.0	26.0	20.0	26.0	20.0	26.0
FD&C Blue No. 2	—	—	0.008	0.009	0.07	0.09	0.07	0.09	0.30	0.39	0.35	0.46	0.50	0.65	0.60	0.78
FD&C Green No. 3	0.01	0.01	0.01	0.01	10.0	13.0	20.0	26.0	20.0	26.0	20.0	26.0	20.0	26.0	20.0	26.0
FD&C Red No. 3	—	—	0.01	0.01	0.6	0.78	0.80	1.04	1.0	1.3	1.0	1.3	8.0	10.4	8.0	10.4
FD&C Red No. 4	—	—	0.01	0.01	0.30	0.39	0.40	0.53	1.0	1.3	1.1	1.4	1.4	1.8	2.0	2.6
FD&C Red No. 40	0.001	0.001	0.05	0.07	0.2	0.26	0.9	1.17	1.3	1.69	5.5	7.15	9.5	12.4	22.0	28.6
FD&C Yellow No. 5	—	—	0.001	0.001	1.1	1.4	1.2	1.56	4.0	5.2	8.4	10.9	12.0	15.6	17.0	22.1
FD&C Yellow No. 6	—	—	0.001	0.001	0.3	0.39	0.3	0.39	3.0	3.9	4.0	5.2	10.0	13.0	15.0	19.5

TABLE 7 GLYCERINE SOLUBILITY OF FD&C COLORS

Federal Name	100% Glycerine				75% Glycerine				50% Glycerine				25% Glycerine			
	25°C		60°C		25°C		60°C		25°C		60°C		25°C		60°C	
	g/100 ml	oz/gal	g/100 ml	oz/gal	g/100 ml	oz/gal	g/100 ml	oz/gal	g/100 ml	oz/gal	g/100 ml	oz/gal	g/100 ml	oz/gal	g/100 ml	oz/gal
FD&C Blue No. 1	20.0	26.0	20.0	26.0	20.0	26.0	20.0	26.0	20.0	26.0	20.0	26.0	20.0	26.0	20.0	26.0
FD&C Blue No. 2	1.0	1.3	1.0	1.3	1.0	1.3	1.0	1.3	1.0	1.3	1.5	2.0	1.0	1.3	1.5	2.0
FD&C Green No. 3	20.0	26.0	20.0	26.0	20.0	26.0	20.0	26.0	20.0	26.0	20.0	26.0	20.0	26.0	20.0	26.0
FD&C Red No. 3	20.0	26.0	20.0	26.0	20.0	26.0	20.0	26.0	16.0	20.8	16.0	20.8	14.0	18.2	19.0	24.7
FD&C Red No. 4	5.8	7.54	5.8	7.54	4.2	5.46	4.2	5.46	4.2	5.46	4.2	5.46	6.0	7.8	6.0	7.8
FD&C Red No. 40	3.0	3.9	8.0	10.4	4.5	5.85	8.8	11.5	12.0	15.6	14.0	18.2	20.0	26.0	20.0	26.0
FD&C Yellow No. 5	18.0	23.4	18.0	23.4	20.0	26.0	20.0	26.0	20.0	26.0	20.0	26.0	20.0	26.0	20.0	26.0
FD&C Yellow No. 6	20.0	26.0	20.0	26.0	18.0	23.4	18.0	26.0	20.0	26.0	20.0	26.0	20.0	26.0	20.0	26.0

TABLE 8 PROPYLENE GLYCOL SOLUBILITY OF FD&C COLORS

Federal Name	100% Glycol				75% Glycol				50% Glycol				25% Glycol			
	25°C		60°C		25°C		60°C		25°C		60°C		25°C		60°C	
	g/100 ml	oz/gal	g/100 ml	oz/gal	g/100 ml	oz/gal	g/100 ml	oz/gal	g/100 ml	oz/gal	g/100 ml	oz/gal	g/100 ml	oz/gal	g/100 ml	oz/gal
FD&C Blue No. 1	20.0	26.0	20.0	26.0	20.0	26.0	20.0	26.0	20.0	26.0	20.0	26.0	20.0	26.0	20.0	26.0
FD&C Blue No. 2	0.1	0.13	0.1	0.13	0.4	0.52	0.4	0.52	0.4	0.52	0.4	0.52	0.6	0.78	2.0	2.6
FD&C Green No. 3	20.0	26.0	20.0	26.0	20.0	26.0	20.0	26.0	20.0	26.0	20.0	26.0	20.0	26.0	20.0	26.0
FD&C Red No. 3	20.0	26.0	20.0	26.0	15.6	20.3	15.6	20.3	6.6	8.6	9.2	12.0	6.6	8.6	9.4	12.2
FD&C Red No. 4	2.0	2.6	2.0	2.6	1.6	2.1	1.6	2.1	2.6	3.38	2.6	3.38	4.4	5.7	4.4	5.7
FD&C Red No. 40	1.5	2.0	1.7	2.2	2.0	2.6	3.2	4.1	7.5	9.8	10.0	13.0	18.0	23.4	22.0	28.6
FD&C Yellow No. 5	7.0	8.1	7.0	9.1	10.4	13.5	13.0	16.9	12.4	16.1	20.0	26.0	20.0	26.0	20.0	26.0
FD&C Yellow No. 6	2.2	2.86	2.2	2.86	2.2	2.86	2.6	3.38	7.0	9.1	12.8	16.6	20.0	26.0	20.0	26.0

TABLE 9 pH STABILITY OF FD&C COLORS

Federal Name	Common Name	pH = 3	pH = 5	pH = 7	pH = 8
FD&C Blue No. 1	Brill. Blue FCF	Slight fade after 1 week	Very slight fade after 1 week	Very slight fade after 1 week	Very slight fade after 1 week
FD&C Blue No. 2	Sod. Indigo Dis.	Appreciable fade after 1 week	Appreciable fade after 1 week	Considerable fade after 1 week	Fades completely
FD&C Green No. 3	Fast Green FCF	Slight fade after 1 week	Very slight fade after 1 week	Very slight fade after 1 week	Slight fade and appreciably bluer
FD&C Red No. 3	Erythrosine	Insoluble	Insoluble	No appreciable change	No appreciable change
FD&C Red No. 4	Ponceau SX	No appreciable change	No appreciable change	No appreciable change	No appreciable change
FD&C Red No. 40	Allura®[a] Red AC	No appreciable change	No appreciable change	No appreciable change	No appreciable change
FD&C Yellow No. 5	Tartrazine	No appreciable change	No appreciable change	No appreciable change	No appreciable change
FD&C Yellow No. 6	Sunset Yellow FCF	No appreciable change	No appreciable change	No appreciable change	No appreciable change

[a] Registered trademark of Buffalo Color Corporation.

TABLE 10 STABILITY OF FD&C COLORS IN THE PRESENCE OF VARIOUS ACIDS

Federal Name	Common Name	10% Citric Acid	10% Acetic Acid	10% Malic Acid	10% Tartaric Acid
FD&C Blue No. 1	Brill. Blue FCF	No appreciable change	No appreciable change	No appreciable change	No appreciable change
FD&C Blue No. 2	Sod. Indigo Dis.	Completely faded after 1 week	Completely faded after 1 week	Considerably faded after 1 week	Considerably faded after 1 week
FD&C Green No. 3	Fast Green FCF	No appreciable change	No appreciable change	Slight fade after 1 week	Slight fade after 1 week
FD&C Red No. 3	Erythrosine	Insoluble	Insoluble	Insoluble	Insoluble
FD&C Red No. 4	Ponceau SX	No appreciable change	No appreciable change	No appreciable change	No appreciable change
FD&C Red No. 40	Allura[a] Red AC	No appreciable change	No appreciable change	No appreciable change	No appreciable change
FD&C Yellow No. 5	Tartrazine	No appreciable change	No appreciable change	No appreciable change	No appreciable change
FD&C Yellow No. 6	Sunset Yellow FCF	No appreciable change	No appreciable change	No appreciable change	No appreciable change

[a]Registered trademark of Buffalo Color Corporation.

TABLE 11 STABILITY OF FD&C COLORS IN THE PRESENCE OF VARIOUS ALKALIS

Federal Name	Common Name	10% Sodium Bicarbonate	10% Sodium Carbonate	10% Ammonium Hydroxide	10% Sodium Hydroxide
FD&C Blue No. 1	Brill. Blue FCF	Slight fade after 1 week	Fades completely	Considerable fade	Fades completely
FD&C Blue No. 2	Sod. Indigo Dis.	Fades completely	Fades completely	Fades Completely	Yellower
FD&C Green No. 3	Fast Green FCF	No appreciable change	Considerable fade and appreciably bluer	Considerable fade and appreciably bluer	Fades completely
FD&C Red No. 3	Erythrosine	No appreciable change	Slight fade after 1 week	Slight fade after 1 week	Fades completely
FD&C Red No. 4	Ponceau SX	Slight fade after 1 week	No appreciable change	No appreciable change	Slightly yellower
FD&C Red No. 40	Allura® ^{a} Red AC	Slightly bluer after 1 week	Appreciably bluer after 1 week	Appreciably bluer after 1 week	Much bluer after 1 week
FD&C Yellow No. 5	Tartrazine	No appreciable change	No appreciable change	No appreciable change	Considerable fade after 1 week
FD&C Yellow No. 6	Sunset Yellow FCF	No appreciable change	No appreciable change	No appreciable change	Slight fade after 1 week

^{a}Registered trademark of Buffalo Color Corporation.

41

TABLE 12 STABILITY OF FD&C COLORS IN THE PRESENCE OF VARIOUS SUGARS

Federal Name	Common Name	10% Cerelose	10% Dextrose	10% Sucrose	10% Cerelose in 2.5% Citric Acid
FD&C Blue No. 1	Brill. Blue FCF	No appreciable change	No appreciable change	No appreciable change	No appreciable change
FD&C Blue No. 2	Sod. Indigo Dis.	Considerable fade after 1 week	Considerable fade after 1 week	Slight fade after 1 week	Considerable fade after 1 week
FD&C Green No. 3	Fast Green FCF	No appreciable change	No appreciable change	No appreciable change	No appreciable change
FD&C Red No. 3	Erythrosine	No appreciable change	No appreciable change	No appreciable change	Insoluble
FD&C Red No. 4	Ponceau SX	No appreciable change	No appreciable change	No appreciable change	No appreciable change
FD&C Red No. 40	Allura®[a] Red AC	No appreciable change	No appreciable change	No appreciable change	
FD&C Yellow No. 5	Tartrazine	No appreciable change	No appreciable change	No appreciable change	No appreciable change
FD&C Yellow No. 6	Sunset Yellow FCF	No appreciable change	No appreciable change	No appreciable chnage	No appreciable change

[a]Registered trademark of Buffalo Color Corporation.

TABLE 13 STABILITY OF FD&C COLORS IN OTHER MEDIA

Federal Name	Common Name	1% Sodium Benzoate	1% Ascorbic Acid	25 ppm Sulfur Dioxide	250 ppm Sulfur Dioxide
FD&C Blue No. 1	Brill. Blue FCF	No appreciable change	Slight fade after 1 week	No appreciable change	Very slight fade after 1 week
FD&C Blue No. 2	Sod. Indigo Dis.	Slight fade after 1 week	Considerable fade after 1 week	Fades completely	Fades completely
FD&C Green No. 3	Fast Green FCF	No appreciable change	Slight fade after 1 week	No appreciable change	Very slight fade after 1 week
FD&C Red No. 3	Erythrosine	Very slight fade after 1 week	Insoluble	Insoluble	Insoluble
FD&C Red No. 4	Ponceau SX	No appreciable change	Considerable fade after 1 week	No appreciable change	No appreciable change
FD&C Red No. 40	Allura®[a] Red AC	No appreciable change	No appreciable change	No appreciable change	No appreciable change
FD&C Yellow No. 5	Tartrazine	No appreciable change	Appreciable fade after 1 week	Appreciable fade after 1 week	Appreciable fade after 1 week
FD&C Yellow No. 6	Sunset Yellow FCF	No appreciable change	Considerable fade after 1 week	Appreciable fade after 1 week	Appreciable fade after 1 week

[a] Registered trademark of Buffalo Color Corporation.

TABLE 14   SOLUBILITIES OF D&C AND EXT. D&C COLORANTS

	H$_2$O	Glycerol	MeOH	EtOH	Petroleum Jelly
D&C Blue No.4	S	S	S	S	C
D&C Blue No. 6	IU	D	I	I	D
D&C Blue No. 9	IU	ID	Ia	I	D
D&C Green No. 5	S	S	S	SS	IE
D&C Green No. 6	I	Ia	SS	SS	M
D&C Green No. 8	SF	SSF	SSF	SSF	Ia
D&C Orange No. 4	S	S	S	M	IE
D&C Orange No. 5	IB	SS	S	M	D
D&C Orange No. 10	IB	SS	S	M	D
D&C Orange No. 11	Similar to those of FD&C Red No. 3				
D&C Orange No. 17	I	D	Ia	Ia	D
D&C Red No. 6	S	S	SS	Ia	I
D&C Red No. 7	I	D	Ia	Ia	D
D&C Red No. 8	I	D	Ia	Ia	D
D&C Red No. 9	I	D	Ia	Ia	D
D&C Red No. 17	I	SS	SS-M	SS	S
D&C Red No. 19	SF	SF	SF	SF	I
D&C Red No. 21	IBF	Da	SS	SS	D
D&C Red No. 22	SF	SF	SF	SF	IE
D&C Red No. 27	IB	Da	SS	SS	D
D&C Red No. 28	S	S	S	S	IE
D&C Red No. 30	IU	D	I	I	I
D&C Red No. 31	M	SS	SS	SS	I
D&C Red No. 33	S	S	SS	SS	I
D&C Red No. 34	I	I	Ia	I	D
D&C Red No. 36	I	D	Ia	Ia	D
D&C Red No. 37	Ia	SS	SF	SF	IE
D&C Red No. 39	Ia	M	M-S	S	I
D&C Violet No. 2	I	Ia	SS	SS	S
D&C Yellow No. 7	IBF	SSF	SF	SS	D
D&C Yellow No. 8	SF	SF	SF	M	IE
D&C Yellow No. 10	S	S	M	SS	I
D&C Yellow No. 11	I	SS	S	S	S
Ext. D&C Violet No. 2	S	S	SS	SS	I
Ext. D&C Yellow No. 7	S	S	M	SS	I

ABBREVIATIONS FOR TABLES 14 AND 15

a—May bleed or stain, very sparingly soluble.

B—Insoluble in water, soluble in aqueous alkaline solution.

b—Turns much bluer in hue.

C—Practically insoluble, but useful in nearly neutral or slightly acid emulsions.

D—Practically insoluble, but may be dispersed by grinding and homogenizing; solid mediums (waxes) should be softened or melted before or during the grinding.

d—Hue becomes duller or darker.

44

TABLE 14 Continued

Toluene	Stearic Acid	Oleic Acid	Mineral Oil	Mineral Wax	Et_2O	Me_2CO	AcOBu
I	C	C	C	C	I	Ia	I
Ia	D	D	D	D	I	I	I
Ia	D	D	D	D	I	I	I
I	IE	IE	IE	IEW	I	SS	I
S	M	M	M	M	SS	SS	S
I	Ia	Ia	I	I	Ia	Ia	Ia
I	IE	IE	IE	IE	I	Ia	I
I	D	D	D	D	M	S	I
I	D	D	D	D	M	S	I
Similar to those of FD&C Red No. 3							
I	D	D	D	D	Ia	Ia	I
I	I	I	I	I	I	Ia	I
I	D	D	D	D	I	Ia	I
I	D	D	D	D	I	Ia	I
I	D	D	D	D	I	I	I
S	S	S	S	S	SS	SS	M
I	IC	SS	IC	IC	S*	SF	I
I	D	D	D	D	M*	S	I
I	IE	IE	IE	IE	Ia	SS	I
I	D	D	D	D	Ia	SS	I
I	IE	IE	IE	IE	Ia	SS	I
Ia	D	D	D	D	Ia	Ia	Ia
I	I	I	I	I	I	Ia	Ia
I	I	I	I	I	I	I	I
I	D	D	D	D	I	D	D
I	D	D	D	D	I	Ia	D
S	S	S	IEG	IEG	S*	SF	SS
Ia	I	SS	I	Ia	S	S	SS
S	S	S	S	SW	SS	SS	S
I	D	D	D	D	SS*	S	I
I	IE	IE	IE	IE	Ia	kIa	I
I	I	I	I	I	Ia	SS	I
S	S	S	S	S	S	S	S
I	I	I	I	I	I	SS	I
I	I	I	I	I	I	M	I

E—Practically insoluble in the fatty acid, oil, or wax, but useful in coloring slightly alkaline aqueous emulsions.

F—Solution usually fluorescent.

G—Soluble or dispersible in oils and waxes in presence of 10-25% of a fatty acid.

I—Insoluble.

J—Tends to thicken or gel the solution.

k—Turns brownish in hue.

L—Turns orange in hue.

M—Moderately soluble (<1%).

m—Turns scarlet in hue.

TABLE 15   FASTNESS PROPERTIES OF D&C AND EXT. D&C COLORANTS

	Light	10% AcOH	10% HCl	10% NaOH
D&C Blue No. 4	3	5	5	4
D&C Blue No. 6	6	7l	5l	L6U
D&C Blue No. 9	7	7l	5l	6lU
D&C Green No. 5	5	5	5	5
D&C Green No. 6	4	5L	5l	6l
D&C Green No. 8	2	l	l	5
D&C Orange No. 4	5	5	5	2m
D&C Orange No. 5	2	4al	4l	Sr
D&C Orange No. 10	2	4al	4l	Sr
D&C Orange No. 11	Similar to D&C Red No. 3			
D&C Orange No. 17	5	5l	ld	ldr
D&C Red No. 6	5	5	4	4d
D&C Red No. 7	6	5l	4l	5l
D&C Red No. 8	6	6l	4ld	4ld
D&C Red No. 9	6	6l	4l	4l
D&C Red No. 17	3	5L	4ld	5l
D&C Red No. 19	3	5	5	2p
D&C Red No. 21	2	3l	3l	5Sr
D&C Red No. 22	2	2py	1py	5
D&C Red No. 27	2	3l	3	5Sr
D&C Red No. 28	3	2p	4p	6
D&C Red No. 30	6	7l	l	6lU
D&C Red No. 31	5	5	4	5
D&C Red No. 33	5	6	3z	5
D&C Red No. 34	4	5l	4	4l
D&C Red No. 36	6	6l	4d	4d
D&C Red No. 37	3	6l	5l	la
D&C Red No. 39	2	Sv	Sv	6Sx
D&C Violet No. 2	4	5l	5l	5l
D&C Yellow No. 7	2	l	l	S6
D&C Yellow No. 8	3	3p	3p	6
D&C Yellow No. 10	3	5	5	4r
D&C Yellow No. 11	2	l	5l	lw
Ext. D&C Violet No. 2	5	5	5	5
Ext. D&C Yellow No. 7	4	5	5	5

p—Dye precipitated as heavy-metal salt or color acid.

r—Turns redder in hue.

S—Dissolves (solubility $\geq$ 1%).

SS—Sparingly soluble (<0.25%).

U—In alkaline-reducing vats a soluble leuco compound forms.

v—Turns violet in hue.

W—Not fast to prolonged storage in some waxes.

w—Becomes tinctorially weaker.

x—Turns yellow in hue.

y—Turns yellower in hue.

TABLE 15 Continued

0.9% Physiol Salt Soln.	5% FeSO₄	5% Alum	Oxidizing Agents	Reducing Agents
6	4	4	2	1
I	I	I	6	U
I	I	I	6	U
5	4	4	3	2
I	I	I	3	2
6	4d	4d	3	3
6	J-p	J-p	3	3
I	I	I	3	3
I	I	I	3	3
Similar to D&C Red No. 3				
I	I	I	3	2
6	p	p	3	1
I	4ld	4l	3	1
I	4ld	4l	3	1
I	4ld	4l	3	1
I	4ld	4l	3	1
6	6	6	3	5
I	Id	4l	4	4
6	3d	2y	4	4
1	I	I	4	4
6	z	p	4	4
I	I	I	5	u
6	p	p	3	1
6	4	4	3	1
I	I	I	3	1
I	4d	4	3	1
la	I	la	3	5
I	4ald	I	3	3
6l	4l	4l	2	1
I	I	I	3	3
6	z-p	p	3	3
6	z	4	2	5
I	I	I	2	5
6	4z	4	3	2
6	zd	4	3	3

z—Hazy or cloudy.

*—Practically colorless.

1—Very poor fastness.

2—Poor fastness.

3—Fair fastness.

4—Moderate fastness.

5—Good fastness.

6—Very good fastness.

7—Excellent fastness.

TABLE 16 POUNDS OF COLOR ADDITIVES CERTIFIED BY FDA DURING FISCAL YEAR.

	1970	1971	1972
FD&C—PRIMARIES			
FD&C Blue No. 1	83,309	92,928	94,796
FD&C Blue No. 2	39,974	43,391	39,218
FD&C Green No. 3	5,005	7,082	5,864
FD&C Red No. 2	1,463,753	1,283,367	729,461
FD&C Red No. 3	154,288	166,744	238,658
FD&C Red No. 4	23,352	21,289	28,640
FD&C Red No. 40	—	263	892,282
FD&C Violet No. 1	8,897	3,200	66,684
FD&C Yellow No. 5	956,681	1,033,464	1,092,724
FD&C Yellow No. 6	939,641	1,016,456	887,444
	3,674,900	3,668,184	4,075,771
FD&C—LAKES			
FD&C Blue No. 1	32,433	24,303	34,279
FD&C Blue No. 2	6,553	11,021	9,988
FD&C Green No. 3	379	114	—
FD&C Red No. 2	53,229	50,676	22,110
FD&C Red No. 3	102,160	110,165	120,690
FD&C Red No. 40	—	—	9,463
FD&C Violet No. 1	14,970	16,591	20,925
FD&C Yellow No. 5	291,750	331,989	391,972
FD&C Yellow No. 6	108,837	111,079	139,358
	610,311	655,938	748,785
D&C—PRIMARIES			
D&C Blue No. 6	514	2,095	1,847
D&C Blue No. 9	—	—	—
D&C Green No. 5	3,492	2,053	1,307
D&C Green No. 6	1,385	3,954	1,400
D&C Green No. 8	15,086	4,879	4,998
D&C Orange No. 4	—	873	1,616
D&C Orange No. 5	8,046	4,342	5,959
D&C Orange No. 10	—	2,430	—
D&C Orange No. 17	—	—	546
D&C Red No. 6	—	—	—
D&C Red No. 8	516	545	778
D&C Red No. 10	—	—	—
D&C Red No. 17	304	—	10
D&C Red No. 19	2,830	709	2,991
D&C Red No. 21	7,028	4,957	3,165
D&C Red No. 22	3,310	4,935	361
D&C Red No. 27	—	—	953
D&C Red No. 28	511	1,747	601
D&C Red No. 30	—	—	—
D&C Red No. 31	868	869	—
D&C Red No. 33	2,341	1,284	1,324
D&C Red No. 36	4,749	4,172	4,161
D&C Red No. 37	516	1,193	547
D&C Red No. 39	2,024	—	—
D&C Violet No. 2	2,821	1,264	1,648
D&C Yellow No. 7	—	400	1,200

TABLE 16 Continued

1973	1974	1975	1976	1977
121,108	159,135	158,539	105,013	184,115
63,135	88,581	84,840	66,046	98,936
4,049	5,180	9,157	6,438	4,111
982,528	902,812	1,377.944	239,257	Delisted
228,436	285,567	337,144	249,581	548,557
15,912	28,261	35,037	4,169	7,822[a]
565,354	729,359	788,147	1,151,065	1,520,648
35,953	Delisted	Delisted	Delisted	Delisted
1,030,987	1,289,878	1,391,325	1,220,459	1,165,528
1,011,164	995,813	1,084,284	770,507	991,347
4,058,626	4,484,586	5,266,417	3,812,535	4,521,064
47,198	67,141	57,637	29,236	96,821
18,890	34,118	35,817	32,667	78,584
—	—	—	—	—
34,888	50,041	40,776	3,871	Delisted
182,913	169,692	181,712	203,967	269,784
27,351	33,482	32,966	46,159	71,206
17,400	Delisted	Delisted	Delisted	Delisted
435,246	492,764	502,832	375,694	727,776
172,622	188,267	235,610	118,159	447,108
936,508	1,035,505	1,087,350	809,753	1,691,279
252	3,042	234	977	1,836
—	—	—	66	—
4,331	13,432	18,670	5,593	3,124
1,989	2,475	—	2,960	660
8,112	11,793	7,529	8,025	10,153
3,827	4,284	1,024	474	2,691
7,082	3,932	5,034	8,324	4,889
—	1,529	525	—	687
—	2,500	6,161	—	1,500
—	—	—	—	543
—	2,073	978	957	454
—	516	—	1,600	Delisted
84	413	707	—	1,316
2,750	6,281	4,272	3,526	2,400
3,390	8,559	4,378	7,826	2,825
2,386	3,182	3,918	2,159	2,857
367	1,935	—	615	1,731
708	1,439	922	—	—
—	638	—	—	2,416
—	—	—	—	—
956	2,666	1,732	6,271	4,469
6,067	7,919	6,858	1,053	3,839
2,134	—	1,223	—	586
2,000	—	—	—	1,720
2,030	500	—	4,836	971
400	476	1,988	924	—

49

TABLE 16   Continued

	1970	1971	1972
D&C Yellow No. 8	4,548	1,581	3,861
D&C Yellow No. 10	4,458	4,376	9,009
D&C Yellow No. 11	1,200	2,179	2,000
	66,547	50,837	50,282
D&C—LAKES			
D&C Blue No. 1	2,429	2,370	457
D&C Blue No. 4	—	705	—
D&C Green No. 5	—	—	—
D&C Green No. 6	—	—	—
D&C Green No. 8	—	—	—
D&C Orange No. 4	628	3,887	1,335
D&C Orange No. 5	7,563	9,360	3,095
D&C Orange No. 17	12,220	16,206	9,164
D&C Red No. 2	693	—	1,483
D&C Red No. 3	6,477	5,743	7,812
D&C Red No. 6	9,192	12,739	20,747
D&C Red No. 7	22,605	2,817	20,929
D&C Red No. 8	6,300	—	—
D&C Red No. 9	27,860	39,091	35,101
D&C Red No. 10	2,056	2,667	1,308
D&C Red No. 11	1,875	1,685	1,030
D&C Red No. 12	2,166	2,663	2,640
D&C Red No. 13	2,017	1,393	746
D&C Red No. 19	12,607	7,376	5,590
D&C Red No. 21	14,049	17,600	6,971
D&C Red No. 27	1,701	7,216	7,617
D&C Red No. 30	6,608	20,209	7,070
D&C Red No. 33	—	—	—
D&C Red No. 34	867	1,551	3,195
D&C Red No. 36	570	2,709	—
D&C Yellow No. 5	20,811	13,460	25,829
D&C Yellow No. 6	5,636	2,674	2 245
D&C Yellow No. 10	2,054	774	1,592
	168,984	174,895	165,956
EXT. D&C—PRIMARIES			
Ext. D&C Green No. 1	256	389	294
Ext. D&C Violet No. 2	—	—	—
Ext. D&C Yellow No. 1	4,141	8,769	8,404
Ext. D&C Yellow No. 7	1,032	—	1,046
	5,429	9,158	9,744
EXT. D&C—LAKES			
Ext. D&C Yellow No. 7	1,764	1,275	1,555
	1,764	1,275	1,555
OTHER COLORANTS			
Citrus Red No. 2	4,612	—	2,300
Orange B	34,017	28,653	20,409
Phthalocyaninato Copper	524	—	—
	39,153	28,653	22,709

[a]Delisted for ingested use.   [b]Delisted.

TABLE 16 Continued

1973	1974	1975	1976	1977
2,367	6,400	6,100	3,800	2,800
14,125	15,787	17,508	12,351	18,751
1,971	3,450	511	5,326	4,906
67,328	105,221	90,272	77,663	78,124
—	1,951	2,417	884	—
—	—	—	—	—
—	—	—	—	—
—	—	—	—	—
—	—	—	—	—
4,147	—	2,238	1,000	1,000
5,667	3,177	12,459	5,494	4,669
14,904	21,405	12,560	16,687	27,191
1,665	—	522	—	Delisted
9,116	4,819	10,891	8,199	6,590
32,850	34,732	17,003	35,809	49,811
32,961	49,150	37,665	30,115	52,875
1,864	3,202	644	1,686	3,623
65,391	52,398	83,338	35,385	87,050
4,520	4,809	7,390	—	Delisted
1,886	5,666	4,010	1,137	Delisted
3,257	8,048	5,879	2,686	2,908[b]
4,268	4,365	5,264	7,621	1,256[b]
16,655	14,553	10,387	11,666	13,217
18,680	13,889	17,887	13,958	3,643
9,337	6,622	10,280	7,283	2,176
11,092	13,501	31,921	16,086	22,040
—	—	—	475	—
2,756	13,048	7,863	6,284	—
3,216	959	4,111	—	—
14,868	30,201	22,454	11,689	27,014
4,253	7,016	2,702	5,866	2,460
1,104	6,566	1,355	7,289	4,449
264,457	300,077	311,240	277,299	311,972
559	1,777	1,981	—	Delisted
—	—	782	884	—
9,869	10,685	15,613	9,648	7,170[b]
1,853	—	—	2,102	—
12,281	12,462	18,376	12,634	7,170
1,516	1,395	730	2,427	—
1,516	1,395	730	2,427	—
1,638	496	12,172	1,752	—
31,211	19,043	31,161	20,464	38,909
—	154	—	—	—
32,849	19,693	43,333	22,216	38,907

TABLE 17 POUNDS OF PRIMARY COLORS USED IN FOODS, DRUGS, AND COSMETICS[a]

Category	FD&C Blue No. 1	FD&C Blue No. 2	FD&C Green No. 3	Orange B	FD&C Red No. 2	FD&C Red No. 3	FD&C Red No. 4	FD&C Violet No. 1	FD&C Yellow No. 5	FD&C Yellow No. 6	Total
Candy, confections	6,632	2,499	124	0	67,637	11,665	0	1,459	59,903	52,770	202,689
Beverages	15,800	2,375	301	0	282,695	1,056	0	985	78,933	181,292	563,437
Dessert powders	3,270	1,659	14	0	62,363	8,616	0	0	59,961	51,622	187,505
Cereals	843	99	0	0	15,558	1,421	0	0	52,496	35,464	105,881
Maraschino cherries	597	0	98	0	8,104	3,469	11,308	0	5,644	4,830	34,050
Pet food	1,473	6,764	0	0	67,058	1,023	0	1,278	101,743	23,226	202,565
Bakery goods	3,680	673	7	0	43,522	9,560	0	369	77,885	42,203	177,899
Ice cream, sherbet, dairy products	2,599	179	7	0	29,697	621	0	45	35,048	23,868	92,064
Sausage	647	0	0	16,890	36,084	4,970	0	0	6,502	99,605	164,698
Snack foods	305	0	0	0	3,623	766	0	2	18,456	11,409	34,561
Meat-stamping inks	11	0	0	0	12	10	0	2,223	15	0	2,271
Miscellaneous	5,345	1,990	1,298	0	46,219	18,200	398	1,134	44,841	29,134	148,559
Subtotal (food use)	41,202	16,238	1,849	16,890	662,572	61,377	11,706	7,495	541,427	555,423	1,916,179
Pharmaceuticals	3,250	593	220	0	21,179	12,168	1,186	347	17,275	15,938	72,156
Cosmetics	397	30	27	0	3,417	903	630	96	3,125	2,148	10,773
Totals	44,849	16,861	2,096	16,890	687,168	74,448	13,522	7,938	561,827	573,509	1,999,108

[a]Figures represent sales for the first 9 months of 1967, but do not include exports or sales to jobbers and other manufacturers.

TABLE 18   MAJOR CATEGORIES OF PROCESSED FOOD MANUFACTURED USING CERTIFIED COLORS AND LEVELS OF COLOR USED

Category	Level of Color Used	
	Range (ppm)	Average (ppm)
Candy and confections	10-400	100
Beverages (liquid and powdered)	5-200	75
Dessert powders	5-600	140
Cereals	200-500	350
Maraschino cherries	100-400	200
Pet foods	100-400	200
Bakery goods	10-500	50
Ice cream and sherbets	10-200	30
Sausage (surface)	40-250	125
Snack foods	25-500	200
Meat-stamping inks	—	—
Miscellaneous (nuts, salad dressing, gravy, spices, jams, jellies, food packaging, etc.)	5-400	—

they are the official FDA designations and thus are given here. The structures shown are, in general, taken from the Colour Index.

FD&C Blue No. 1

Synonyms: Brilliant Blue FCF; CI Food Blue 2 (42090).

Chemical Structure:

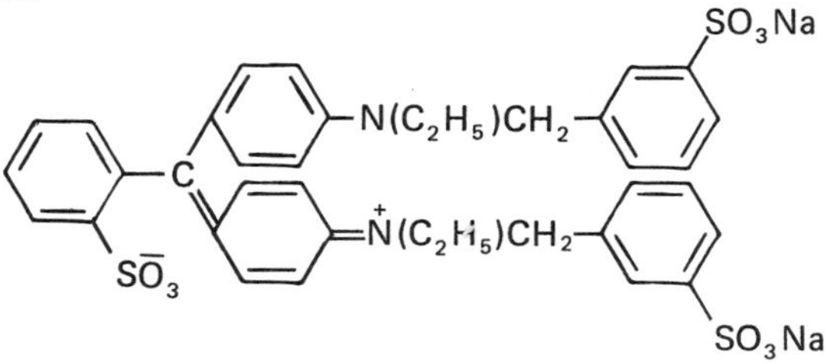

Identity: Principally the disodium salt of ethyl [4-[p-[ethyl(m-sulfobenzyl)-amino]-α-(o-sulfophenyl)benzylidene]-2,5-cyclohexadien-1-yldiene] (m-sulfo-benzyl) ammonium hydroxide inner salt with smaller amounts of the isomeric disodium salts of ethyl[4-[p-[ethyl(p-sulfobenzyl)amino]-α-(o-sulfophenyl)-benzylidene]-2,5-cyclohexandien-1-ylidene](p-sulfobenzyl) ammonium hydroxide inner salt and ethyl[4-[p-[ethyl(o-solfobenzyl)amino]-α-(o-sulfo-phenyl)benzylidene]-2,5-cyclohexadien-1-ylidene](o-sulfobenzyl) ammonium hydroxide inner salt.

Empirical Formula: $C_{37}H_{34}N_2O_9S_3Na_2$.

Molecular Weight: 792.84.

Dye Classification: Triphenylmethane.

Manufacturing Process: Condense benzaldehyde-o-sulfonic acid with α-(N-ethylanilino)m-toluenesulfonic acid ("benzylethylaniline sulfonic acid").

FD&C Blue No. 2

Synonyms: Indigotine, Indigo Carmine; CI Food Blue 1 (73015).

Chemical Structure:

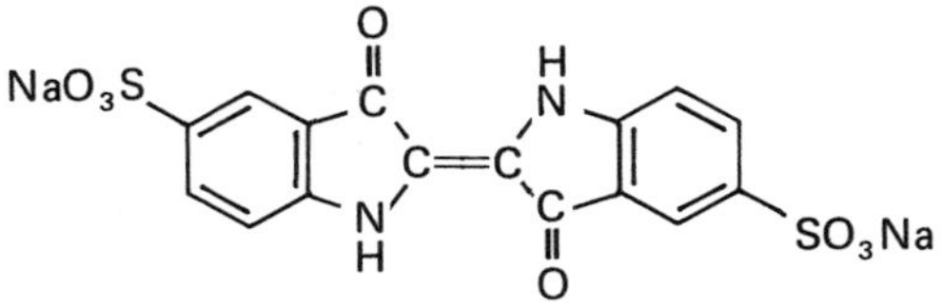

Identity: Principally the disodium salt of 5,5′-disulfo-3,3′-dioxo-$\Delta^{2,2'}$-biindoline with smaller amounts of the isomeric disodium salt of 5,7′-disulfo-3,3′-dioxo-$\Delta^{2,2'}$-biindoline.

Empirical Formula: $C_{16}H_8N_2O_8S_2Na_2$.

Molecular Weight: 466.35.

Dye Classification: Indigoid.

Manufacturing Process: Sulfonation of indigo.

FD&C Green No. 3

Synonyms: Fast Green FCF; CI Food Green 3 (42053).

Chemical Structure:

Identity: Disodium salt of 4-[[4-(N-ethyl-m-sulfobenzylamino)phenyl](4-hydroxy-2-sulfoniumphenyl)methylene]-[1-(N-ethyl-N-m-sulfobenzyl)-$\Delta^{2,5}$-cyclohexadienimine].

Empirical Formula: $C_{37}H_{34}O_{10}N_2S_3Na_2$.

Molecular Weight: 808.84.

Dye Classification: Triphenylmethane.

Manufacturing Process: Condense p-hydroxybenzaldehyde-o-sulfonic acid with α-(N-ethylanilino)-m-toluenesulfonic acid.

FD&C Red No. 3

Synonyms: Erythrosine, Erythrosine Bluish; CI Food Red 14 (45430).
Chemical Structure:

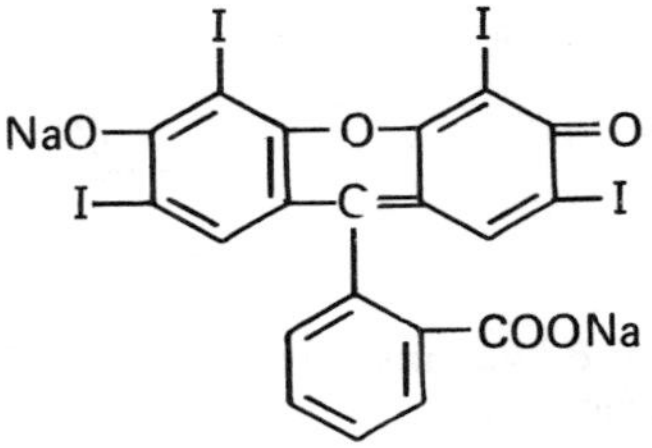

Identity: Principally the monohydrate of 9(o-carboxyphenyl)-6-hydroxy-2,4,5,7-tetraiodo-3H-xanthen-3-one, disodium salt, with smaller amounts of lower iodinated fluoresceins.
Empirical Formula: $C_{20}H_6O_5I_4Na_2$.
Molecular Weight: 879.86.
Dye Classification: Xanthene.
Manufacturing Process: Iodination of fluorescein (D&C Yellow No. 7).

FD&C Red No. 4

Synonyms: Ponceau SX; CI Food Red 1 (14700).
Chemical Structure:

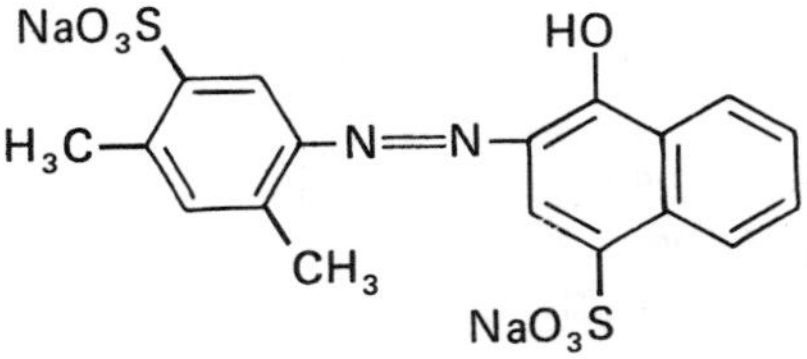

Identity: Principally the disodium salt of 3-[(2,4-dimethyl-5-sulfophenyl)-azo]-4-hydroxy-1-naphthalenesulfonic acid.
Empirical Formula: $C_{18}H_{14}N_2O_7S_2Na_2$.
Molecular Weight: 480.42.
Dye Classification: Monoazo.
Manufacturing Process: Couple diazotized 1-amino-2,4-dimethylbenzene-5-sulfonic acid with 1-naphthol-4-sulfonic acid.

FD&C Red No. 40

Synonyms: Allura® Red AC; CI Food Red 17 (16035).

Chemical Structure:

Identity: Disodium salt of 6-hydroxy-5-[(2-methoxy-5-methyl-4-sulfophenyl)-azo]-2-naphthalenesulfonic acid.

Empirical Formula: $C_{18}H_{14}N_2O_8S_2Na_2$.

Molecular Weight: 496.42.

Dye Classification: Monoazo.

Manufacturing Process: Couple diazotized 5-amino-4-methoxy-2-toluenesulfonic acid with 6-hydroxy-2-naphthalenesulfonic acid.

FD&C Yellow No. 5

Synonyms: Tartrazine; CI Food Yellow 4 (19140).

Chemical Structure:

Identity: Trisodium salt of 5-Oxo-1-(p-sulfophenyl)-4-[(p-sulfophenyl)azo]-2-pyrazoline-3-carboxylic acid.

Empirical Formula: $C_{16}H_9N_4O_9S_2Na_3$.

Molecular Weight: 534.36.

Dye Classification: Pyrazolone.

Manufacturing Processes: (a) Condense phenylhydrazine-p-sulfonic acid with oxalacetic ester, couple the product with diazotized sulfanilic acid, then hydrolyze the ester with sodium hydroxide or (b) condense phenylhydrazine-p-sulfonic acid with dihydroxytartaric acid.

FD&C Yellow No. 6

Synonyms: Sunset Yellow; CI Food Yellow 3 (15985).

Chemical Structure:

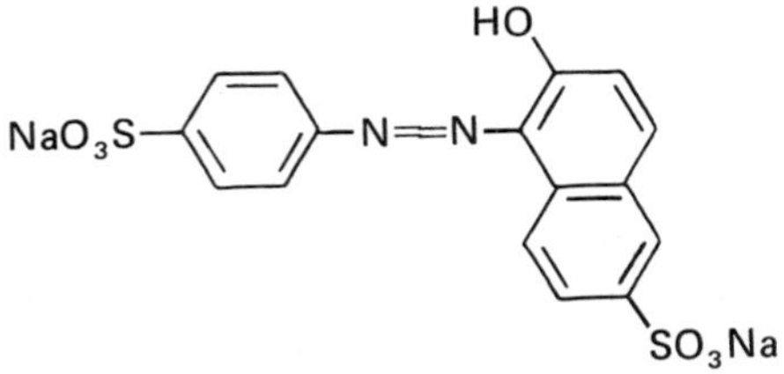

Identity: Disodium salt of 1-p-sulfophenylazo-2-naphthol-6-sulfonic acid.
Empirical Formula: $C_{16}H_{10}N_2O_7S_2Na_2$.
Molecular Weight: 452.36.
Dye Classification: Monoazo.
Manufacturing Process: Couple diazotized sulfanilic acid with 2-naphthol-6-sulfonic acid.

Citrus Red 2

Synonyms: CI Solvent Red 80 (12156).
Chemical Structure:

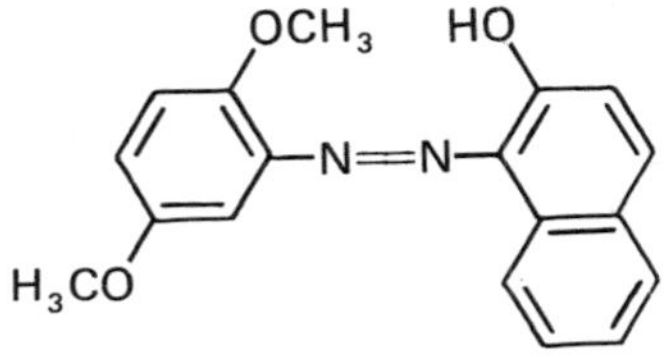

Identity: 1-(2,5-Dimethoxyphenylazo)-2-naphthol.
Empirical Formula: $C_{18}H_{16}N_2O_3$.
Molecular Weight: 308.34.
Dye Classification: Monoazo.
Manufacturing Process: Couple diazotized 2,5-dimethoxyaniline with 2-naphthol.

Orange B

Synonyms: CI Acid Orange 137 (19235).
Chemical Structure:

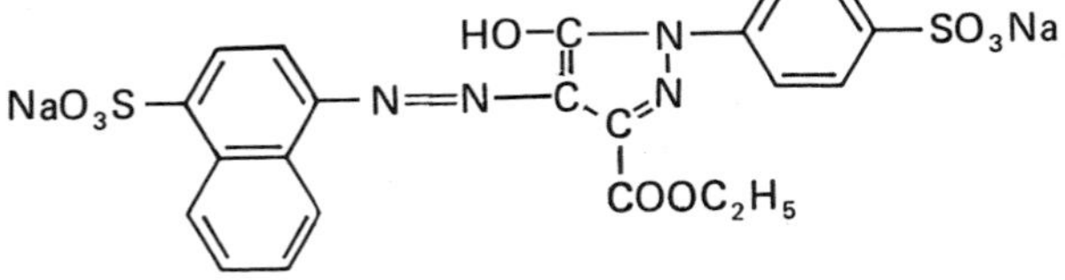

Identity: Disodium salt of 1-(4-sulfophenyl)-3-ethylcarboxy-4-(4-sulfonaphthylazo)-5-hydroxypyrazole.

Empirical Formula: $C_{22}H_{16}N_4O_9S_2Na_2$.

Molecular Weight: 590.49.

Dye Classification: Pyrazolone.

Manufacturing Process: React phenylhydrazine-*p*-sulfonic acid with the sodium derivative of diethyl hydroxymaleate; partially hydrolyze, to remove one ethyl group; then couple with diazotized naphthionic acid.

D&C Blue No. 4

Synonyms: Alphazurine FG, Erioglaucine; CI Acid Blue 9 (42090).

Chemical Structure:

Identity: NH_4 Salt corresponding to FD&C Blue No. 1.

Empirical Formula: $C_{37}H_{42}N_4O_9S_3$.

Molecular Weight: 782.94.

Dye Classification: Triphenylmethane.

Manufacturing Process: Same as FD&C Blue No. 1.

D&C Blue No. 6

Synonyms: Indigo; CI Vat Blue 1 (73000).

Chemical Structure:

Identity: Indigotin.

Empirical Formula: $C_{16}H_{10}N_2O_2$.

Molecular Weight: 262.27.

Dye Classification: Indigoid.

Manufacturing Processes: (a) Convert *N*-phenylglycine into pseudoindoxyl by fusion with sodium amide (or sodium and a current of ammonia) in presence of a mixture of potassium and sodium hydroxides and sodium cyanide. Oxidize the pseudoindoxyl with air; (b) convert phenylglycine-*o*-carboxylic acid[*N*-(carboxymethyl)-anthranilic acid] into indoxylic acid by fusion with alkalis and follow by air oxidation in alkaline solution.

D&C Blue No. 9

Synonyms: Indanthrene Blue, Carbanthrene Blue; CI Vat Blue 6 (69825).

Chemical Structure:

Identity: Principally 7,16-dichloro-6,15-dihydro-5,9,14,18-anthrazinetetrone.

Empirical Formula: $C_{28}H_{12}N_2O_4Cl_2$.

Molecular Weight: 511.32.

Dye Classification: Anthraquinone vat.

Manufacturing Process: Chlorinate indanthrene.

D&C Brown No. 1

Synonyms: Resorcin Brown; CI Acid Orange 24 (20170).

Identity: A mixture of the sodium salts of 4[[5-[(dialkylphenyl)azo]-2,4-dihydroxyphenyl]azo]-benzenesulfonic acid. The alkyl group is principally the methyl group.

Empirical Formula: $C_{20}H_{17}N_4O_5S_1Na_1$.

Molecular Weight: 448.43.

Dye Classification: Diazo.

Manufacturing Process: Couple diazotized sulfanilic acid and diazotized crude xylidine with resorcinol.

D&C Green No. 5

Synonyms: Alizarin Cyanine Green F; CI Acid Green 25 (61570).

Chemical Structure:

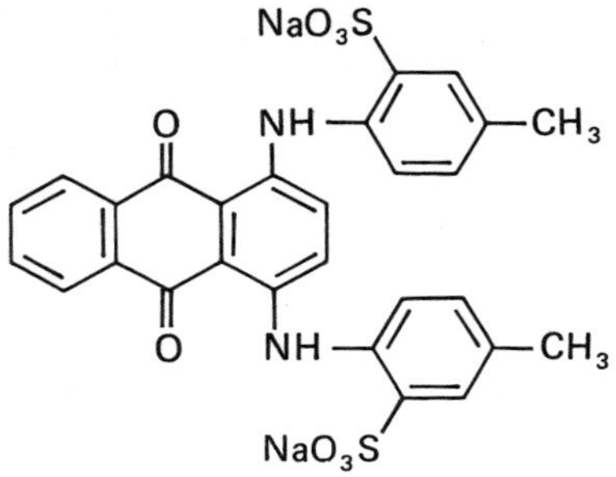

Identity: Principally the disodium salt of 2,2′-[(9,10-dihydro-9,10-dioxo-1,4-anthracenediyl)diimino]bis[5-methylbenzenesulfonic acid].

Empirical Formula: $C_{28}H_{20}N_2O_8S_2Na_2$.

Molecular Weight: 622.58.

Dye Classification: Anthraquinone.

Manufacturing Process: Condense leucoquinizarin with *p*-toluidine, then sulfonate.

D&C Green No. 6

Synonyms: Quinizarin Green SS; CI Solvent Green 3 (61565).

Chemical Structure:

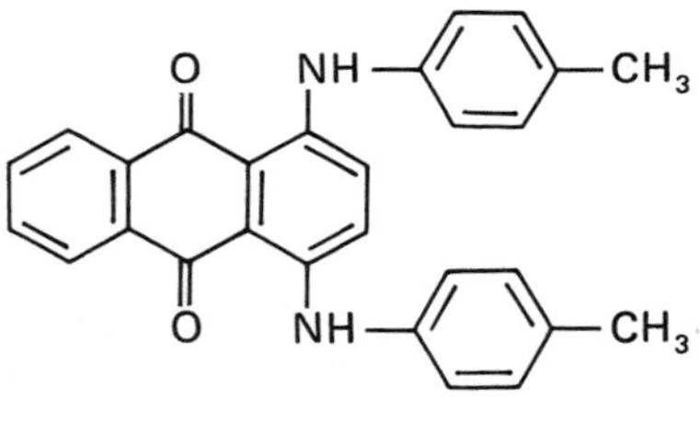

Identity: 1,4-di(*p*-toluidino)anthraquinone.

Empirical Formula: $C_{28}H_{22}N_2O_2$.

Molecular Weight: 418.50.

Dye Classification: Anthraquinone.

Manufacturing Process: Condense leucoquinizarin with *p*-toluidine.

D&C Green No. 8

Synonyms: Pyranine Concentrated; CI Solvent Green 7 (59040).

Chemical Structure:

Identity: Principally the trisodium salt of 8-hydroxy-1,3,6-pyrenetrisulfonic acid.

Empirical Formula: $C_{16}H_7O_{10}S_3Na_3$.

Molecular Weight: 524.37.

Dye Classification: Pyrene.

Manufacturing Process: Sulfonate pyrene to tetrasulfonic acid, salt out with sodium chloride, hydrolyze in sodium hydroxide solution, add formic acid, and salt out with sodium chloride.

D&C Orange No. 4

Synonyms: Orange II; CI Acid Orange 7 (15510).

Chemical Structure:

Identity: Monosodium salt of 1-p-sulfophenylazo-2-naphthol.

Empirical Formula: $C_{16}H_{11}N_2O_4S_1Na_1$.

Molecular Weight: 350.32.

Dye Classification: Monoazo.

Manufacturing Process: Couple diazotized sulfanilic acid with 2-naphthol.

D&C Orange No. 5

Synonyms: Dibromofluorescein; CI Solvent Red 72 (45370:1).

Chemical Structure:

Identity: 4,5-Dibromo-3,6-fluorandiol.
Empirical Formula: $C_{20}H_{10}O_5Br_2$.
Molecular Weight: 490.07.
Dye Classification: Fluoran.
Manufacturing Process: Brominate fluorescein (D&C Yellow No. 7).

D&C Orange No. 10

Synonyms: Diiodofluorescein; CI Solvent Red 73 (45425:1).
Chemical Structure:

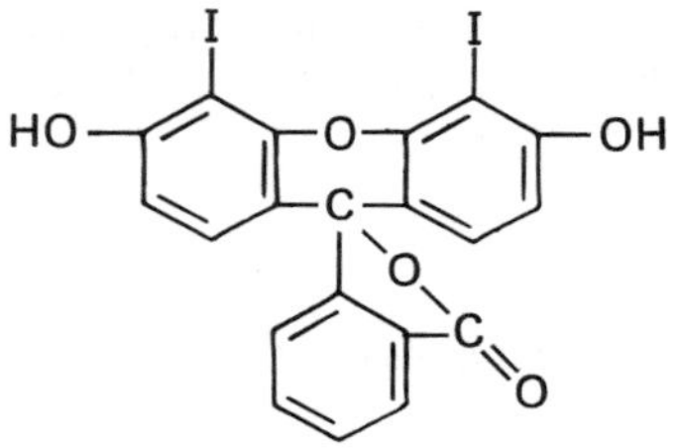

Identity: 4,5-Diiodo-3,6-fluorandiol.
Empirical Formula: $C_{20}H_{10}O_5I_2$.
Molecular Weight: 584.07.
Dye Classification: Fluoran.
Manufacturing Process: Iodinate fluorescein (D&C Yellow No.7).

D&C Orange No. 11

Synonyms: Erythrosine Yellowish Na; CI Acid Red 95 (45425).
Chemical Structure:

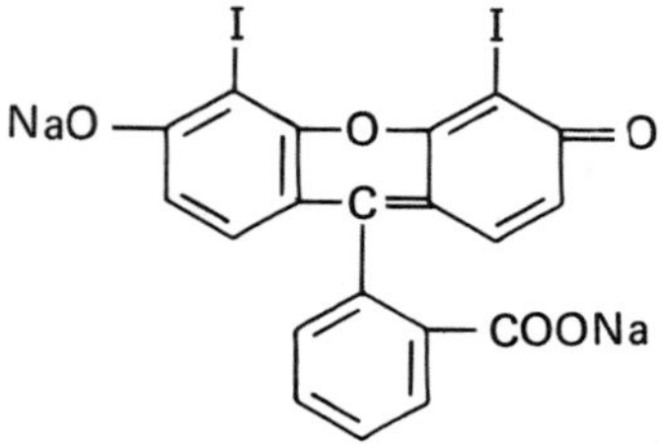

Identity: Disodium salt of 9-o-carboxyphenyl-6-hydroxy-4,5-diiodo-3-iso-xanthone.
Empirical Formula: $C_{20}H_8N_2O_5I_2$.
Molecular Weight: 628.07.

Dye Classification: Xanthene.
Manufacturing Process: Convert D&C Orange No. 10 to the Na salt.

D&C Orange No. 17

Synonyms: Permatone Orange; CI Pigment Orange 5 (12075).
Chemical Structure:

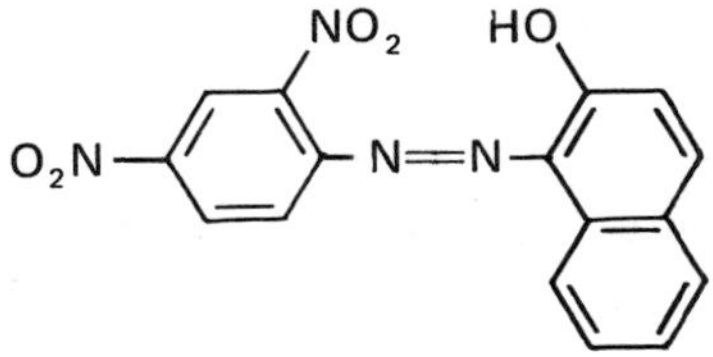

Identity: 1-(2,4-Dinitrophenylazo)-2-naphthol.
Empirical Formula: $C_{16}H_{10}N_4O_5$.
Molecular Weight: 338.28.
Dye Classification: Monoazo.
Manufacturing Process: Couple diazotized 2,4-dinitroaniline with 2-naphthol.

D&C Red No. 6

Synonyms: Lithol Rubin B; CI Pigment Red 57 (15850).
Chemical Structure:

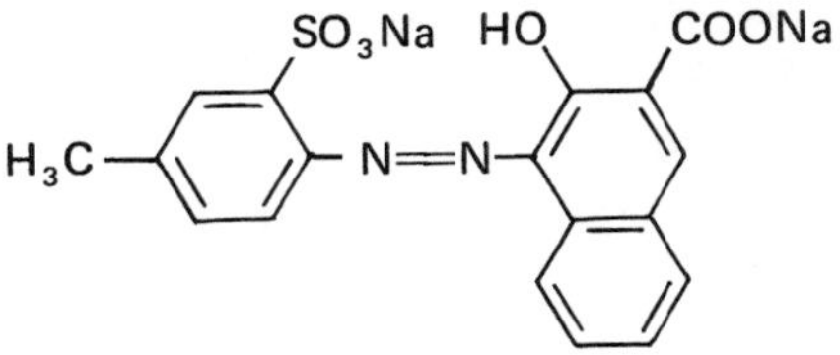

Identity: Disodium salt of 4-(o-sulfo-p-tolylazo)-3-hydroxy-2-naphthoic acid.
Empirical Formula: $C_{18}H_{12}N_2O_6S_1Na_2$.
Molecular Weight: 430.34.
Dye Classification: Monoazo.
Manufacturing Process: Couple diazotized 6-amino-m-toluenesulfonic acid with 3-hydroxy-2-naphthoic acid.

D&C Red No. 7

Synonyms: Lithol Rubin B Ca; CI Pigment Red 57:1 (15850:1).

Chemical Structure:

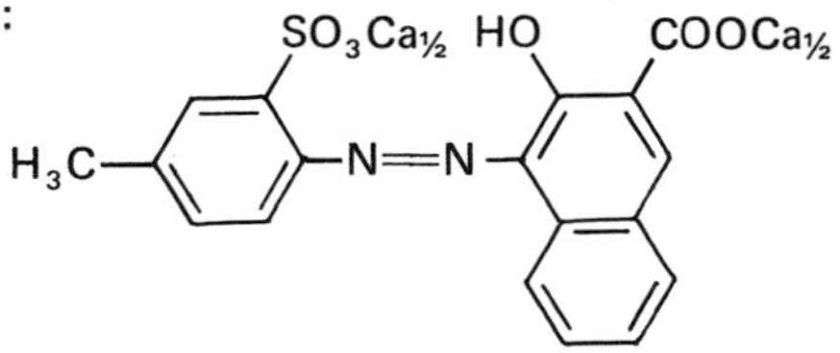

Identity: Calcium salt of 4-(o-sulfo-p-tolylazo)-3-hydroxy-2-naphthoic acid.
Empirical Formula: $C_{18}H_{12}N_2O_6S_1Ca_1$.
Molecular Weight: 424.44.
Dye Classification: Monoazo.
Manufacturing Process: Heat D&C Red No. 6 with $CaCl_2$.

D&C Red No. 8

Synonyms: Lake Red C; CI Pigment Red 53 (15585).
Chemical Structure:

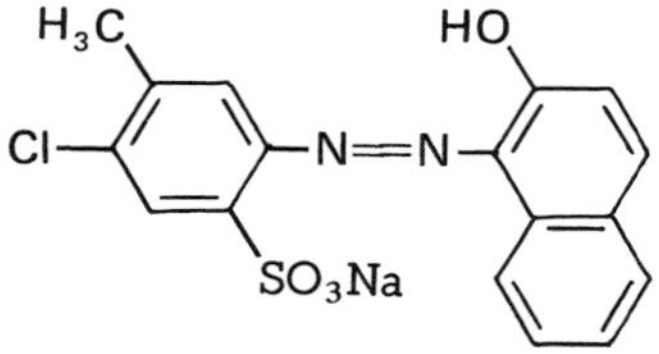

Identity: Monosodium salt of 1-(4-chloro-o-sulfo-5-tolylazo)-2-naphthol.
Empirical Formula: $C_{17}H_{12}N_2O_4S_1Na_1Cl_1$.
Molecular Weight: 398.79.
Dye Classification: Monoazo.

Manufacturing Process: Couple diazotized 2-amino-5-chloro-p-toluenesulfonic acid with 2-naphthol.

D&C Red No. 9

Synonyms: Lake Red C Ba; CI Pigment Red 53:1 (15585:1).
Chemical Structure:

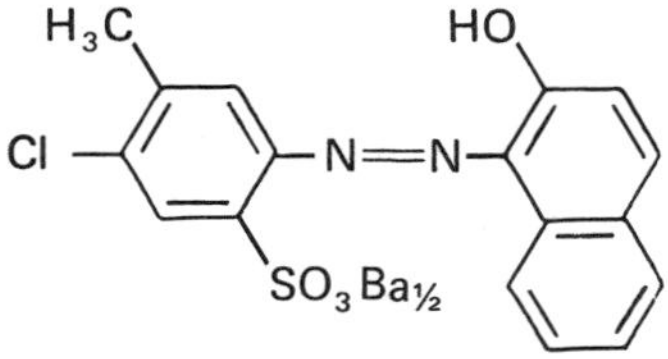

Identity: Monobarium salt of 1-(4-chloro-o-sulfo-5-tolylazo)-2-naphthol.

Empirical Formula: $C_{17}H_{12}N_2O_4S_1Ba_{1/2}Cl_1$.
Molecular Weight: 444.47.
Dye Classification: Monoazo.
Manufacturing Process: Boil D&C Red No. 8 with $BaCl_2$.

D&C Red No. 17

Synonyms: Toney Red, Sudan III; CI Solvent Red 23 (26100).
Chemical Structure:

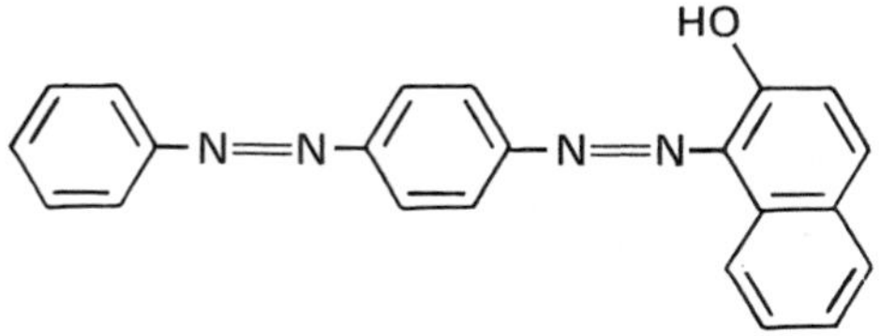

Identity: Principally 1-[[4-(phenylazo)phenyl]azo]-2-naphthalenol.
Empirical Formula: $C_{22}H_{16}N_4O_1$.
Molecular Weight: 352.40.
Dye Classification: Diazo.
Manufacturing Process: Couple diazotized aminoazobenzene with 2-naphthol.

D&C Red No. 19

Synonyms: Rhodamine B; CI Basic Violet 10 (45170).
Chemical Structure:

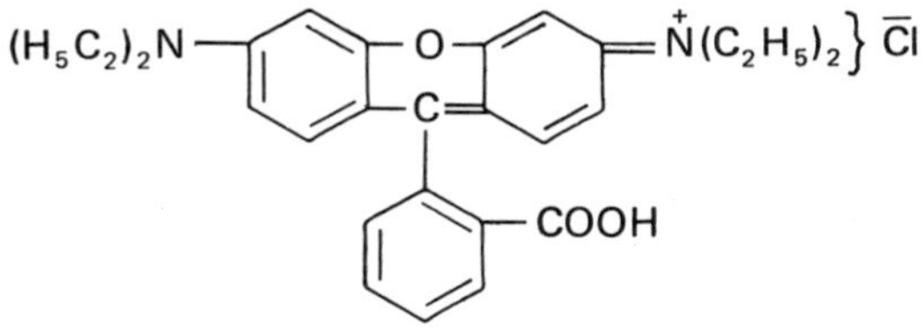

Identity: 3-Ethochloride of 9-o-carboxyphenyl-6-diethylamino-3-ethylimino-3-isoxanthene.
Empirical Formula: $C_{28}H_{31}N_2O_3Cl_1$.
Molecular Weight: 479.02.
Dye Classification: Xanthene.

Manufacturing Processes: Fuse m-diethylaminophenol with phthalic anhydride, then treat the base with dilute HCl; or treat fluorescein chloride under pressure with diethylamine.

D&C Red No. 21

Synonyms: Tetrabromofluorescein; CI Solvent Red 43 (45380:2).

Chemical Structure:

Identity: 2,4,5,7-Tetrabromo-3,6-fluorandiol.

Empirical Formula: $C_{20}H_8O_5Br_4$.

Molecular Weight: 647.90.

Dye Classification: Fluoran.

Manufacturing Process: Brominate fluorescein (D&C Yellow No. 7).

D&C Red No. 22

Synonyms: Eosin Y; CI Acid Red 87 (45380).

Chemical Structure:

Identity: Disodium salt of 2,4,5,7-tetrabromo-9-o-carboxyphenyl-6-hydroxy-3-isoxanthone.

Empirical Formula: $C_{20}H_6O_5Na_2Br_4$.

Molecular Weight: 691.86.

Dye Classification: Xanthene.

Manufacturing Process: Convert D&C Red No. 21 to the Na salt.

D&C Red No. 27

Synonyms: Tetrabromotetrachlorofluorescein; CI Solvent Red 48 (45410:1).

Chemical Structure:

Identity: 2,4,5,7-Tetrabromo-12,13,14,15-tetrachloro-3,6-fluorandiol.

Empirical Formula: $C_{20}H_4O_5Cl_4Br_4$.

Molecular Weight: 785.68.

Dye Classification: Fluoran.

Manufacturing Process: Condense resorcinol with tetrachlorophthalic anhydride, then brominate.

D&C Red No. 28

Synonyms: Phloxine B; CI Acid Red 92 (45410).

Chemical Structure:

Identity: Disodium salt of 2,4,5,7-tetrabromo-9-(3,4,5,6-tetrachloro-o-carboxyphenyl)-6-hydroxy-3-isoxanthone.

Empirical Formula: $C_{20}H_2O_5Na_2Cl_4Br_4$.

Molecular Weight: 829.64.

Dye Classification: Xanthene.

Manufacturing Process: Convert D&C Red No. 27 to the Na salt.

D&C Red No. 30

Synonyms: Helindone Pink CN; CI Vat Red 1 (73360).

Chemical Structure:

Identity: 5,5'-Dichloro-3,3'-dimethylthioindigo.

Empirical Formula: $C_{18}H_{10}O_2S_2Cl_2$.

Molecular Weight: 393.30.

Dye Classification: Indigoid.

Manufacturing Process: Oxidize 6-chloro-4-methyl-thioindoxyl, or chlorinate 4,4'-dimethylthioindigo.

D&C Red No. 31.

Synonyms: Brilliant Lake Red R; CI Pigment Red 64:1 (15800:1).

Chemical Structure:

Identity: Principally the calcium salt of 3-hydroxy-4-(phenylazo)-2-naphthalencarboxylic acid.

Empirical Formula: $C_{17}H_{11}N_2O_3Ca_{1/2}$.

Molecular Weight: 311.33.

Dye Classification: Monoazo.

Manufacturing Process: Couple diazotized aniline with 3-hydroxy-2-naphthoic acid and then convert to the Ca salt.

D&C Red No. 33

Synonyms: Acid Fuchsine; CI Acid Red 33 (17200).

Chemical Structure:

Identity: Disodium salt of 8-amino-2-phenylazo-1-naphthol-3,6-disulfonic acid.

Empirical Formula: $C_{16}H_{11}N_3O_7S_2Na_2$.

Molecular Weight: 467.38.

Dye Classification: Monoazo.

Manufacturing Process: Couple diazotized aniline with 8-amino-1-naphthol-3,6-disulfonic acid in alkaline solution.

D&C Red No. 34

Synonyms: Deep Maroon, Fanchon Maroon, Lake Bordeaux B; CI Pigment Red 63:1 (15880:1).

Chemical Structure:

Identity: Principally the calcium salt of 3-hydroxy-4-(1-sulfo-2-naphthylazo)-2-naphthalenecarboxylic acid.

Empirical Formula: $C_{21}H_{12}N_2O_6S_1Ca_1$.

Molecular Weight: 460.47.

Dye Classification: Monazo.

Manufacturing Process: Couple diazotized 2-naphthylamine-1-sulfonic acid with 3-hydroxy-2-naphthoic acid and then convert to the Ca salt.

D&C Red No. 36

Synonyms: Flaming Red; CI Pigment Red 4 (12085).

Chemical Structure:

Identity: 1-(o-Chloro-p-nitrophenylazo)-2-naphthol.

Empirical Formula: $C_{16}H_{10}N_3O_3Cl_1$.

Molecular Weight: 327.73.

Dye Classification: Monoazo.

Manufacturing Process: Couple diazotized 2-chloro-4-nitroaniline with 2-naphthol.

D&C Red No. 37

Synonyms: Rhodamine B—Stearate.

Identity: 3-Ethosterate of 9-o-carboxyphenyl-6-diethylamino-3-ethylimino-3-isoxanthene.

Empirical Formula: $C_{46}H_{66}N_2O_5$.

Molecular Weight: 727.04.

Dye Classification: Xanthene.

Manufacturing Process: Same as for D&C Red No. 19, except treat the base with stearic acid.

D&C Red No. 39

Synonyms: Alba Red; CI Pigment Red 100 (13058).

Chemical Structure:

Identity: o-[p-(β,β'-Dihydroxy-diethylamino)phenylazo]-benzoic acid.

Empirical Formula: $C_{17}H_{19}N_3O_4$.

Molecular Weight: 329.36.

Dye Classification: Monoazo.

Manufacturing Process: Couple diazotized anthranilic acid with 2,2'-(phenylimino)diethanol.

D&C Violet No. 2

Synonyms: Alizurol Purple SS; CI Solvent Violet 13 (60725).

Chemical Structure:

Identity: Principally 1-hydroxy-4-[(4-methylphenyl)amino]-9,10-anthracenedione.

Empirical Formula: $C_{21}H_{15}N_1O_3$.

Molecular Weight: 329.35.

Dye Classification: Anthraquinone.

Manufacturing Process: Condense quinizarin with p-toluidine, or condense 1-hydroxy-4-halogenoanthraquinone with p-toluidine.

D&C Yellow No. 7

Synonyms: Fluorescein; CI Solvent Yellow 94 (45350:1).

Chemical Structure:

Identity: Principally fluorescein.

Empirical Formula: $C_{20}H_{12}O_5$.

Molecular Weight: 332.31.

Dye Classification: Fluoran.

Manufacturing Process: Condense resorcinol with phthalic anhydride in the presence of $ZnCl_2$ or H_2SO_4.

D&C Yellow No. 8

Synonyms: Uranine; CI Acid Yellow 73 (45350).

Chemical Structure:

Identity: Principally the disodium salt of fluorescein.

Empirical Formula: $C_{20}H_{10}O_5Na_2$.

Molecular Weight: 376.27.

Dye Classification: Xanthene.

Manufacturing Process: Convert D&C Yellow No. 7 to the Na salt.

D&C Yellow No. 10

Synonyms: Quinoline Yellow; CI Acid Yellow 3 (47005).

Identity: Disodium salt of the disulfonic acid of 2-(2-quinolyl)-1,3-indan-dione.

Empirical Formula: $C_{18}H_9N_1O_8S_2Na_2$.

Molecular Weight: 477.37.

Dye Classification: Quinoline.

Manufacturing Process: Sulfonate D&C Yellow No. 11.

D&C Yellow No. 11

Synonyms: Quinoline Yellow SS, Quinoline Yellow Spirit Soluble; CI Solvent Yellow 33 (47000).

Chemical Structure:

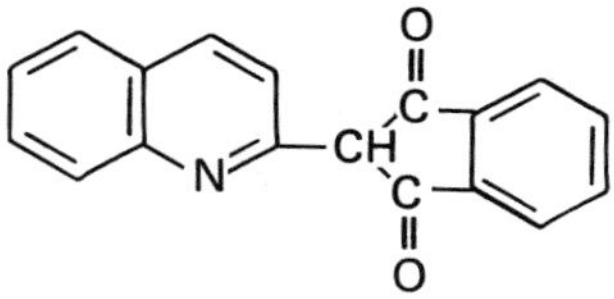

Identity: 2-(2-Quinolyl)-1,3-indandione.

Empirical Formula: $C_{18}H_{11}N_1O_2$.

Molecular Weight: 273.29.

Dye Classification: Quinoline.

Manufacturing Process: Condense quinaldine with phthalic anhydride in the presence of $ZnCl_2$.

Ext. D&C Violet No. 2

Synonyms: Alizarine Violet; CI Acid Violet 43 (60730).

Chemical Structure:

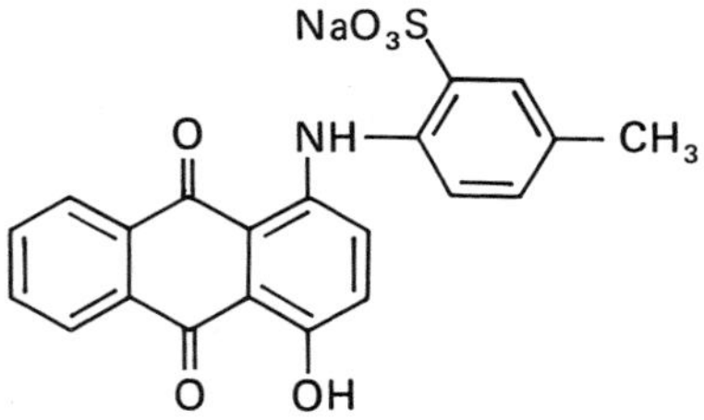

Identity: Principally the monosodium salt of 2-[(9,10-dihydro-4-hydroxy-9,10-dioxo-1-anthracenyl)amino]-5-methylbenzesulfonic acid.

Empirical Formula: $C_{21}H_{14}N_1O_6S_1Na_1$.

Molecular Weight: 431.39.

Dye Classification: Anthraquinone.

Manufacturing Process: Sulfonate D&C Violet No. 2 and then convert to the sodium salt.

Ext. D&C Yellow No. 7

Synonyms: Naphthol Yellow S; CI Acid Yellow 1 (10316).

Chemical Structure:

$$NaO_3S - \text{[naphthalene ring with ONa at top, NO}_2\text{ at right, NO}_2\text{ at bottom]}$$

Identity: Principally the disodium salt of 8-hydroxy-5,7-dinitro-2-naphthalenesulfonic acid.

Empirical Formula: $C_{10}H_4N_2O_8S_1Na_2$.

Molecular Weight: 358.19.

Dye Classification: Nitro.

Manufacturing Process: Nitrate the di- or trisulfonic acids of 1-naphthol or the nitroso compound of the 2,7-disulfonic acid.

[Phthalocyaninato(2−)] Copper

Synonyms: Copper Phthalocyanine; CI Pigment Blue 15 (74160).

Chemical Structure:

Identity: [Phthalocyaninato(2−)] copper.

Empirical Formula: $C_{32}H_{16}Cu_1N_8$.

Molecular Weight: 576.08.

Dye Classification: Phthalocyanine.

Manufacturing Processes: (a) Heat phthalonitrile with cuprous chloride at 180-200°C. (b) Heat phthalic anhydride, phthalimide, or phthalamide with a copper salt and urea, cyanoguanidine or p-toluenesulfonamide and cuprous (or cupric) chloride in the presence of ammonium molybdate or arsenic oxide (phthalic anhydride/urea process).

Chapter 5 Colorants Exempt from Certification

Colorants exempt from certification are those that, as provided for by law, have been exempted by the Commissioner of Food and Drug from the batch-certification procedures imposed on synthetic aromatic organic colorants because, after careful consideration of their composition, method of manufacture, toxic potential, impurities, and so on, he has concluded that they pose no threat to the public health. Although by definition these colors do not require certification prior to their sale they are, nevertheless, subject to surveillance by FDA to ensure that they meet current government specifications and are used in accordance with the law.

With the passage of the 1960 amendments all exempt colorants then in use were provisionally listed pending completion of the studies needed to obtain their "permanent" listing. Since this time many of them as well as several completely new colors have achieved this status. Exempt color additives now in use and their status are shown in Tables 1-3. Specifications have not as yet been prepared for all these colors but those which have can be found in Appendix A.

As can be seen, these lists are comprised of a wide variety of organic and inorganic compounds representing the animal, vegetable, and mineral kingdoms. Some, like β-carotene and zinc oxide, are essentially pure, factory-produced chemicals of definite and known composition. Others, including caramel, annatto extract, and carmine, are mixtures obtained from natural sources and have somewhat indefinite composition. Many of the materials included on the lists are relatively unimportant as colorants per se and are only classified as such because of the loose definition of a color additive included in the 1960 amendments. Only the more important of the colorants are considered in detail here.

ANNATTO EXTRACT

The annatto tree (*Bixa orellana*) is a large, fast-growing shrub cultivated in tropical climates, including parts of South America, India, East Africa, and the Caribbean. The tree produces large clusters of brown or crimson capsular fruit containing seeds coated with a thin, highly colored resinous coating or mark that serves as the raw material for the preparation of the colorant known as annatto extract.

The colorant is prepared by leaching the annatto seeds with an extractant prepared from one or more approved, food-grade materials taken from a list that includes various solvents, edible vegetable oils and fats, and alkaline aqueous and alcoholic solutions. Depending on the use intended, the alkaline extracts are often treated with food-grade acids to precipitate the annatto pigments, which in turn may or may not be further purified by recrystallization from an approved solvent. Annatto extract is one of the oldest known dyes, used since antiquity for the coloring of food, textiles, and cosmetics. It has been used in the United States and Europe for over 100 years as a color additive for butter and cheese.

The chief coloring principle found in the oil or fat extracts of annatto seeds is the carotenoid bixin (CI Natural Orange 4, CI No. 75120):

BIXIN: $C_{25}H_{30}O_4$ (mw 394.52)

The major colorant in alkaline aqueous extracts is norbixin:

NORBIXIN: $C_{24}H_{28}O_4$ (mw 380.48)

Annatto extract is sold in several physical forms, including dry powders, oil solutions and suspensions, and alkaline aqueous solutions, all containing 1-15% active colorant calculated as bixin. It is used in products at levels of 0.5-10 ppm as pure color, resulting in hues ranging from butter-yellow to peach, depending on the type of color preparation employed and the product colored. Annatto extract's chief use is in foods such as butter, margarine, cooking oils, salad dressings, cereals, ice cream, ice cream cones, and spices.

The chemistry and performance of annatto extract is essentially that of bixin, a brownish-red crystalline material that melts at 198°C. It is moderately stable toward light and has good stability toward oxidation, change in pH, and microbiological attacks. Bixin is very stable toward heat up to 100°C, fairly stable at 100-125°C and unstable above 125°C, where it tends to form 13-carbomethoxy-4,8-dimethyltridecahexane-oic acid.

β-CAROTENE

β-Carotene is an isomer of the naturally occurring carotenoid, carotene (CI Natural Yellow 26, CI No. 75130). It is the pigment largely responsible for the color of various products obtained from nature, including butter, cheese, carrots, alfalfa, and certain cereal grains. The colorant is synthetically produced from acetone, using the process developed in the 1950s by Hoffman-LaRoche, which results in the formation of the optically inactive all-*trans* form. It is this synthesis that made β-carotene so important in the history of the use of color additives since it was one of the first "natural" colorants synthetically produced on a commercial scale and the one that eventually raised the question as to whether factory-produced analogues of natural colorants should require certification by FDA such as "coal-tar dyes" do, and whether such compounds could continue to be referred to as "natural colors." This controversey eventually led to the creation of the category of colorants called "colorants exempt from certification."

β-CAROTENE: $C_{40}H_{56}$ (mw 536.89)

β-Carotene forms reddish-violet platelets that melt at 183°C. It is insoluble in water, ethanol, glycerine, and propylene glycol, and only slightly soluble in boiling organic solvents such as ether (0.05%), benzene (0.2%), carbon disulfide (1%), and methylene chloride (0.5%). Its solubility in edible oils is about 0.08% at room temperature, 0.2% at 60°C, and 0.8% at 100°C. β-Carotene is sensitive to alkali and very sensitive to air and light, particularly at high temperatures. Pure, crystalline β-carotene remains unchanged for long periods of time when stored under CO_2 below 20°C but is almost completely destroyed after only 6 weeks when stored in air at 45°C. Vegetable fat and oil solutions and suspensions are quite stable under normal handling conditions. β-Carotene is a rarity among color additives in that it is one of the few with nutritional value since it is converted biologically by humans into vitamin A; 1 g of β-carotene = 1,666,666 USP units of vitamin A.

β-Carotene is marketed as dry crystals packed under nitrogen, as liquid and semisolid suspensions in edible oils including vegetable, peanut, and butter oils, as water-dispersible beadlets composed of colorant plus vegetable oil, sugar, gelatin, and carbohydrate, and as emulsions.

The colorant is used at levels ranging from 2-50 ppm as pure color to shade margarine, shortening, butter, cheese, baked goods, confections, juices, and beverages. Its chief advantages over other colorants are its nutritional value and its ability to duplicate natural shades.

β-APO-8'-CAROTENAL

This colorant (CI Food Orange 6, CI No. 40820) is an aldehydic carotenoid widely distributed in nature; it is isolated from numerous items, including spinach, oranges, grass, tangerines, and marigolds. It is synthetically produced as the crystalline all-*trans* stereoisomer, which is a purplish-black powder that melts in the range 136-140°C (corrected).

β-APO-8'-CAROTENAL: $C_{30}H_{40}O$ (mw 416.65)

β-Apo-8'-carotenal has provitamin activity with one gram of the colorant equal to 1,200,000 IU of vitamin A. Like all crystalline carotenoids, it slowly decomposes in air through oxidation of its conjugated double bonds and thus must be stored in sealed containers under an atmosphere of inert gas, preferably under refrigeration. Also like other carotenoids, β-apo-8'-carotenal readily isomerizes to a mixture of its *cis* and *trans* stereoisomers when its solutions are heated to about 60°C or exposed to ultraviolet light.

In general, its solubility characteristics are similar to those of β-carotene except that it is slightly more soluble in the usual solvents. In addition, because of its aldehydic group, β-apo-8'-carotenal is slightly soluble in polar solvents such as ethanol. Its solubility in various solvents is:

Solvent	Solubility at 24°C in Weight Percent
Vegetable oils	0.7-1.5
Orange oil	1.5-2.0
Ethanol	~0.2
Propylene glycol	Trace
Cyclohexane	~0.7
Chloroform	>1.0

Vegetable-oil solutions of the colorant are orange to red, depending on their concentration. Aqueous dispersions range in hue from orange to orange-red.

β-Apo-8′-carotenal is sold as a dry powder, as 1-1.5% vegetable oil solutions, as 20% suspensions in vegetable oil, and as 10% dry beadlets. The vegetable-oil suspensions are purplish-black fluids at room temperatures that set to thick pastes when refrigerated. The dry beadlets are colloidal dispersions of colorant in a matrix of gelatin, vegetable oil, sugar, starch, and antioxidants.

β-Apo-8′-carotenal is used wherever an orange to reddish-orange shade is desired. The dry beadlets are water-dispersible and can be used to color aqueous based foods and beverages such as juices, fruit drinks, soups, jams, jellies, and gelatins. The vegetable-oil solutions and suspensions are most useful in fat base or fat containing foods including process cheese, margarine, fats, and oils. Use levels typically range within 1-20 ppm as pure color.

CANTHAXANTHIN

The newest of the synthetically produced carotenoid color additives, canthaxanthin (β-carotene-4,4′-dione), became commercially available about 1969. Its CI designation is Food Orange 8, CI No. 40850.

Unknown until 1950 when F. Haxo isolated it from an edible mushroom (*Cantharellus cinnabarinus*), canthaxanthin has since been identified in sea trout, algae, daphnia, salmon, brine shrimp, and several species of flamingo. Crystalline canthaxanthin is prepared synthetically from acetone or β-ionone using procedures similar to those used for β-carotene and β-apo-8′-carotenal.

CANTHAXANTHIN: $C_{40}H_{52}O_2$ (mw 564.85)

Canthaxanthin crystallizes from various solvents as brownish-violet, shiny leaves that melt with decomposition at 210°C. As is the case with carotenoids in general, the crystals are sensitive to light and oxygen and, when heated in solution or exposed to ultraviolet light or iodine, form a mixture of *cis* and *trans* stereoisomers. Consequently, crystalline canthaxanthin should be stored under inert gas at low temperatures. Unlike its cousin carotenoid colorants β-carotene and β-apo-8′-carotenal, canthaxanthin has no vitamin A activity. It is chemically stable at pH 2-8 (the range normally encountered in foods) and unaffected by heat in systems with a minimal oxygen content.

The solubility of canthaxanthin in most solvents is low compared to β-carotene and β-apo-8'-carotenal. Some representative values follow:

Solvent	Solubility at 25°C in Weight Percent
Vegetable oils	0.02
Orange oil	2.0
Ethanol	Insoluble
Acetone	0.03
Propylene glycol	Trace
Benzene	0.2
Chloroform	10

Oil solutions of canthaxanthin are red at all concentrations. Aqueous dispersions are orange or red depending on the type of emulsion prepared.

Besides as a dry powder, canthaxanthin is commercially available as a water-dispersible, dry beadlet composed of 10% colorant, gelatin, vegetable oil, sugar, and starch. Canthaxanthin is used at levels of 5-60 ppm as pure color to produce a tomato red. The colorant is useful in coloring tomato products such as tomato soup, spaghetti sauce, and pizza sauce, Russian and French dressings, fruit drinks, sausage products, and baked goods. It has also been found to be an effective broiler pigmenter when added to poultry feeds as a supplement to the natural carotenoids present.

CARAMEL

Officially, "The color additive caramel is the dark-brown liquid or solid material resulting from the carefully controlled heat treatment of the following food-grade carbohydrates: dextrose, invert sugar, lactose, malt syrup, molasses, starch hydrolysates and fractions thereof, or sucrose." Practically speaking, caramel is burned sugar.

Caramel is most often made from liquid corn syrup with a reducing sugar content of 85% or more, expressed as dextrose. Sucrose (canesugar) is rarely used as the starting material because of its relatively high cost and process difficulties frequently encountered since, after inversion, the dextrose and levulose present react at different rates, making the burning process difficult to control and resulting in a product inferior to that made from corn sugar. In most cases a small amount of an approved acid, alkali, or salt is used to expedite the reaction and to obtain products with specific properties for specific applications.

To prepare the colorant the liquid corn sugar, and the appropriate reactants are cooked at about 250°F for several hours or until the proper tinctorial power has been obtained. The product is then filtered and stored cool to minimize further caramelization. Often it is dehydrated to produce powdered colorant.

Caramel coloring is freely soluble in water and insoluble in most organic solvents. Its solubility in solutions containing 50-70% alcohol varies with the type of caramel. In concentrated form the colorant has a distinctive burned taste that is unnoticeable at the typical levels of use. The specific gravity of caramel coloring syrups ranges from 1.25 to 1.38, whereas the total solids content varies from 50% to 75%. The pH of the acid-proof caramels used for carbonated beverages and acidified solutions is normally 2.8-3.0. Most bakers' caramels, which are a less refined grade of colorant used for cookies, cakes, bread, and so on, have a slightly higher pH due to differences in their manufacturing processes.

In aqueous solution caramel coloring exhibits colloidal properties, with the particles carrying small positive or negative electrical charges, depending on the method used in its manufacture and the pH of the product being colored. The nature of this charge is most important in using caramel since it must be the same as that of the product it is added to, or else mutual attraction will occur causing flocculation or precipitation. A good beverage caramel should carry a strong negative charge and have an isoelectric point at pH = 1.5 or less.

A major use for caramel is in soft drinks, particularly root beers and colas. It is also used extensively to standardize the hue of blended whiskeys and beer. Other uses include the coloring of baked goods, syrups, preserves, candies, pet foods, gravies, canned meat products, cough syrups, and pharmaceuticals. Where the use of liquid coloring is impractical, such as in cake mixes and other dry products, powdered caramel is added. Typical use levels are high (1000-5000 ppm), but the colorant is relatively inexpensive and shows good stability in most products.

CARMINE; COCHINEAL EXTRACT

Among the more interesting of the color additives in use today are cochineal extract and its related colorant, carmine. They are interesting not only because of their characteristics, but also because of the part their source,

CARMINIC ACID: $C_{22}H_{20}O_{13}$ (mw 492.39)

cochineal, played in the political and economic history of the New World and those who settled in it.

Cochineal extract (CI No. 75470) is the concentrated solution obtained after removing the alcohol from an aqueous-alcoholic extract of cochineal, which is the dried bodies of the femal insect *Coccus cacti*, a variety of shield louse. The coloring principle of the extract is believed to be carminic acid, an anthraquinone comprising approximately 10% of cochineal and 2% of its extract.

Carmine is the aluminum or calcium-aluminum lake on an aluminum hydroxide substrate of the coloring principle (again, chiefly carminic acid) obtained by the aqueous extraction of cochineal. Carmine is normally 50% or more carminic acid.

Cochineal itself is an insect that lives on a species of cactus, *Nopalea coccinelliferna*, and was once known only in Mexico. The Azetecs cultivated it for its color value and often exacted it as tribute. It is believed that Cortez discovered native Mexicans using cochineal when he arrived there in 1518 and at first believed it to be kermes, one of the most ancient dyestuffs that was also widely known in Europe at the time. The eventual discovery that cochineal was in fact a new colorant, and one 10 times stronger than kermes, gave the Spaniards an exclusive on what was to become an important and lucrative article of commerce that they jealously controlled until Mexico finally freed itself from Spain. By the end of the 16th century as much as 500,000 pounds of cochineal were being shipped from Mexico to Spain each year—a rather astounding figure when one realizes that it requires about 70,000 hand-gathered insects to make a single pound of cochineal. Numerous attempts were made to grow cochineal in other areas of the world, but most failed due partly to the specialized climates needed for its cultivation and partly to the Spaniards doggedness in guarding what they considered a good thing. In spite of these obstacles, cochineal was eventually cultivated elsewhere, including the Canary Islands, Spain, the East and West Indies, Palestine, and parts of Central and South America. The cochineal trade peaked about 1870 then declined rapidly due to the introduction of synthetic colors in 1856.

Cochineal extract is typically acid (pH = 5-5.3) and has a total solids content of about 6%. It frequently contains sodium benzoate as a preservative. Cochineal extract varies in shade from orange to red, depending on pH. It is insoluble in typical solvents including water, glycerine, and propylene glycol but can be dispersed in water. It exhibits good stability toward light and oxidation but poor stability toward pH and microbiological attack. Its tinctorial strength is only moderate. Use levels are within the range 25-1000 ppm.

Carmine is a pigment and thus exhibits little solubility in most solvents. Since it is also an aluminum lake, it can be solubilized by strong acids and bases that cause degradation of the substratum and release of the color. Carmine is useful for producing pink shades in retorted protein products, candy, confections, rouge, eye shadow, and pill coatings.

CHROMIUM OXIDE GREENS

These are essentially pure Cr_2O_3, in either anhydrous or hydrate form. Chrome oxide greens are usually prepared by one of two methods:

Method 1: Fuse potassium bichromate and boric acid, drown the product in water, then dry it at high temperature. Precipitate chrome alum with sodium hydroxide, then roast the chromous hydroxide; extract, wash, and dry at a high temperature.

This colorant is essentially Cr_2O_3 and is a yellowish green pigment of good strength and opacity and excellent stability.

Method 2: Paste potassium bicromate with three times its weight of boric acid, then roast the mixture at $500°C$ in a muffle in an oxidizing atmosphere. Hydrolyze the melt by quenching it in hot water, treat it with superheated steam, grind it, and wash it with hot water.

This product (Veridian, Guignet's Green) is essentially $Cr_2O_3 \cdot 2H_2O$ and is a more bluish and brilliant green than chrome oxide green. It is very transparent, and has good strength and excellent stability.

Both colorants are used in eye makeup and soap.

GUANINE (PEARL ESSENCE)

This colorant is one of a variety of natural pearl-essence colorants available and is essentially crystalline guanine (2-aminohypoxanthine) obtained from the scales of various fish, including menhaden, herring, and alewives. To prepare the colorant, scales are scraped from the fish, levigated and washed with water, then made into one or more commercial forms, depending on the intended end use. Typically, guanine is supplied as a paste or suspension in water, castor oil, or nitrocellulose. Guanine is not a colorant in the strict sense but instead is used to produce iridescence in a product.

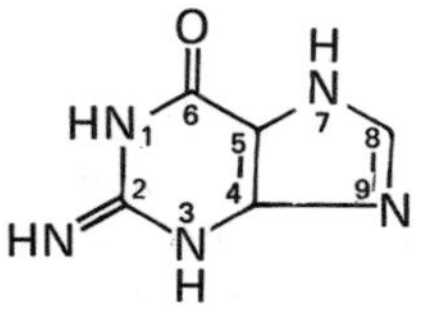

GUANINE: $C_5 H_5 N_5 O$ (mw 151.13)

The hue of natural pearl essence varies greatly with the amount and type of pigment found in the fish scales. Carotenoids produce reds and yellows, melanin results in blacks, and combinations of guanine and melanin produce greens and blues. Only when guanine is found alone is the product silvery or pearly white.

Much of the pearl essence used today as a color additive is synthetic and produced from bismuth oxychloride. Synthetic pearl essence has the advantages that its properties are more easily controlled during its manufacture and that its production is not dependent on the supply of fish and their condition, environment, and so on as the natural product is. Synthetic pearl essence has good heat stability up to 230-260°C, but only fair light and sulfide stability. Both products are used in lipsticks, nail polishes, and eye makeup.

IRON OXIDES

This colorant is one or a combination of various synthetically prepared iron oxides, including the hydrated forms. The naturally occurring oxides are unacceptable as a color additive because of the difficulties frequently encountered in purifying them.

Iron oxide is recognized under various names, including CI Pigment Black 11 and CI Pigment Browns 6 and 7 (CI No. 77499), CI Pigment Yellows 42 and 43 (CI No. 77492), and CI Pigment Reds 101 and 102 (CI No. 77491). The chemical composition and hence the empirical formula of the colorant varies greatly with the method of manufacture used but can generally be represented as $FeO \cdot xH_2O$, $Fe_2O_3 \cdot xH_2O$ or some combination thereof. Most are made from copperas (ferrous sulfate, $FeSO_4 \cdot 7H_2O$). The commonly used forms are the yellow hydrated oxide (ochre) and the brown, red, and black oxides.

The yellow oxides are prepared by precipitating hydrated ferric oxide from a ferrous salt using an alkali, followed by oxidation. The shades obtained range from light lemon yellow to orange, depending on the conditions used for the precipitation and oxidation. Yellow oxides contain about 85% Fe_2O_3 and 15% water of hydration.

Brown oxides are manufactured either by blending mixtures of the red, yellow, and black oxides or by precipitation of an iron salt with alkali followed by partial oxidation of the precipitate. The result is a mixture of red Fe_2O_3 and black Fe_3O_4 ($FeO \cdot Fe_2O_3$).

Red iron oxides are usually prepared by calcining the yellow oxides to form Fe_2O_3. The shade of the red oxide depends on the characteristics of the original yellow pigment, and the conditions of calcination and ranges from light to dark red. The product is 96-98.5% Fe_2O_3.

The black oxides are prepared by the controlled precipitation of Fe_3O_4 (treat $FeSO_4 \cdot 7H_2O$ with NaOH and O_2) to form a mixture of ferrous and ferric oxides.

Iron oxides are stable pigments insoluble in most solvents but usually soluble in hydrochloric acid. Those not soluble in HCl can be fused with potassium hydrogen sulfate ($KHSO_4$) and then dissolved in water.

The major use of iron oxide as a colorant is in cosmetics, particularly eye makeup and face powders. It is also permitted in dog and cat food at levels not exceeding 0.25% by weight of the finished food, and in drugs.

PAPRIKA AND PAPRIKA OLEORESINS

Paprika is the deep red, sweet, pungent powder prepared from the ground, dried pod of mild capsicum (*Capsicum annum*). It is one of the two principal kinds of red pepper; the other is cayenne pepper or cayenne. Paprika is produced in large quantities in Hungary and is also available from many warm-climate areas, including Africa, Spain, and the American tropics. The chief classifications of paprika are Hungarian paprika, which has the pungency and flavor characteristics of that produced in Hungary (Rosen-paprika and Koenigspaprika), and Spanish paprika (pimenton, pimiento), which has the characteristics of paprika produced in Spain.

Paprika oleoresin is the combination of flavor and color principles obtained by extracting paprika with any one or a combination of approved solvents: acetone, ethyl alcohol, ethylene dichloride, hexane, isopropyl alcohol, methyl alcohol, methylene chloride, and trichloroethylene. Depending on their source, paprika oleoresins are brown-red, slightly viscous, homogeneous liquids, pourable at room temperature and containing 2-5% sediment.

The oleoresins are avaiable in various standardized forms in which 1 pound of oleoresin is equal to 10-30 pounds of paprika. Paprika oleoresins are typically standardized by dilution with vegetable oil or mono- or diglycerides.

Paprika and its oleoresin are approved for use in foods in general where its application as a color additive frequently overlaps its use as a spice. Both products have good tinctorial strength and are used at levels of 0.2-100 ppm to produce orange to bright red shades.

SAFFRON

Saffron, known also as CI Natural Yellow 6 (CI No. 75100), *safran, crocine, crocétine*, and *crocus*, is the dried stigma of *Crocus sativus*, a plant indigenous to the Orient but also grown in North Africa, Spain, Switzerland, Greece, Austria, and France. It is a reddish brown or golden yellow odoriferous powder having a slightly bitter taste. The stigmas of approximately 165,000 blossoms are required to make 1 kg of colorant.

The coloring principles of saffron are crocin and crocetin.

$$H_{22}C_{18} \bigg\langle \begin{array}{l} COOC_{12}H_{21}O_{11} \\ COOC_{12}H_{21}O_{11} \end{array}$$

CROCIN: $C_{44}H_{64}O_{26}$ (mw 1,008.99)

$$\begin{array}{c} \text{CH}_3 \qquad \text{CH}_3 \\ | \qquad\quad | \\ \text{HOOC}-\text{C}\quad\text{CH}\quad\text{C}\quad\text{CH CH CH CH} \end{array}$$

$$\text{CH CH CH CH C}\quad\text{CH}\quad\text{C}-\text{COOH}$$
$$\begin{array}{cc} | & | \\ \text{CH}_3 & \text{CH}_3 \end{array}$$

CROCETIN: $C_{20}H_{24}O_4$ (mw 328.41)

Crocin is a yellow-orange glycoside that is freely soluble in hot water, slightly soluble in absolute alcohol, glycerine, and propylene glycol, and insoluble in vegetable oils. Crocin melts with decomposition at about $186°C$ and has absorption maxima in methanol at about 464 nm and 434 nm.

Crocetin is a dicarboxylic acid that forms brick-red rhombs from acetic anhydride that melt with decomposition at about $285°C$. It is very sparingly soluble in water and most organic solvents but soluble in pyridine and similar organic bases as well as in dilute sodium hydroxide.

As a food colorant, saffron shows good overall performance. In general, it is stable toward light, oxidation, microbiological attack, and changes in pH. Its tinctorial strength is relatively high, resulting in use levels of 1-260 ppm.

TALC

Talc, CI No. 77019, is finely powdered, native, hydrous magnesium silicate ($3 MgO \cdot 4SiO_2 \cdot H_2O$, "soapstone") sometimes containing a small amount of aluminum silicate. It is produced in many parts of the world, including France, Italy, India, and the United States. The typical composition of USP talc is:

Silicon dioxide (SiO_2)	60.13%
Magnesium oxide (MgO)	32.14%
Calcium oxide (CaO)	0.39%
Aluminum oxide (Al_2O_3)	1.84%
Ferric oxide (Fe_2O_3)	0.15%
Acid solubles	<2.0%
Water solubles	<0.1%
Loss on ignition	4.90%
Lead (Pb)	<5 ppm
Arsenic (As)	<1 ppm

Theoretically, talc is a pure white, odorless, unctuous powder rated as among the softest materials available, assigned a hardness of No. 1-1.5 on the Mohs Mineralogical Scale. Actually, it is a white-gray powder possessing varying amounts of softness and slip depending on its origin. The best grades of talc are very white crystalline powders with a lamellar structure, a greasy feel, and a particle size of 74 μ or less. Micronized talcs are often 40 μ or less in size. The specific gravity of talc is about 2.70.

TITANIUM DIOXIDE

Titanium dioxide (TiO_2; mol. wt. 79.90; Titanic Earth; CI No. 77891) is the whitest, brightest pigment known today, with a hiding power four to five times greater than that of its closest rival, zinc oxide.

Titanium dioxide exists in nature in three crystalline forms: anatase, brookite, and rutile, with anatase as the commonly available form. Anatase has a high refractive index (2.52) and excellent stability toward light, oxidation, changes in pH, and microbiological attack. Titanium dioxide is virtually insoluble in all common solvents.

Titanium dioxide is permitted in foods at levels up to 1% and is used to color such products as confectionary panned goods, cheeses, and icings. It is also widely used in tableted drug products and in numerous cosmetics such as lipsticks, nail enamels, face powders, eye makeup, and rouges.

The colorant's chief disadvantages are its inability to blend well with the other ingredients usually found in powder formulations, its tendency to produce blue undertones, and its ability to catalyze the oxidation of perfumes.

TURMERIC AND TURMERIC OLEORESIN

Turmeric (CI Natural Yellow 3, CI No. 75300) is the dried and ground rhizome or bulbous root of *Curcuma longa,* a perennial herb of the Zingiberaceae family native to southern Asia and cultivated in China, India, and the East Indies. It is a yellow powder with a characteristic odor and a sharp taste.

Turmeric oleoresin is the combination of flavor and color principles obtained from turmeric by extracting it with one or a combination of the following

CURCUMIN: $C_{21}H_{20}O_6$ (mw 368. 39)

solvents: acetone, ethyl alcohol, ethylene dichloride, hexane, isopropyl alcohol, methyl alcohol, methylene chloride, and trichloroethylene.

The principal coloring matter in turmeric and its oleoresin is curcumin, an orange-yellow, crystalline powder, insoluble in water and ether but soluble in ethanol and glacial acetic acid. It has a reported melting point of 180-183°C.

Turmeric is available as a powder and as a suspension in a variety of carriers, including edible vegetable oils and fats and mono- and diglycerides. Turmeric oleoresin is most often sold as solutions in propylene glycol with or without added emulsifying agents. Both products exhibit poor to moderate stability to light, oxidation, and change in pH but good tinctorial strength. Turmeric is typically used at levels of 0.2-60 ppm, whereas use levels for its oleoresin are 2-640 ppm. Both are used alone or in combination with other colorants such as annatto to shade pickles, mustard, spices, margarine, cooking oils, and salad dressings.

ULTRAMARINES

The ultramarines are synthetic, inorganic pigments of somewhat indefinite composition. Basically, they are sodium aluminosulfosilicates with crystal structures related to the zeolites and empirical formulas that can be approximated as $Na_7Al_6Si_6O_{24}S_3$. They are intended as the duplicate of the colorants produced from the naturally occuring semiprecious gem, lazurite (*Lapis lazuli*). Their color is believed due to polysulfide linkages in a highly resonant state.

Ultramarines are manufactured by the heat-treating and then very slow cooling of various combinations of kaolin (China clay) silica, sulfur, soda ash, and sodium sulfate plus a carbonaceous reducing agent such as rosin or charcoal pitch. The formulation of ingredients, temperature, time, cooling rate, subsequent treatment, and other variables determines the resultant color. Firing temperatures range from 700-800°C, whereas firing times vary from a few to as many as 150 hr.

The basic product of the ignition is Ultramarine Green (CI Pigment Green 24, CI No. 77013). This is converted into Ultramarine Blue (CI Pigment Blue 29, CI No. 77007) by further heat treatment in the presence of sulfur, or into Ultramarine Violet (CI Pigment Violet 15, CI No. 77007) by heating with 5% ammonium chloride for 4 days at 200-250°C, or into Ultramarine Red (CI No. 77007) by treatment with gaseous hydrochloric acid at 70-200°C for 4 days or by reaction with gaseous nitric acid at higher temperatures.

Ultramarines are insoluble in water and organic solvents but soluble in acids, which cause their discoloration and the liberation of hydrogen sulfide. They have excellent permanency and resistance to alkali but poor tinting and hiding power.

Ultramarine Blue is used in salt intended for animal feed (≤0.5% w/w). All ultramarines are used in the cosmetic field in such products as mascara, eyebrow pencils, and soaps.

ZINC OXIDE

Of all the white pigments used in the cosmetic field, zinc oxide ranks among the most important. Although it does not have the hiding power of colorants such as titanium dioxide, zinc oxide has certain advantages, including its brightness, ability to provide opacity without blue undertones, adhesiveness or "stick," and therapeutic properties, as it is mildly antiseptic and has drying and healing effects on the skin.

Zinc oxide (Mol. wt. 81.38; CI No. 77947) is a white or yellowishwhite amorphous, odorless powder with pH = 6.95-7.37. It is practically insoluble in water but soluble in dilute acetic acid, mineral acids, ammonia, ammonium carbonate, and alkali hydroxides.

As a colorant, zinc oxide is used in face powders, rouges, and eye makeups at levels of 5-30%.

MISCELLANEOUS COLORANTS

Other colorants not requiring certification have been defined in the Code of Federal Regulations. Most of these are of only minor to moderate importance and have only limited usage.

Alumina—A white, odorless, tasteless, amorphous powder consisting essentially of aluminum hydroxide, $Al_2O_3 \cdot XH_2O$.

Aluminum powder—Finely divided particles of aluminum prepared from virgin aluminum. It is free from admixture with other substances.

Bismuth oxychloride—A synthetically prepared white or nearly white amorphous or finely crystalline, odorless powder consisting principally of BiOCl. Bismuth oxychloride is synthetic pearl essence. It is used in lipstick, nail polish, eye makeup, and other cosmetics to produce a lustrous, pearly effect. See Guanine.

Bronze powder—A very fine metallic powder prepared from alloys consisting principally of virgin electrolytic copper and zinc with small amounts of the virgin metals aluminum and tin. It contains small amounts of stearic or oleic acid as a lubricant.

Calcium carbonate—A fine, white synthetically prepared powder consisting essentially of precipitated calcium carbonate, $CaCO_3$.

Carrot oil—The liquid or the solid portion of the mixture, or the mixture itself obtained by the hexane extraction of edible carrots (*Daucus carota* L.) with subsequent removal of the hexane by vacuum distillation. The resultant mixture of solid and liquid extractives consists chiefly of oils, fats, waxes, and carotenoids naturally occurring in carrots.

Chromium-cobalt-aluminum oxide—A blue-green pigment obtained by calcining a mixture of chromium oxide, cobalt carbonate, and aluminum oxide. It may contain small amounts (<1% each) of oxides of barium, boron, silicon, and nickel.

Copper powder—A very fine free-flowing metallic powder prepared from virgin electrolytic copper. It contains small amounts of stearic or oleic acid as a lubricant.

Corn endosperm oil—A reddish brown liquid composed chiefly of glycerides, fatty acids, sitosterols, and carotenoid pigments obtained by isopropyl alcohol and hexane extraction from the gluten fraction of yellow corn grain.

Dehydrated beets (beet powder)—A dark red powder prepared by dehydrating sound, mature, good quality, edible beets. The active ingredient in beet powder is Betanin. Alone, the colorant produces hues resembling raspberry or cherry. In combination with water-soluble annatto, strawberry shades are obtained. Beet powder dissolves readily in water and water-based products. It is stable when used from pH = 4 to pH = 7. Exposure to air, light, and heat should be avoided: maximum storage of 3 months is recommended. Beet powder is used in yogurt, sherbert, ice cream, beverages, desserts, frosting, candy, and gelatins in concentrations of 0.1-1% of the final product.

Dihydroxyacetone—this colorant is 1,3-dihydroxy-2-propanone.

Disodium EDTA-copper—Disodium [[N,N'-1,2-ethanediylbis[N-(carboxymethyl)glycinato]](4-)-N,N',O,O',O^N,O$^{N'}$]cuprate(2-).

Dried algae meal—A dried mixture of algae cells (genus *Spongiococcum*, separated from its culture broth), molasses, cornsteep liquor, and a maximum of 0.3% ethoxyquin. The algae cells are produced by suitable fermentation, under controlled conditions, from a pure culture of the genus *Spongiococcum*.

Ferric ammonium citrate—A mixture of complex chelates prepared by the interaction of ferric hydroxide with citric acid in the presence of ammonia. The chelates occur in brown and green forms, are deliquescent in air, and are reducible by light.

Ferric ammonium ferrocyanide—The blue pigment obtained by oxidizing under acidic conditions with sodium dichromate the acid-digested precipitate resulting from mixing solutions of ferrous sulfate and sodium ferrocyanide in the presence of ammonium sulfate. The oxidized product is filtered, washed, and dried. The pigment consists principally of ferric ammonium ferrocyanide with small amounts of ferric ferrocyanide and ferric sodium ferrocyanide.

Ferrous gluconate—Fine yellowish gray or pale greenish yellow powder or granules having a slight odor resembling that of burned sugar. One gram dissolves in about 10 ml of water with slight heating. It is practically insoluble in alcohol. A 1:20 solution is acid to litmus.

$$\left[HOCH_2CH\underset{OH}{-}CH\underset{OH}{C}HCH\underset{OH}{C}HC\overset{O}{\underset{O^-}{\diagup}} \right]_2 Fe\cdot 2H_2O$$

$$C_{12}H_{22}FeO_{14}\cdot 2H_2O \qquad mw\ 482.18$$

Fruit juice—The concentrated or unconcentrated liquid expressed from mature varieties of fresh, edible fruits; or, a water infusion of the dried fruit.

Grape skin extract (enocianina)—A purplish red liquid prepared by the aqueous extraction (steeping) of the fresh deseeded marc remaining after grapes have been pressed to produce grape juice or wine. It contains the common components of grape juice (anthocyanins, tartaric acid, tannins, sugars, minerals, etc.), but not in the same proportions as found in grape juice. During the steeping process, sulfur dioxide is added and most of the extracted sugars are fermented to alcohol. The extract is concentrated by vacuum evaporation, during which practically all of the alcohol is removed. A small amount of sulfur dioxide may be present.

Guaiazulene—Principally 1,4-dimethyl-7-isopropyl-azulene.

Henna—The dried leaf and petiole of *Lawsonia alba* Lam. (*Lawsonia inermis* L.).

Logwood extract—A reddish brown-black solid material extracted from the heartwood of the leguminous tree *Haematoxylon campechianum*. The active colorant substance is principally hematein. The latent coloring material is the unoxidized or leuco form of hematein called *hematoxylin*. The leuco form is oxidized by air.

Manganese violet—A violet pigment obtained by reacting phosphoric acid, ammonium dihydrogen orthophosphate, and manganese dioxide at temperatures above $450°F$. The pigment is a manganese ammonium pyrophosphate complex having the approximate formula: $Mn(III)NH_4P_2O_7$.

Mica—A white powder obtained from the naturally occurring mineral, muscovite mica, consisting predominantly of a potassium aluminum silicate, $K_2Al_4(Al_2Si_6O_{20})(OH_4)$ or, alternatively, $H_2KAl_3(SiO_4)_3$. Mica may be identified and semiquantitatively determined by its characteristic X-ray-diffraction pattern and by its optical properties.

Potassium sodium copper chlorophyllin (chlorophyllin-copper complex)—A green-black powder obtained from chlorophyll by replacing the methyl and phytyl ester groups with alkali and replacing the magnesium with copper. The source of the chlorophyll is dehydrated alfalfa.

Pyrogallol—This colorant is 1,2,3-trihydroxybenzene.

Pyrophyllite—A naturally occurring mineral substance consisting predominantly of a hydrous aluminum silicate, $Al_2O_3 \cdot 4SiO_2 \cdot H_2O$, intimately mixed with lesser amounts of finely divided silica, SiO_2. Small amounts (usually $<3\%$) of other silicates, such as potassium aluminum silicate, may be present. Pyrophyllite may be identified and semiquantitatively determined by its characteristic X-ray powder-diffraction pattern and by its optical properties.

Riboflavin—A yellow-orangish yellow crystalline powder having a slight odor. It melts at about $280°C$, and its saturated solution is neutral to litmus. When dry, it is not affected by diffused light, but when in solution, light induces deterioration. One gram dissolves in about 3000-20,000 ml of water, depending on the internal crystalline structure. It is less soluble in alcohol than in water. It is insoluble in ether and in chloroform but is very soluble in dilute solutions of alkalies. A solution of 1 mg in 100 ml of water is pale greenish yellow by transmitted light and has an intense yellowish green fluorescence that disappears on the addition of mineral acids or alkalies.

$$\text{(chemical structure)}$$

$$C_{17}H_{20}N_4O_6 \quad \text{mw } 376.37$$

Tagetes meal and extract—Tagetes (Aztec marigold) meal is the dried, ground flower petals of the Aztec marigold (*Tagetes erecta* L.) mixed with not more than 0.3% ethoxyquin. Tagetes extract is a hexane extract of the flower petals of the Aztec marigold. It is mixed with an edible vegetable oil, or with an edible vegetable oil and a hydrogenated edible vegetable oil, and not more than 0.3% ethoxyquin. It may also be mixed with soy flour or corn meal as a carrier.

Toasted partially defatted cooked cottonseed flour—This product is prepared by delinting and decorticating food-quality cottonseed. The meats are screened, aspirated, and rolled; moisture is adjusted, the meats heated, and the oil expressed; the cooked meats are cooled, ground, and reheated to obtain a product varying in shade from light to dark brown.

Vegetable juice—The concentrated or unconcentrated liquid expressed from mature varieties of fresh, edible vegetables.

BIBLIOGRAPHY

ANDREU, R. F. Farmacognosia *17*, 145-224 (1957). A Drug Which is Gradually Disappearing from the Medical Armamentarium: Saffron (Historical Study). An extensive review of saffron.

Annatto Food Colors. Charles Hansen's Laboratory, 9015 West Maple St., Milwaukee, Wisc. 53214. A brief description of what annatto is and how it is used.

BAUERNFEIND, J. C., BUNNELL, R. H. Food Technol. *16*, 76-82 (1962). β-Apo-8′-Carotenal—A New Food Color. Describes the properties, market forms, uses, stability, and other characteristics of the colorant.

BAUERNFEIND, J. C., OSADCA, M., BUNNELL, R. H. Food Technol. *16*, 101-107 (1962). β-Carotene, Color and Nutrient for Juices and Beverages. A general discussion of the use of β-carotene as a color additive for juices and beverages.

BUNNELL, R. H., BORENSTEIN, B. Food Technol. *21*, 13A-16A (1967). Canthaxanthin, A Potential New Food Color. A brief review of the history, natural occurrence, properties, market forms, and stability of canthaxanthin.

BUNNELL, R. H., DRISCOLL, W., BAUERNFEIND, J. C. Food Technol. *12*, 536 (1958). Coloring water-Base Foods With β-Carotene.

DENDY, D. A. V. East Afr. Agric. Forest. J. *32*, 126-132 (1966). Annatto, The Pigment of Bixa Orellana. The manufacture of annatto.

EICHENBERGER, W. R. Paper presented at the ACS Meeting, August 29, 1972. Caramel Colors: Manufacture, Properties and Food Applications.

GORDON, H. T. Food Technol. (May) 64-66 (1972). Coloring Foods With Carotenoids. A brief description of the properties, commercial forms, and uses of β-carotene, β-apo-8'-carotenal, and canthaxanthin.

ISLER, O., RUEGG, R., SCHUDEL, P. Chimia *15*, 208-226 (1961). Synthetic Carotenoids for Food Coloring. Includes a discussion of β-carotene, β-apo-8'-carotenal, and canthaxanthin from the standpoint of preparation, toxicity, analysis, and application.

ISLER, O., RUEGG, R., SCHWIETER, U. Pure Appl. Chem. *14*, 245-264 (1967). Carotenoids as Food Colors. Describes the preparation and analysis of various carotenoids including β-carotene, canthaxanthin, and β-apo-8'-carotenal.

KAMPFER, W., STIEG, F., Jr., Color Eng. *44*, 35-40 (1967). Titanium Dioxide as a Colorant. A description of the manufacture, properties, and uses of titanium dioxide as a colorant for paint, food, plastics, and other materials.

LINNER, R. T. Baker's Digest, April 1965. Caramel Coloring—Production, Composition and Functionality.

MARCUS, F. K. Ger. 1,156,529, October 31, 1963. Fabrication of Oil and Water Soluble Coloring from Annatto Seeds for Coloring of Margarine and Cheese.

MAYER, F., COOK, A. H. *The Chemistry of Natural Coloring Matters.* ACS Monograph, Reinhold, New York, 1943.

NORTH, R. Canner Packer, May 1969. Add a Pinch of Burnt Sugar for Color. A description of caramel, and how it is made and used.

PECK, F. W. Food Eng. (March) 94 (1955). Caramel—Its Properties and Its Uses.

RATH, F. Ger. 927,305, May 5, 1955. Dyeing of Food and Drugs with Natural Dyes. Natural dyes like norbixin, crocetin and carminic acid are discussed from an applications standpoint.

REITH, J. F., GIELEN, J. W. J. Food Sci. *36*, 861-864 (1971). Properties of bixin, norbixin, and annatto extracts.

SATO, T., SUZUKI, H. Nippon Shokuhin Kogyo Gakkaishi *13*, 488-491 (1966). Coloring of Vienna Sausage with Water-Soluble Annatto. A study of the coloring of sausages with annatto and zanthene-type pigments from the standpoint of fading, penetration, and other variables.

SCHWARZ, G., MUMM, H., WOERNER, F. Molkerei Käserei—Ztg. 9, 1430-1433 (1958). Coloring Cheeses With Annatto and Carotene Dyes and their Detection.

TODD, P. H., Jr. U.S. 3,162,538, December 22. 1964. Vegetable Base Food Coloring. Describes the use of bixin and turmeric for coloring butter, margarine, cheese, and other fatty and oily foods.

USOVA, E. M., VOROSHIN, E. M., ROSTOVSKII, V. S., MOROZ, A. M., YAKHINA, F. Kh. Izv. Vysshikh Uchebn, Zavedenii, Pishchevaya Tekhnol. 4, 151-153 (1966). Food Dyes from Mountain Ash Berries and Nettles. Describes the use of natural colorants as replacements for tartrazine, indigo carmine, and annatto.

VISHNEVETSKAYA, S. G. Maslob.—Zhir. Prom. 28, 30-32 (1962). Properties and Applications of Henna.

Appendix **A** Colorant Specifications

The specifications cited here are based on the 1978 edition of the Code of Federal Regulations (21 CFR 1.1) and revisions to 21 CRF 1.1 that appeared in the Federal Register throughout 1978. All are maxima unless indicated otherwise.

In addition to these requirements, color additives must be free of all impurities other than those named to the extent that such impurities can be avoided by good manufacturing practice.

FD&C Blue No. 1

Sum of volatile matter (at 135°C) and chlorides and sulfates (calculated as sodium salts)—15.0%, total
Water-insoluble matter—0.2%
Leuco base—5%
Sum of o-, m-, and p-sulfobenzaldehydes—1.5%
N-Ethyl-N-(m-sulfobenzyl)sulfanilic acid—0.3%
Subsidiary colors—6.0%
Chromium (as Cr)—50 ppm
Arsenic (as As)—3 ppm
Lead (as Pb)—10 ppm
Total color—85.0% min.

FD&C Blue No. 2

Sum of volatile matter (at 135°C) and chlorides and sulfates (calculated as sodium salts)—15.0%, total
Water-insoluble matter—0.4%
Isatin-4-sulfonic acid—0.4%
Isomeric colors—18.0%
Lower sulfonated subsidiary colors—5.0%
Lead (as Pb)—10 ppm
Arsenic (as As)—3 ppm
Total color—85.0% min.

FD&C Green No. 3

Volatile matter (at 135°C)–10.0%
Water-insoluble matter–0.5%
Ether extracts–0.4%
Chlorides and sulfates of sodium–5.0%
Mixed oxides–1.0%
Subsidiary dyes–5.0%
Pure dye (as determined by titration with titanium trichloride)–85.0% min.

FD&C Red No. 3

Volatile matter (at 135°C) and chlorides and sulfates (calculated as the
    sodium salts)–13%, total
Water-insoluble matter–0.2%
Sodium iodide–0.4%
Lead (as Pb)–10 ppm
Arsenic (as As)–3 ppm
Unhalogenated intermediates–0.1%
Triiodoresorcinol–0.2%
2(2′,4′-dihydroxy-3′,5′-diiodobenzoyl) benzoic acid–0.2%
Monoiodofluoresceins–1.0%
Other lower iodinated fluoresceins–9.0%
Total color–87.0% min.

FD&C Red No. 4

Sum of volatile matter (at 135°C) and chlorides and sulfates (calculated as
    sodium salts)–13%, total
Water-insoluble matter–0.2%
5-Amino-2,4-dimethyl-1-benzenesulfonic acid, sodium salt–0.2%
4-Hydroxy-1-naphthalenesulfonic acid, sodium salt–0.2%
Subsidiary colors–2%
Lead (as Pb)–10 ppm
Arsenic (as As)–3 ppm
Mercury (as Hg)–1 ppm
Total color–87% min.

FD&C Red No. 40

Sum of volatile matter (at 135°C) and chlorides and sulfates (calculated as
    sodium salts)–14.0% total
Water-insoluble matter–0.2%
Higher-sulfonated subsidiary colors (as sodium salts)–1.0%
Lower-sulfonated subsidiary colors (as sodium salts)–1.0%

Disodium salt of 6-hydroxy-5-[(2-methoxy-5-methyl-4-sulfophenyl)azo]-8-(2-methoxy-5-methyl-4-sulfophenoxy)-2-naphthalenesulfonic acid—1.0%
Sodium salt of 6-hydroxy-2-naphthalenesulfonic acid—0.3%
4-Amino-5-methoxy-o-toluenesulfonic acid—0.2%
Disodium salt of 6,6'-oxybis(2-naphthalenesulfonic acid)—1.0%
Lead (as Pb)—10 ppm
Arsenic (as As)—3 ppm
Total color—85.0% min.

FD&C Yellow No. 5

Volatile matter (at 135°C) and chlorides and sulfates (calculated as the sodium salts)—13.0%, total
Water-insoluble matter—0.2%
Lead (as Pb)—10 ppm
Arsenic (as As)—3 ppm
Phenylhydrazine-p-sulfonic acid—0.2%
Other uncombined intermediates—0.2% each
Subsidiary dyes—1.0%
Total color—87.0% min.

FD&C Yellow No. 6

Volatile matter (as 135°C)—10.0%
Water-insoluble matter—0.5%
Ether extracts—0.2%
Chlorides and sulfates of sodium—5.0%
Mixed oxides—1.0%
Subsidiary dyes—5.0%
Pure dye (as determined by titration with titanium trichloride)—85.0% min.

FD&C Lakes

Prepared from previously certified FD&C colors.
Soluble chlorides and sulfates (as sodium salts)—2.0%
Inorganic matter, insoluble in hydrochloric acid—0.5%

Citrus Red No. 2

Volatile matter (at 100°C)—0.5%
Water-soluble matter—0.3%
Matter insoluble in carbon tetrachloride—0.5%
Uncombined intermediates—0.05%
Subsidiary dyes—2.0%
Lead (as Pb)—10 ppm

Arsenic (as As)—1 ppm
Pure dye—98% min.

Orange B

Volatile matter (at $135°C$)—6.0%
Chlorides and sulfates (calculated as the sodium salts)—7.0%, total
Water insoluble matter—0.2%
1-(4-sulfophenyl)-3-ethylcarboxy-5-hydroxypyrazolone and 1-(4-sulfophenyl)-
  3-carboxy-5-hydroxypyrazolone—0.7%
Naphthionic acid—0.2%
Phenylhydrazine-p-sulfonic acid—0.2%
Trisodium salt of 1-(4-sulfophenyl)-3-carboxy-4-(4-sulfonaphthylazo)-5-hy-
  droxypyrazole—6.0%
Other subsidiary dyes—1.0%
Lead (as Pb)—10 ppm
Arsenic (as As)—1 ppm
Pure dye—87.0% min.

D&C Blue No. 4

Sum of volatile matter (at $135°C$) and chlorides and sulfates (calculated as
  sodium salts)—15%, total
Water-insoluble matter—0.2%
Leuco base—5%
Sum of o-, m-, and p-sulfobenzaldehydes, ammonium salts—1.5%
N-Ethyl, N-(m-sulfobenzyl) sulfanilic acid, ammonium salt—0.3%
Subsidiary colors—6%
Chromium (as Cr)—50 ppm
Lead (as Pb)—20 ppm
Arsenic (as As)—3 ppm
Mercury (as Hg)—1 ppm
Total color—85% min.

D&C Blue No. 6

Volatile matter (at $135°C$)—3%
Matter insoluble in N,N-dimethylformamide—1%
Isatin—0.3%
Anthranilic acid—0.3%
Indirubin—1%
Lead (as Pb)—10 ppm
Mercury (as Hg)—1 ppm
Arsenic (as As)—3 ppm
Pure color—95% min.

D&C Blue No. 9

Volatile matter (at 135°C)—3%
Matter extractable by alcoholic HCl (0.1 ml of concentrated hydrochloric acid per 50 ml of 95% ethyl alcohol)—1%
2-Aminoanthraquinone—0.2%
Organically combined chlorine in pure dye—13.0%-14.8%
Lead (as Pb)—20 ppm
Arsenic (as As)—3 ppm
Pure Dye—97% min.

D&C Brown No. 1

Sum of volatile matter (at 135°C) and chlorides and sulfates (calculated as sodium salts)—16% , total
Water-insoluble matter—0.2%
Sulfanilic acid, sodium salt—0.2%
Resorcinol—0.2%
Xylidines—0.2%
Disodium salt of 4[[5-[(4-sulfophenyl)azo]-2,4-dihydroxyphenyl]azo]benzenesulfonic acid—3%
Monosodium salt of 4[[5-[(2,4-dimethylphenyl)azo]-2,4-dihydroxyphenyl]-azo]benzenesulfonic acid—29% min., 39% max.
Monosodium salt of 4[[5-[(2,5-dimethylphenyl)azo]-2,4-dihydroxyphenyl]-azo]benzenesulfonic acid—12% min., 17% max.
Monosodium salt of 4[[5-[(2,3-dimethylphenyl)azo]-2,4-dihydroxyphenyl]-azo]benzenesulfonic acid—6% min., 13% max.
Monosodium salt of 4[[5-[(2-ethylphenyl)azo]2,4-dihydroxyphenyl]azo]-benzenesulfonic acid—5% min., 12% max.
Monosodium salt of 4[[5-[(3,4-dimethylphenyl)azo]-2,4-dihydroxyphenyl]-azo]benzenesulfonic acid—3% min., 9% max.
Monosodium salt of 4[[5-[(2,6-dimethylphenyl)azo]-2,4-dihydroxyphenyl]-azo]benzenesulfonic acid—3% min., 8% max.
Monosodium salt of 4[[5-[(4-ethylphenyl)azo]-2,4-dihydroxyphenyl]azo]-benzenesulfonic acid—2% min., 8% max.
Lead (as Pb)—20 ppm
Arsenic (as As)—3 ppm
Mercury (as Hg)— 1 ppm
Total color—84% min.

D&C Green No. 5

Sum of volatile matter (at 135°C) and chlorides and sulfates (calculated as sodium salts)—20.0%, total
Water-insoluble matter —0.2%
1,4-Dihydroxyanthraquinone—0.2%

2-Amino-*m*-toluenesulfonic acid—0.2%
Subsidiary colors—5.0%
Lead (as Pb)—10 ppm
Arsenic (as As)—3 ppm
Total color 80.0% min.

D&C Green No. 6

Volatile matter (at 135°C)—2.0%
Water-soluble matter—0.3%
Matter insoluble in carbon tetrachloride—1.5%
Intermediates—0.5%
Lead (as Pb)—10 ppm
Arsenic (as As)—1 ppm
Pure Color—96.0% min.

D&C Green No. 8

Volatile matter (at 135°C)—15%
Water-insoluble matter—0.2%
Chlorides and sulfates (calculated as sodium salts)—20%
Trisodium salt of 1,3,6-pyrenetrisulfonic acid—6%
Tetrasodium salt of 1,3,6,8-pyrenetetrasulfonic acid—1%
Pyrene—0.2%
Lead (as Pb)—20 ppm
Arsenic (as As)—3 ppm
Mercury (as Hg)—1 ppm
Total color—65% min.

D&C Orange No. 4

Sum of volatile matter (at 135°C) and chlorides and sulfates (calculated as sodium salts)—13%, total
Water-insoluble matter—0.2%
2-Naphthol—0.4%
Sulfanilic acid, sodium salt—0.2%
Subsidiary colors—3%
Lead (as Pb)—20 ppm
Arsenic (as As)—3 ppm
Mercury (as Hg)—1 ppm
Total color—87% min.

D&C Orange No. 5

Volatile matter (at 135°C)—5.0%
Insoluble matter (alkaline solution)—1.0%

Ether extracts (from alkaline solution)—0.5%
Sodium chloride—3.0%
Mixed oxides—1.0%
Free bromine—0.02%
Organically combined bromine in pure dye—31.0-35.0%
Pure dye (as determined gravimetrically)—90.0% min.

D&C Orange No. 10

Volatile matter (at 135°C)—5.0%
Insoluble matter (alkaline solution)—1.0%
Ether extracts (from alkaline solution)—0.5%
Sodium chloride—3.0%
Mixed oxides—1.0%
Free iodine—0.05%
Organically combined iodine in pure dye—41.5-45.5%
Pure dye (as determined gravimetrically)—85.0% min.

D&C Orange No. 11

Volatile matter (at 135°C)—10.0%
Water-insoluble matter—1.0%
Ether extracts—0.5%
Chlorides and sulfates of sodium—3.0%
Mixed oxides—1.0%
Free iodine—0.05%
Organically combined iodine in pure dye—38.5-42.5%
Pure dye (as determined gravimetrically)—85.0% min.

D&C Orange No. 17

Volatile matter (at 135°C)—5.0%
Sulfated ash—1.0%
Matter insoluble in toluene—1.5%
2,4-Dinitroaniline—0.2%
β-Naphthol—0.2%
Pure dye (as determined by titration with titanium trichloride)—90.0% min.

D&C Red No. 6

Volatile matter (at 135°C)—10.0%
Water-insoluble matter—1.0%
Ether extracts (isopropyl ether)—0.5%
Chlorides and sulfates of sodium—6.0%
Mixed oxides—1.0%
Pure dye (as determined by titration with titanium trichloride)—82.0% min

D&C Red No. 7

Volatile matter (at 135°C)—8.0%
Ether extracts (isopropyl ether)—0.5%
Chlorides and sulfates (as calcium salts)—6.0%
Oxides of iron and aluminum—1.0%
Pure dye (as determined by titration with titanium trichloride)—85.0% min.

D&C Red No. 8

Volatile matter (at 135°C)—10.0%
Ether extracts (isopropyl ether)—0.5%
Lake Red C Amine (2-chloro-5-aminotoluene-4-sulfonic acid)—0.2%
β-Naphthol—0.2%
Chlorides and sulfates of sodium—5.0%
Mixed oxides—1.0%
Pure dye (as determined by titration with titanium trichloride)—85.0% min.

D&C Red No. 9

Volatile matter (at 135°C)—5.0%
Ether extracts (isopropyl ether)—0.5%
Lake Red C Amine (2-chloro-5-aminotoluene-4-sulfonic acid)—0.2%
β-Naphthol—0.2%
Chlorides and sulfates of sodium—6.0%
Oxides of iron and aluminum—1.0%
Pure dye (as determined by titration with titanium trichloride)—87.0% min.

D&C Red No. 17

Volatile matter (at 135°C)—5%
Matter insoluble in both toluene and water (the color additive is mixed in toluene and the resultant residue is isolated and mixed with water to obtain the matter insoluble in both toluene and water)—0.5%
2-Naphthol—0.2%
1-(Phenylazo)-2-naphthol—3%
1-[[2-(Phenylazo)phenyl]azo]-2-naphthalenol—2%
Lead (as Pb)—20 ppm
Arsenic (as As)—3 ppm
Mercury (as Hg)—1 ppm
Chlorides and sulfates (calculated as sodium salts)—3%
Aniline—0.2%
4-aminoazobenzene—0.1%
Total color—90% min.

D&C Red No. 19

Volatile matter (at 135°C)—5.0%
Water-insoluble matter—1.0%
Ether extracts (from acid solution)—0.5%
Diethyl-*m*-aminophenol—0.2%
Chlorides and sulfates of sodium—2.0%
Mixed oxides—1.0%
Pure dye (as determined by titration with titanium trichloride)—92.0% min.

D&C Red No. 21

Volatile matter (at 135°C)—6.0%
Insoluble matter (alkaline solution)—1.0%
Ether extracts (from alkaline solution)—0.5%
Chlorides and sulfates of sodium—2.0%
Mixed oxides—1.0%
Free bromine—0.02%
Organically combined bromine in pure dye—47.5-51.5%
Pure dye (as determined gravimetrically)—93.0% min.

D&C Red No. 22

Volatile matter (at 135°C)—10.0%
Water-insoluble matter—1.0%
Ether extracts—0.5%
Chlorides and sulfates of sodium—5.0%
Mixed oxides—1.0%
Free bromine—0.02%
Organically combined bromine in pure dye—44.5-48.5%
Pure dye (as determined gravimetrically)—85.0% min.

D&C Red No. 27

Volatile matter (at 135°C)—5.0%
Insoluble matter (alkaline solution)—1.0%
Ether extracts (from alkaline solution)—0.5%
Sodium chloride—3.0%
Mixed Oxides—1.0%
Free Halogens—0.02%
Organically combined bromine in pure dye—38.5-42.5%
Organically combined chlorine in pure dye—16.0—19.0%
Pure dye (as determined gravimetrically)—90.0%

D&C Red No. 28

Volatile matter (at $135°C$)—10.0%
Water-insoluble matter—1.0%
Ether extracts—0.5%
Chlorides and sulfates of sodium—5.0%
Mixed oxides—1.0%
Free halogens—0.02%
Organically combined bromine in pure dye—36.5-40.5%
Organically combined chlorine in pure dye)16.0-18.0%
Pure dye (as determined gravimetrically)—85.0%

D&C Red No. 30

Volatile matter (at $135°C$)—5.0%
Matter insoluble in xylene—1.0%
Sodium chloride—3.0%
Mixed oxides—1.0%
Pure dye (as determined by titration with titanium trichloride)—90.0% min.

D&C Red No. 31

Sum of volatile matter (at $135°C$) and chlorides and sulfates (calculated as
    sodium salts)—10%, total
Aniline—0.2%
3-Hydroxy-2-naphthoic acid, calcium salt—0.4%
Subsidiary colors—1%
Lead (as Pb)—20 ppm
Arsenic (as As)—3 ppm
Mercury (as Hg)—1 ppm
Total Color—90% min.

D&C Red No. 33

Volatile matter (at $135°C$)—6.0%
Water-insoluble matter—1.0%
Ether extracts—0.5%
Aniline—0.2%
Chlorides and sulfates of sodium—10.0%
Mixed oxides—1.0%
Pure dye (as determined by titration with titanium trichloride)—82.0% min.

D&C Red No. 34

Sum of volatile matter (at $135°C$) and chlorides and sulfates (calculated as
    sodium salts)—15%, total

2-Amino-1-naphthalenesulfonic acid—0.2%
3-Hydroxy-2-naphthoic acid—0.4%
Subsidiary colors—4%
Lead (as Pb)—20 ppm
Arsenic (as As)—3 ppm
Mercury (as Hg)— 1 ppm
Total Color—85% min.

D&C Red No. 36

Volatile matter (at 135°C)—5.0%
Sulfated ash—1.0%
Matter insoluble in toluene—1.0%
o-Chloro-p-nitroaniline—0.2%
β-Naphthol—0.2%
Pure dye (as determined by titration with titanium trichloride)—90.0% min.

D&C Red No. 37

Volatile matter (at 80°C)—2.0%
Sulfated ash—3.0%
Matter insoluble in benzene—0.5%
Diethyl-m-aminophenol—0.2%
Stearic acid (not part of the dye)—50.0%
Pure dye (as determined by titration with titanium trichloride)—50.0% min.

D&C Red No. 39

Volatile matter (at 100°C)—2.0%
Matter insoluble in acetone—1.0%
Anthranilic acid—0.2%
N,N-(β-β'-Dihydroxy-diethyl) aniline—0.2%
Subsidiary colors—3.0%
Lead (as Pb)—20 ppm
Arsenic (as As)—3 ppm
Pure color—95.0% min.

D&C Violet No. 2

Volatile matter (at 135°C)—2.0%
Matter insoluble in both carbon tetrachloride and water—0.5%
p-Toluidine—0.2%
1-Hydroxy-9,10-anthracenedione—0.5%
1,4-Dihydroxy-9,10-anthracenedione-0.5%
Subsidiary colors—1.0%
Lead (as Pb)—20 ppm

Arsenic (as As)—3 ppm
Total color—96.0% min.

D&C Yellow No. 7

Sum of water and chlorides and sulfates (calculated as sodium salts)—6%, total
Matter insoluble in alkaline water—0.5%
Resorcinol—0.5%
Phthalic acid—0.5%
2-(2,4-Dihydroxybenzoyl)benzoic acid—0.5%
Lead (as Pb)—20 ppm
Arsenic (as As)—3 ppm
Mercury (as Hg)—1 ppm
Total color—94% min.

D&C Yellow No. 8

Sum of water and chlorides and sulfates (calculated as sodium salts)—15%, total
Matter insoluble in alkaline water—0.3%
Resorcinol—0.5%
Phthalic acid—1%
2-(2,4-Dihydroxybenzoyl)benzoic acid—0.5%
Lead(as Pb)—20 ppm
Arsenic (as As)—3 ppm
Mercury (as Hg)—1 ppm
Total color—85% min.

D&C Yellow No. 10

Volatile matter (at $135°C$)—10.0%
Water-insoluble matter—1.0%
Ether extracts—0.5%
Quinaldine—0.2%
Chlorides and sulfates of sodium—6.0%
Mixed oxides—1.0%
Pure dye (as calculated from organically combined nitrogen)—82.0% min.

D&C Yellow No. 11

Volatile matter (at $135°C$)—1%
Ethyl alcohol-insoluble matter—0.4%
Phthalic acid—0.3%
Quinaldine—0.2%
Subsidiary colors—5%
Lead (as Pb)—20 ppm

Arsenic (as As)—3 ppm
Mercury (as Hg)—1 ppm
Total color—96% min.

D&C Lakes

Ether extracts—0.5%
Soluble chlorides and sulfates (as sodium salts)—3.0%
Intermediates—0.2%

Ext. D&C Violet No. 2

Sum of volatile matter (at 135°C) and chlorides and sulfates (calculated as
  sodium salts)—18%, total
Water-insoluble matter—0.4%
1-Hydroxy-9,10-anthracenedione—0.2%
1,4-Dihydroxy-9,10-anthracenedione—0.2%
p-Toluidine—0.1%
p-Toluidinesulfonic acids, sodium salts—0.2%
Subsidiary colors—1%
Lead (as Pb)—20 ppm
Arsenic (as As)—3 ppm
Mercury (as Hg)—1 ppm
Total color—80% min.

Ext. D&C Yellow No. 7

Sum of volatile matter (at 135°C) and chlorides and sulfates (calculated as
  sodium salts)—15%, total
Water-insoluble matter—0.2%
1-Naphthol—0.2%
2,4-Dinitro-1-naphthol—0.03%
Lead (as Pb)—20 ppm
Arsenic (as As)—3 ppm
Mercury (as Hg)—1 ppm
Total color—85% min.

Ext. D&C Lakes

Ether Extracts—0.5%
Soluble chlorides and sulfates (as sodium salts)—3.0%
Intermediates—0.2%

[Phthalocyanınato(2-)] Copper

Volatile matter (at 135°C)—0.3%

Salt content (as NaCl)—0.3%
Alcohol-soluble matter—0.5%
Organic chlorine—0.2%
Aromatic amines—0.05%
Lead (as Pb)—40 ppm
Arsenic (as As)—3 ppm
Mercury (as Hg)—1 ppm
Total color—98.5% min.

Alumina

Acidity or alkalinity: agitate 1 g of colorant with 25 ml of water and filter; the filtrate shall be neutral to litmus paper
Matter insoluble in dilute HCl—0.5%
Lead (as Pb)—10 ppm
Arsenic (as As)—1 ppm
Mercury (as Hg)—1 ppm
Aluminum oxide (Al_2O_3)—50% min.

Aluminum Powder

Fineness, 100% shall pass through a 200-mesh screen and 95% shall pass through a 325-mesh screen
Mercury—1 ppm
Arsenic—3 ppm
Lead—20 ppm
Aluminum—99% min.

Annatto Extract (and Pigments Precipitated Therefrom)

Arsenic (as As)—3 ppm
Lead (as Pb)—10 ppm
Solvent residue—no more than that permitted for corresponding solvent in spice oleoresins

β-Apo-8'-Carotenal

Physical state—solid
1% Solution in chloroform—clear
Melting point (decomposition)—136-140°C (corrected)
Loss of weight on drying—0.2%
Residue on ignition—0.2%
Lead (as Pb)—10 ppm
Arsenic (as As)—1 ppm
Assay (spectrophotometric)—96-101%

Bismuth Oxychloride

Volatile matter—0.5%
Lead (as Pb)—20 ppm
Arsenic (as As)—3 ppm
Mercury (as Hg)—1 ppm
Bismuth oxychloride—98% min.

Bronze Powder

Stearic or oleic acid—5%
Cadmium (as Cd)—15 ppm
Lead (as Pb)—20 ppm
Arsenic (as As)—3 ppm
Mercury (as Hg)—1 ppm
Aluminum (as Al)—0.5%
Tin (as Sn)—0.5%
Copper (as Cu)—70% min.; 95% max.
Zinc (as Zn)—30%
Maximum particle size 45 μ (95% min.)
Al, Zn, Sn, and Cu content shall be based on the weight of the dried powder
   after thorough washing with ether

Calcium Carbonate

Loss on drying (200°C for 4 hr)—2%
Acid insolubles—0.2%
Heavy metals—30 ppm
Magnesium and alkali salts—1.0%
Barium—no green color when a platinum wire is dipped in a 2.5% acidified
   sample solution and held in a nonluminous flame
Assay (dry basis)—98% min.

Canthaxanthin

Physical state—solid
1% Solution in chloroform—complete and clear
Melting range (decomposition)—207-212°C (corrected)
Loss on drying—0.2%
Residue on ignition—0.2%
Total carotenoids other than transcanthaxanthin—5%
Lead—10 ppm
Arsenic—3 ppm
Mercury—1 ppm
Assay—96-101%

Caramel

Lead (as Pb)—10 ppm
Arsenic (as As)—3 ppm
Mercury (as Hg)—0.1 ppm

Carmine

Volatile matter ($135°$C for 3 hr)—20.0%
Ash—12.0%
Lead (as Pb)—10 ppm
Arsenic (as As)—1 ppm
Carminic acid—50.0% min.
Viable salmonella microorganisms—none

β-Carotene

Physical state—solid
1% Solution in chloroform—clear
Loss on drying—0.2%
Residue on ignition—0.2%
Lead (as Pb)—10 ppm
Arsenic (as As)—3 ppm
Assay (spectrophotometric)—96-101%

Carrot Oil

Hexane—25 ppm

Chromium-Cobalt-Aluminum Oxide

Chromium (as Cr_2O_3)—34-37%
Cobalt (as CaO)—29-34%
Aluminum (as Al_2O_3)—29-35%
Lead (as Pb)—30 ppm
Arsenic (as As)—3 ppm
Total oxides of Al, Cr, and Co—97% min.
Lead and arsenic shall be determined in the solution obtained by boiling 10 g of the colorant for 15 min. in 50 ml of $0.5N$ HCl

Chromium Hydroxide Green

Water-soluble matter—2.5%
Chromium (as Cr_2O_3) in 2% NaOH extract—0.1% (based on sample weight)
Boron (as B_2O_3)—8%

Total volatile matter at $1000°C-20\%$
$Cr_2O_3-75\%$ min.
Lead (as Pb)-20 ppm
Arsenic (as As)-3 ppm
Mercury (as Hg)-1 ppm

Chromium Oxide Greens

Chromium (as Cr_2O_3) in 2% NaOH extract-0.075% (based on sample weight)
Arsenic (as As)-3 ppm
Lead (as Pb)-20 ppm
Mercury (as Hg)-1 ppm
$Cr_2O_3-95\%$ min.

Cochineal Extract

pH (at $25°C)-5$-5.5
Protein ($N \times 6.25)-2.2\%$
Total solids-5.7-6.3%
Methyl alcohol-150 ppm
Lead (as Pb)-10 ppm
Arsenic (as As)-1 ppm
Carminic acid-1.8% min.
Viable salmonella microorganisms$-$none

Copper Powder

Stearic or oleic acid-5%
Cadmium (as Cd)-15 ppm
Lead (as Pb)-20 ppm
Arsenic (as As)-3 ppm
Mercury (as Hg)-1 ppm
Copper (as Cu)-95% min.
Maximum particle size 45 μ (95% minimum)

Corn Endosperm Oil

Total fatty acids-85% min.
Iodine value-118-134
Saponification value-165-185
Unsaponifiable matter-14%
Hexane-25 ppm
Isopropyl alcohol-100 ppm

Dehydrated Beets (Beet Powder)

Volatile matter—4%
Acid-insoluble ash—0.5%
Lead (as Pb)—10 ppm
Arsenic (as As)—1 ppm
Mercury (as Hg)—1 ppm

Dihydroxyacetone

Volatile matter (at 34.6°C for 3 hr at ≤30 mm Hg pressure)—0.5%
Residue on ignition—0.4%
Lead (as Pb)—20 ppm
Arsenic (as As)—3 ppm
Iron (as Fe)—25 ppm
1,3-Dihydroxy-2-propanone—98% min.

Disodium EDTA-Copper

Total copper—13.5% min.
Total (ethylenedinitrilo)tetraacetic acid—62.5% min.
Free copper—100 ppm
Free disodium salt of (ethylenedinitrilo) tetraacetic acid—1.0%
Moisture—15%
Water-insoluble matter—0.2%
Lead (as Pb)—20 ppm
Arsenic (as As)—3 ppm

Ferric Ammonium Citrate

Iron (as Fe)—14.5-18.5%
Lead (as Pb)—20 ppm
Arsenic (as As)—3 ppm

Ferric Ammonium Ferrocyanide

Oxalic acid or its salts—0.1%
Water-soluble matter—3%
Water-soluble cyanide—0.5 ppm
Volatile matter—4%
Lead (as Pb)—20 ppm
Arsenic (as As)—3 ppm
Nickel (as Ni)—60 ppm
Cobalt (as Co)—60 ppm
Mercury (as Hg)—1 ppm
Total iron (as Fe)—33-37%

Ferric Ferrocyanide (Proposed Specifications)

Water-soluble cyanide—10 ppm
Lead (as Pb)—20 ppm
Arsenic (as As)—3 ppm
Nickel (as Ni)—200 ppm
Cobalt (as Co)—200 ppm
Mercury (as Hg)—1 ppm
Oxalic acid— .1%
Water-soluble matter—3%
Volatile matter—10%
Total iron (as Fe) corrected for volatile matter—37-45%

Ferrous Gluconate

Assay (as $C_{12}H_{22}FeO_{14}$, dried basis)—95.0% min.
Loss on drying (105°C for 4 hr)—6.5-10%
Arsenic (as As)—3 ppm
Chloride—700 ppm
Ferric iron—2%
Lead—10 ppm
Mercury—3 ppm
Oxalic acid—passes test
Reducing sugars—passes test
Sulfate—0.1%

Grape-skin Extract (Enocianina)

Pesticide residues—not more than permitted in or on grapes by regulations
    promulgated under section 408 of the Federal Food, Drug, and Cosmetic
    Act
Lead (as Pb)—10 ppm
Arsenic (as As)—1 ppm

Guaiazulene

Melting point—30.5-31.5°C
Lead (as Pb)—20 ppm
Arsenic (as As)—3 ppm
Mercury (as Hg)—1 ppm
Total color—99% min.

Guanine

Guanine—75% min.
Hypoxanthine—25%

Ash (ignition at 800°C)—2%
Lead (as Pb)—20 ppm
Arsenic (as As)—3 ppm
Mercury (as Hg)— 1 ppm
Assay (total purines)—96% min.

Henna

Shall contain no more than 10% of plant material from *Lawsonia alba* Lam.
(*Lawsonia inermis* L.) other than the leaf and petiole, and shall be free from
    admixture with material from any other species of plant
Moisture—10%
Total ash—15%
Acid-insoluble ash—5%
Lead (as Pb)—20 ppm
Arsenic (as As)—3 ppm

Logwood Extract

Volatile matter (at 110°C)—15%
Sulfated ash—20%
Hematein—5-20%
Lead (as Pb)—70 ppm
Arsenic (as As)—4 ppm
Mercury (as Hg)—3 ppm

Manganese Violet

Ash (at 600°C)—81% min.
Volatile matter (at 135°C for 3 hr)—1%
Water-soluble substances—6%
pH of filtrate of 10 g of color additive (shaken occasionally for 2 hr with 100
    ml of freshly boiled distilled water)—4.7-2.5
Lead (as Pb)—20 ppm
Arsenic (as As)—3 ppm
Mercury (as Hg)—1 ppm
Total color (based on Mn content of as-is sample)—93% min.

Mica

Fineness
    through a 100-mesh sieve—100% min.
    through a 200-mesh sieve—80% min.
Loss on ignition at 600-650°C—2%
Lead (as Pb)—20 ppm

Arsenic (as As)—3 ppm
Mercury (as Hg)—1 ppm

Paprika Oleoresin

Solvent residue—no more than that permitted for the corresponding solvent in spice oleoresins

Potassium Sodium Copper Chlorophyllin (Chlorophyllin-Copper Complex)

Moisture—5.0%
Nitrogen—5.0%
pH of 1% solution—9-11
Total copper—4-6%
Free copper—0.25%
Iron—0.5%
Lead (as Pb)—20 ppm
Arsenic (as As)—5 ppm
Ratio, absorbance at 405 nm: absorbance at 630 nm—3.4-3.9
Total color—75% min.

Pyrogallol

Melting point—130-133°C
Residue on ignition—0.1%
Lead (as Pb)—20 ppm
Arsenic (as As)—3 ppm

Pyrophyllite

Lead (as Pb)—20 ppm
Arsenic (as As)—3 ppm
Lead and arsenic shall be determined in the solution obtained by boiling 10 g of pyrophyllite for 15 min in 50 ml of $0.5N$ HCl.

Riboflavin

Assay (as $C_{17}H_{20}N_4O_6$, dry basis)—98.0—102.0%
Specific rotation, $[\alpha]_D^{25}$ (dry basis)—between $-112°$ and $-122°$
Loss on drying—1.5%
Lumiflavin—passes test
Residue on ignition—0.3%

Synthetic Iron Oxide (for Dog and Cat Food)

Arsenic (as As)—5 ppm

Lead (as Pb)—20 ppm
Mercury (as Hg)—3 ppm

Synthetic Iron Oxide (for Ingested or Topically Applied Drugs and Cosmetics

Arsenic (as As)—3 ppm
Lead (as Pb)—10 ppm
Mercury (as Hg)—3 ppm

Tagetes (Aztec Marigold) Meal and Extract

Tagetes meal shall be free from admixture with other plant material from *Tagetes erecta* L. and from plant material or flowers of any other species of plant.

Tagetes extract shall be prepared from tagetes petals meeting the above-mentioned specification and, in addition, shall conform to the following requirements:

Melting point—53.5-55.0°C
Iodine value—132-145
Saponification value—175-200
Acid value—0.60-1.20
Titer—35.5-37.0
Unsaponifiable matter—23.0-27.0%
Hexane residue—25 ppm

All determinations except the hexane residue shall be made on the initial extract of the flower petals (after drying in a vacuum oven at 60°C for 24 hr) prior to the addition of oils and ethoxyquin; hexane determination shall be made on the color additive after addition of vegetable oils, hydrogenated vegetable oils and ethoxyquin

Talc

Loss on ignition (at red heat to constant weight)—5%
Acid-soluble substances as sulfate—2%
Reaction and soluble substances—0.1%
Water-soluble iron—passes test

Lead (as Pb)—20 ppm
Arsenic (as As)—3 ppm

Lead and arsenic shall be determined in the solution obtained by boiling 10 g of talc for 15 min. in 50 ml of 0.5N hydrochloric acid

Titanium Dioxide

Lead (as Pb)—10 ppm

Arsenic (as As)—1 ppm
Antimony (as Sb)—2 ppm
Mercury (as Hg)—1 ppm
Loss on ignition at 800°C (after drying for 3 hr at 105°C)—0.5%
Water-soluble substances—0.3%
Acid-soluble substances—0.5%
TiO_2 (After drying for 3 hr at 105°C)—99.0% min.
Lead, arsenic, and antimony shall be determined in the solution obtained by boiling 10 g of the colorant for 15 min in 50 ml of 0.5N hydrochloric acid

Toasted Partially Defatted Cooked Cottonseed Flour

Arsenic (as As)—0.2 ppm
Lead (as Pb)—10 ppm
Free gossypol—450 ppm

Turmeric Oleoresin

Solvent Residue—no more than that permitted for the corresponding solvent in spice oleoresins

Ultramarine Blue (for Coloring Salt Intended for Animal Feed)

Lead (as Pb)—10 ppm
Arsenic (as As)—1 ppm
Mercury (as Hg)—1 ppm

Ultramarines (for Coloring Externally Applied Cosmetics)

Lead (as Pb)—20 ppm
Arsenic (as As)—3 ppm
Mercury (as Hg)—1 ppm

Zinc Oxide

Zinc oxide (as ZnO)—99% min.
Loss on ignition at 800°C—1%
Cadmium (as Cd)—15 ppm
Mercury (as Hg)—1 ppm
Arsenic (as As)—3 ppm
Lead (as Pb)—20 ppm

Appendix B Domestic Suppliers of Color Additives

Buffalo Color Corporation, 1 Garret Mountain Plaza, West Paterson, N. J. 07424 (800-631-0171). Certified FD&C and D&C colors.

Colorcon Inc., Moyer Blvd., West Point, Pa. 19486 (215-646-7080). Certified FD&C lakes.

Crompton & Knowles Corporation, Route 724, Gibraltar, Pa. 19524 (215-582-8765). Certified FD&C and D&C colors, carmine lake, and beet powder.

Food Concentrates, Inc., P. O. Box 1014-A Rahway, N. J. 07065 (201-634-2161). Caramel.

Fritzsche Dodge & Olcott Inc., 76 Ninth Ave., New York, N.Y. 10011 (212-929-4100). Paprika and turmeric.

Chr. Hansen's Laboratory, Inc., 9015 W. Maple St., Milwaukee, Wisc. 53214 (414-476-3630). Beet powder, annatto, turmeric, and paprika.

Hilton-Davis, 2235 Langdon Farm Rd. Cincinnati, Ohio 45237 (513-351-1300). Certified FD&C and D&C colorants, iron oxides, and inorganic colors. Thomasset Colors.

Hoffmann-LaRoche Inc., Nutley, N. J. 07110 (800-631-7276). β-Carotene, β-apo-8'-carotenal, and canthaxanthin.

Kalamazoo Spice Extraction Co., 3711 West Main St., Kalamazoo, Mich. 49007 (616-349-9711). Paprika, turmeric, and annatto.

H. Kohnstamm & Company, 161 Avenue of the Americas, New York, N. Y. 10013 (212-620-4833). Certified FD&C colorants, carmine lake, annatto, caramel, turmeric, grape-skin extract, fruit juice, titanium dioxide, synthetic iron oxides, certified D&C colorants, alumina, hydrous and anhydrous chromium oxides, ultramarines, and ferric ferrocyanide.

The Mearl Corporation, 41 East 42nd St., New York, N. Y. 10017. Copper and bronze powders.

Meer Corporation, 9500 Railroad Ave., North Bergen, N. J. 07047 (201-861-9500). Beet powder, annatto, paprika, turmeric, cochineal, grape-skin extract, caramel, and sarfron.

Miles Laboratories, Inc., Marschall Division, 1127 Myrtle St., Elkhart, Ind. 46514. Annatto.

Neumann-Buslee & Wolfe, Inc., 521 Santa Rosa Dr., Des Plaines, Ill. 60018

(312-827-2153). Certified FD&C colorants.

Pfizer, Inc., 235 East 42nd St., New York, N. Y. 10017. Calcium carbonate, synthetic iron oxides,and talc.

Pylam Products Company, Inc., 95-10 218th St., Queens Village, N. Y. 11429 (212-464-0860). Certified FD&C and D&C colorants and grape-skin extract.

Rona Pearl, Inc., East 21st St., Bayonne, N. J. 07002 (201-437-0800). Natural and synthetic pearl pigments, titanium dioxide-, mica-, and talc-base colorants.

Smith Chemical & Color Company, 104-20 Dunkirk St., Jamaica, L. I., N. Y. 11412 (212-454-9400). Iron oxides, ultramarine blue, chromium oxide, mica, talc, zinc oxide, and calcium carbonate.

Stange Company, 342 N. Western Ave., Chicago, Ill. 60612 (312-733-6945). Certified FD&C colorants, and turmeric.

Sun Chemical Corporation, 441 Tompkins Ave., Rosebank, Staten Island, N. Y. 10305 (212-981-1600). Certified D&C colorants, iron oxides, manganese violet, ultramarine blue, chromium hydroxide green, chromium oxide green, and titanium dioxide.

Warner-Jenkinson, 2526 Baldwin St., St. Louis, Mo. 63106 (314-531-1500). Certified FD&C colorants.

D. D. Williamson & Company, Inc., P. O. Box 6001, Louisville, Ky. 40206 (502-895-2438). Caramel colors.

Whittaker, Clark & Daniels, Inc., 1000 Coolidge St., South Plainfield, N. J. 07080 (201-561-6100). Certified D&C colorants, calcium carbonate, chrome oxide greens, mica, talc, titanium dioxide, ultramarines and zinc oxide.

Appendix C Glossary

NOTE: Some of the following terms have broader meanings than those stated. The definitions given here are as the terms relate to color additives in particular.

ADULTERATE—To render impure, spurious, or inferior by adding extraneous or improper ingredients.

BLEED—Leaching of an impurity or minor constituent from a dyed article or a solid dye.

BLOWOUT—Procedure (or its result) whereby solid colorant is dispersed onto a moist absorbent surface to detect impurities or a physical mixture of colorants.

BRIGHTNESS—The attribute of a color that classifies it as equivalent to some member of the series of achromatic (neutral) color perceptions ranging from very dim to very bright or dazzling. Analogous to "value" in the Munsell system of color notation.

CERTIFICATION—The submission of a sample of color additive to the Food and Drug Administration and, after subsequent analysis, the issuance of a certificate of acceptance or "certification."

CHROMA—In the Munsell system of color notation, that quality of color by which we distinguish a strong one from a weak one; the intensity of a distinctive hue; color intensity.

COAL-TAR DYE—An erroneous name often used to describe certifiable colors in the belief that they are derived from coal tar.

COLOR—That aspect of visual perception by which an observer distinguishes differences between two structure-free fields of view of the same size and shape, such as may be caused by differences in the spectral composition of the radiant energy concerned in the observation.

COLOR ADDITIVE—A dye, pigment, or other substance synthesized, extracted, isolated, or otherwise derived from a vegetable, animal, mineral, or other source and that, when added or applied to a food, drug, cosmetic, or the human body or any part thereof, is capable of imparting color, either alone or through reaction with another substance.

COLORANT—A substance such as a dye or pigment that colors or modifies the color of something else.

COSMETIC—Articles intended to be rubbed, poured, sprinkled, or sprayed on, introduced into, or otherwise applied to the human body or any part thereof for cleansing, beautifying, promoting attractiveness, or altering the appearance; articles (except soap) intended for use as a component of any such articles.

DELANEY CLAUSE—That portion of the Color Additive Amendments of 1960 that forbids the use in foods, drugs, and cosmetics of any color additive that can be shown by reasonable tests to cause cancer in man or other animals.

DILUENT—Any component of a color-additive mixture that is not inherently a color additive and has been intentionally mixed therein to facilitate the use of the mixture in coloring foods, drugs, or cosmetics or in coloring the human body.

DRAW-DOWN—Samples used to judge undertone and masstone, prepared by spreading a blob of pigment onto a white backing with a single stroke of a blade.

DRUG—Articles recognized in the official *United States Pharmacopeia*, official *Homeopathic Pharmacopeia of* the United States, or official *National Formulary*, or any supplement thereto; articles intended for use in the diagnosis, cure, mitigation, treatment, or prevention of disease in man or other animals; articles (other than food) intended to affect the structure or any function of the body of man or other animals; articles intended for use as a component of any articles specified above, but not including devices or their components, parts, or accessories.

DYE—A chemical compound that is capable of imparting color and that is soluble in the vehicle in which it is applied.

EXCIPIENT—An inert substance used as a diluent or vehicle.

FLASHING—The visible effect of individual colors in a color blend separately dissolving when the blend is added to a solvent.

FOOD—Articles (including chewing gum) used for food or drink for man or other animals; items used for components of any such article.

HIDING POWER—The opacity of a colored film, usually measured by observing the amount of black transmitted through equal film thicknesses of color when the colored dispersion is drawn down on a sheet of checkered black-and-white paper.

HOMOLOGOUS COLORS—A series of colorants with similar chemical structures that differ only in their chain lengths or in the number of substituent groups they contain.

HUE—In the Munsell system of color notation, the name of a color. That quality by which we distinguish one color family from another, such as red from yellow or green from blue or purple.

INTERMEDIATE—A compound from which a colorant is directly or indirectly synthesized.

ISOMERIC COLORS—Colorants with the same empirical formula but different structural forms.

LAKE—A pigment prepared by precipitating a soluble dye onto an insoluble reactive or adsorptive substratum or diluent.

LISTED COLORANTS–Sometimes mistakenly called "permanently" listed colorants. Those colorants that have been sufficiently evaluated to convince the Food and Drug Administration of their safety for the application intended. See PROVISIONALLY LISTED COLORANTS.

LOT NUMBER–The identifying number or symbol assigned by the Food and Drug Administration to a batch of color additive after certification.

MASSTONE–The color (without regard to background) of a thick layer of a pigment incorporated into a vehicle.

NATURAL COLORANT–One obtained from natural sources; not man-made.

OLEORESIN–The mixture of color and flavor principles obtained from a spice or herb by extracting it with one or more selected solvents and then removing the solvent.

OPACITY–The quality or state of being opaque (i.e., impenetrable by light).

PIGMENT–A colored or white chemical compound that is capable of imparting color and is insoluble in the solvent in which it is being applied. That which is a pigment in relation to one solvent may be a dye in relation to another solvent.

PLATING–The process by which powdered colorant is uniformly deposited onto the surface of a particulate substrate by dry mixing.

POUR-OUT–The process (or its result) whereby a dye solution or dispersion is uniformly spread as a broad streak on a flat uncolored surface for evaluation.

PRIMARY COLOR–A single colorant containing no diluent. See SECONDARY COLOR.

PROVISIONALLY LISTED COLORANTS–Colorants that are not considered unsafe but that nevertheless have not undergone all the tests required by the Color Additives Amendments of 1960 to establish their eligibility for listing. See LISTED COLORANTS.

PURE COLOR (DYE)–The amount of color contained in a color additive, exclusive of any intermediate, diluent, substratum, or other substance.

SATURATION–That attribute of a color perception that determines the degree of its difference from the achromatic (neutral) color perception most resembling it.

SECONDARY COLORANT–A color additive made by mixing two or more straight colors, or one or more straight colors and one or more diluents, or both.

SHADE–Hue.

SUBSIDIARY COLORS–A structural variant of a dye in which the variation is the position, number, and/or chain length of substituent groups.

SUBSTRATUM–The substance on which the pure color in a lake is extended.

TINCTORIAL STRENGTH–A measure of the potential coloring power of a dye.

TONER—An organic pigment containing no substratum or diluent.

UNDERTONE—Color of a thin layer of a pigment incorporated into a vehicle and drawn down on white paper, or color of a tint of the pigment, sometimes as viewed by transmitted light.

VALUE—In the Munsell system of color notation, the lightness of a color. That quality by which we distinguish a light color from a dark one.

PART **B** COLORANT ANALYSIS

Chapter 6 Identification

The techniques used for the analysis of color additives differ little from those used for the examination of other organic compounds. However, because of the extent to which these colorants are examined, particularly the certified food colors, they are among the most thoroughly analyzed group of organic chemicals available today.

Unlike technical dyestuffs, color additives are usually analyzed on an absolute basis rather than versus a standard sample. The reason for this, of course, is obviously the need to know the exact nature of any compound consumed by man or applied to his body.

As is the case with most branches of chemistry, the methodology used to analyze colorants is experiencing a renaissance with many of the old wet procedures being replaced or supplemented by sophisticated instrumental techniques. The majority of the procedures in use today have been developed by both industry and government (FDA), and many have been collaboratively studied for their precision and accuracy. Although few methods are designated as official, many have been used for so long that they often appear as such.

The most frequently used sounding board for the dissemination of new technology in the field of color-additive analysis is the Journal of the Association of Official Analytical Chemists. Timetested methods can be found in a book entitled *Official Methods of Analysis of the Association of Official Analytical Chemists.*

IDENTIFICATION OF COLOR ADDITIVES

Numerous procedures have been used to identify color additives. The methods employed are usually limited only by the inventiveness of the analyst and the equipment to which he has access. Those described here presume that the chemist has a single colorant and that it is not in a food, drug, or cosmetic matrix. The isolation of colorants from product matrices and the resolution of mixtures of colorants are separate problems and are considered in detail later.

In attempting to identify any unknown, the colorant's physical properties, including its solubility, crystal structure, melting point, if any, and color in solution as well as its color under ultraviolet (UV) light, both as is and in solution, provide important clues.

Migration rates such as a colorant's R_f in thin-layer (TLC) and paper-chromatography systems, its retention time or volume during column chromatography, its ionic mobility in electrophoresis experiments, and its partition coefficient during solvent-solvent extraction are all useful as methods of identification. However, one must realize that such constants are not unique and are only conclusive as means of identification when determined versus knowns and in a variety of media.

The behavior of a colorant when mixed with various reagents, including nitric acid, sulfuric acid, hydrochloric acid, sodium hydroxide, and sodium carbonate can be very informative. Qualitative tests for various functional groups and metals can also be meaningful.

One of the more elegant approaches to the identification of azo colorants involves their reduction followed by the identification of the reduction products. Water-soluble colors are usually reduced in hot water with sodium hydrosulfite. Oil-soluble colors are best cleaved in alcohol and under a stream of inert gas using titanium trichloride as the reducing agent. Usually, the reduction products obtained consist of the amine originally diazotized and used to form the dye (or a reduction product of this amine) plus an amino derivative of the compound to which the diazo component was originally coupled. Basic components obtained can be separated from neutral and acidic materials by stream distillation or by extraction from alkaline solution, whereas neutral components can be steam distilled or extracted from neutral solutions. Acidic materials such as sulfonic acids can be neither steam distilled nor extracted from water using simple liquid/liquid techniques and thus remain behind in either of the preceding schemes. Alternately, the reduction products can be separated using chromatographic procedures similar to those described for the determination of uncombined intermediates in color additives. In any event, the products obtained are best identified using UV spectrophotometry. A major advantage of this approach is the need for only a few milligrams of sample to run the test.

The most widely used and in general the most conclusive procedures for identifying color additives are instrumental in nature. Thermal techniques such as differential thermal analysis (DTA) and differential scanning calorimetry (DSC) are useful since the thermograms produced are "fingerprints" of the compounds examined. Thermal methods, though, are generally more valuable as qualitative tools for the study of purity and stability and thus far have found little application as methods of identification. The real workhorse in this area is spectrometry.

Ultraviolet and visible spectrometry are usually the simplest to perform and require the least amount of sample, often as little as 0.1 mg. Where possible, spectra of the unknown should be compared with those of knowns in several solvents since, although two colorants may have almost identical spectra in any one solvent, it is rare that their spectra will be the same in several of them. The solvents chosen for such comparisons should be as different as possible—aprotic versus protic, acid versus alkaline, polar versus nonpolar, and so on—with due consideration, of course, of the solvent's spectral characteristics. Spectra can usually be adequately compared by visual inspection alone; however, sophisticated electronic equipment is available, particularly for work in the visible region, that can evaluate and compare them mathematically. When working in the UV region one must be careful that the colorant does not contain colorless UV-absorbing excipients that can clutter the spectra and mislead the analyst. Ultraviolet and visible spectrometry are fast, reliable, and relatively simple procedures for identifying dyestuffs and should be used whenever possible. Their use requires only a modest amount of training, whereas the necessary equipment is moderate to expensive in price.

Infrared (IR) has also been used extensively for the identification of unknowns. These techniques are generally a little more complicated and expensive to use than UV and visible spectrometry but usually provide a higher order of certainty when dealing with unknowns. Infrared spectra have been obtained as Nujol mulls, as KBr pellets, in solution, and as complexes in liquid ion exchange resins such as Amberlite LA-2 (Rohm and Haas, Philadelphia, Pa.). The procedure to use, of course, depends on the nature of the dyestuff being examined. Organic diluents such as dextrin and sugar and certain inorganics such as sulfates, present naturally or as deliberately added diluents, complicate IR spectra and should be removed prior to analysis or taken into account when interpreting them. When comparing spectra prepared as mulls or KBr pellets, it's also necessary to remember that differences observed can be due to purely physical reasons including the sample's particle size, the pressure and time used in preparing the KBr pellet, and other variables and may have nothing to do with the structure of the dyestuff itself.

Recently nuclear magnetic resonance (NMR) has been used to identify color additives. Good spectra of the certified watersoluble food colors have already been obtained and published using a mixed, deuterated solvent (water:dimethylsulfoxide; $D_2O:DMSO-d_6$, 2:1 v/v) at 100-105°C, and it can be presumed that more work in this field will soon follow. Nuclear magnetic resonance is one of the least sensitive, most complicated, and most expensive of the spectral techniques in use today but is an excellent tool for identification purposes and for studying the structure of organic compounds. Inorganic salts such as sodium chloride and sodium sulfate do not interfere with NMR spectra, but protonated impurities, such as acetates, sugars, and other dyestuffs, do.

Not much has been reported in the field of color additives using Raman spectroscopy; however, this tool should be excellent for the identification of colorants, particularly the water-soluble ones.

A great many spectra of color additives have been published to date, some of which are referred to in the bibliography at the end of this chapter. Many of these are of high-enough quality that they can be used as standards for comparison purposes; however, comparisons are best made against knowns prepared by the same analyst at the same time and on the same equipment

REACTIONS OF SOME NATURAL COLORING MATERIALS[a]

Coloring Matter	Concentrated Hydrochloric Acid	10% Sodium Hydroxide Solution	Sodium Hyposulfite
Annatto	Remains orange, little change		Little affected
Caramel	Little or no change	Little change or slightly browner	Slightly paler
Carotene and xanthophyll	Little change, perhaps slightly paler	Little or no change	Little affected
Cochineal	Little or no change	Violet	No marked change
Logwood	Deep red with excess of acid	Violet to violet-blue	Almost decolorized, color returning imperfectly by reoxidation
Saffron	Little or no change	Remains yellow	Little affected
Turmeric (solution in ethyl ether or ethanol characterized by pure yellow color and light green fluorescence)	Orange-red or carmine-red on addition of several volumes of concentrated acid	Orange-brown	Little affected

From: Official Methods of Analysis, 11th ed., The Association of Official Analytical Chemists, Washington, D.C., 1970, p. 581.

[a]Procedure: Dissolve the color in a small amount of ethanol and dilute with water. To individual portions of this solution apply reagents as follows:

Hydrochloric acid—Add one or two drops of concentrated hydrochloric acid. Then dilute to three or four times the sample volume with concentrated hydrochloric acid.

Sodium hydroxide solution—Make solution slightly alkaline with one drop of 10% sodium hydroxide.

Sodium hyposulfite—Add $Na_2S_2O_4$ crystals.

0.5% Ferric Chloride Solution	10% Alum Solution	5% Uranium Acetate Solution	Sulfuric Acid on Dry Color
No marked change, perhaps somewhat browner			Blue
No change			
			Blue, reaction obtained with difficulty
Slightly darker		Green	
Dark shades of violet, brown, or black (first hue often fleeting)	Rose red (change rather slow)	Violet, quickly fading	Red, changing to yellow
No marked change, perhaps somewhat browner	Little change	Not affected	Blue
No marked change, perhaps somewhat browner	Little change	Somewhat browner	Red

Ferric chloride solution—Add fresh 0.5% $FeCl_3$ solution dropwise; colors are not always obtained if an excess is used.

Alum solution—Add to the test solution 20% of its volume of 10% potassium alum or ammonium alum solution.

Uranium acetate solution—Add 5% $UO_2(OAc)_2 \cdot 2H_2O$ solution dropwise.

Sulfuric acid on dry color—Dry a small portion of the color in an evaporating dish. Cool. Treat the residue with one or two drops of cold concentrated sulfuric acid. The color formed is sometimes extremely transitory and may be noted only when the acid wets the residue.

used to prepare the unknowns. Selected thermograms and spectra are included here to illustrate the value of the various techniques (see Figs. 1a-k, 2a-j, and 3a-h).

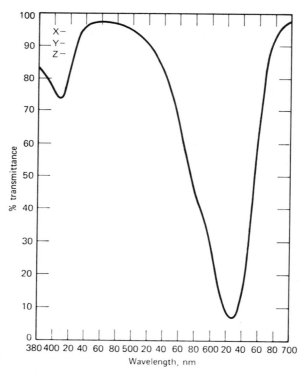

Figure 1a **Visible Spectrum of FD&C Blue No. 1: Concentration, 8 ppm; Cell Thickness, 1 cm; Solvent, Water; Reference, Water** (From *Encyclopedia of Industrial Chemical Analysis*, **Vol. 10, Wiley, New York, 1970)**

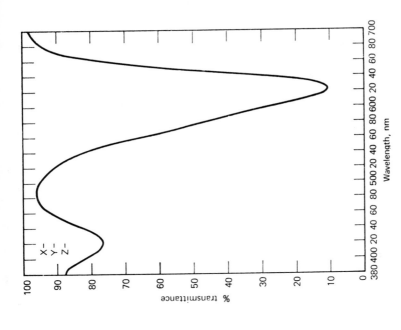

Figure 1*b* Visible Spectrum of FD&C Blue No. 2: Concentration, 22 ppm; Cell Thickness, 1 cm; Solvent, Water; Reference, Water (From *Encyclopedia of Industrial Chemical Analysis*, Vol. 10, Wiley, New York, 1970)

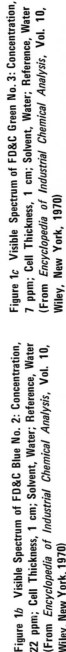

Figure 1*c* Visible Spectrum of FD&C Green No. 3: Concentration, 7 ppm; Cell Thickness, 1 cm; Solvent, Water; Reference, Water (From *Encyclopedia of Industrial Chemical Analysis*, Vol. 10, Wiley, New York, 1970)

131

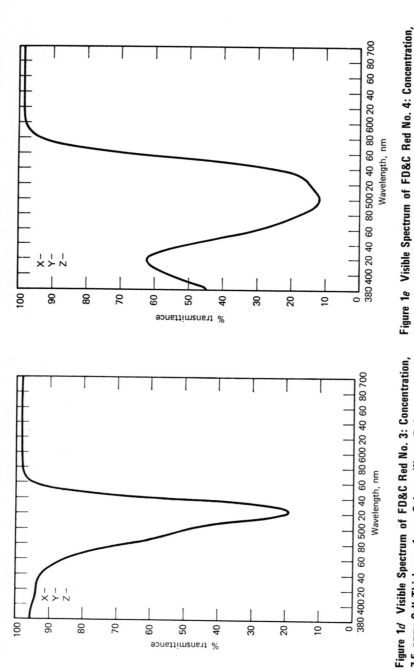

Figure 1d Visible Spectrum of FD&C Red No. 3: Concentration, 7.5 ppm; Cell Thickness, 1 cm; Solvent, Water; Reference, Water (From *Encyclopedia of Industrial Chemical Analysis*, Vol. 10, Wiley, New York, 1970)

Figure 1e Visible Spectrum of FD&C Red No. 4: Concentration, 20 ppm; Cell Thickness, 1 cm; Solvent, Water; Reference, Water (From *Encyclopedia of Industrial Chemical Analysis*, Vol. 10, Wiley, New York, 1970)

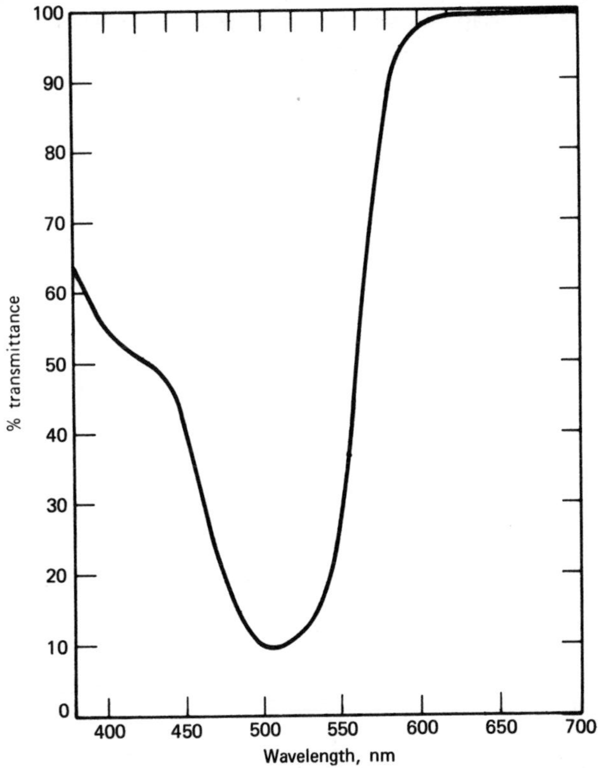

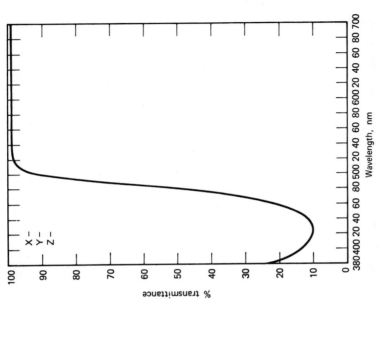

Figure 1g Visible Spectrum of FD&C Yellow No. 5: Concentration, 20 ppm; Cell Thickness, 1 cm; Solvent, Water; Reference, Water (From *Encyclopedia of Industrial Chemical Analysis*, Vol. 10, Wiley, New York, 1970)

Figure 1f Visible Spectrum of FD&C Red No. 40: Concentration, 20 ppm; Cell Thickness, 1 cm; Solvent, Water; Reference, Water

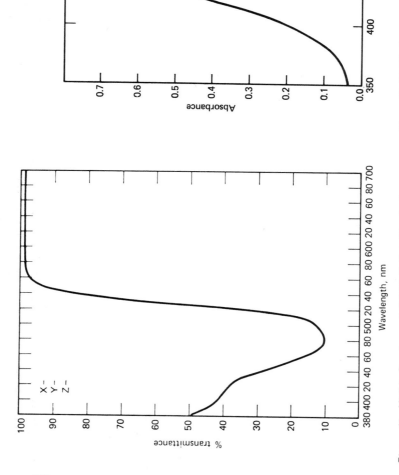

Figure 1h Visible Spectrum of FD&C Yellow No. 6: Concentration, 20 ppm; Cell thickness, 1 cm; Solvent, Water; Reference, Water (From *Encyclopedia of Industrial Chemical Analysis*, Vol. 10, Wiley, New York, 1970)

Figure 1i Visible Spectrum of All-*trans* β-Carotene: Concentration, 2.68 μg/ml; Cell Thickness 1 cm; Solvent, Cyclohexane

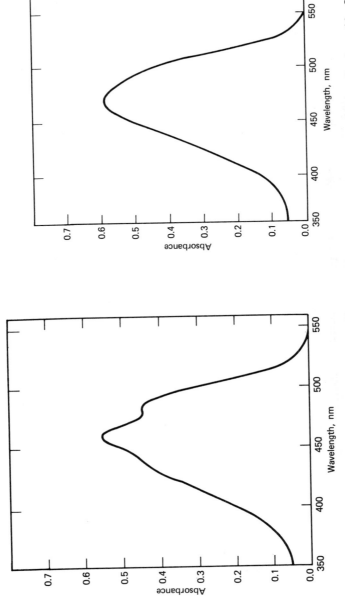

Figure 1j Visible Spectrum of All-*trans* β-apo-8'-Carotenal: Concentration, 2.11 μg/ml; Cell Thickness, 1 cm; Solvent, Cyclohexane

Figure 1k Visible Spectrum of All-*trans* Canthaxanthin: Concentration, 2.67 μg/ml; Cell Thickness, 1 cm; Solvent, Cyclohexane

135

Figure 2 Nuclear Magnetic Resonance Spectra Obtained on Varian A60 Spectro-photometer Equipped with Variable-temperature Probe and Using the Following Operating Conditions: Probe Temperature, 100-105°C; Filter Bandwidth, 0.4 Hz; Radiofrequency field, 0.07 mG; Sweep Time, 500 sec; Sweep Offset, 0 Hz; Solvent, D_2O/DMSO-d_6 (Dimethylsulfoxide) (2 + 1, v/v); Marker TSP (Sodium 3-Trimethylsilylpropionate-2,2,3,3-d_4)

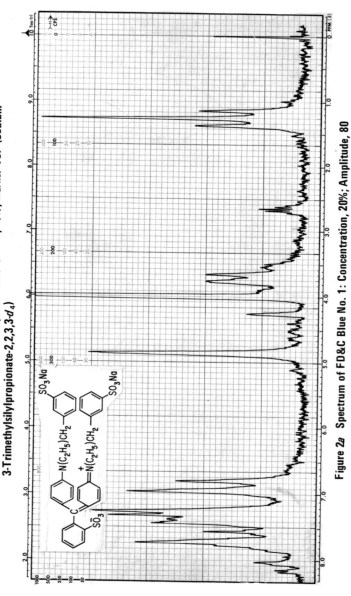

Figure 2a Spectrum of FD&C Blue No. 1: Concentration, 20%; Amplitude, 80

136

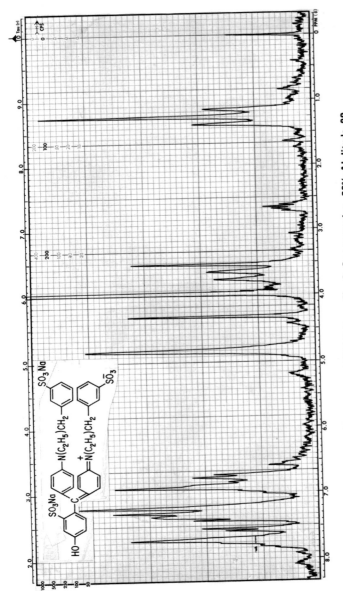

Figure 2b Spectrum of FD&C Green No. 3: Concentration, 20%; Alplitude, 80

137

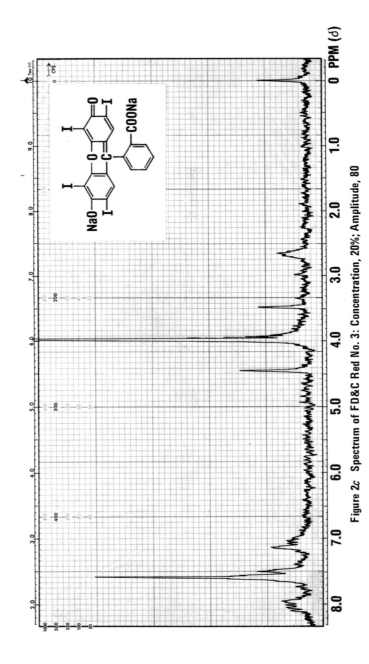

Figure 2c Spectrum of FD&C Red No. 3: Concentration, 20%; Amplitude, 80

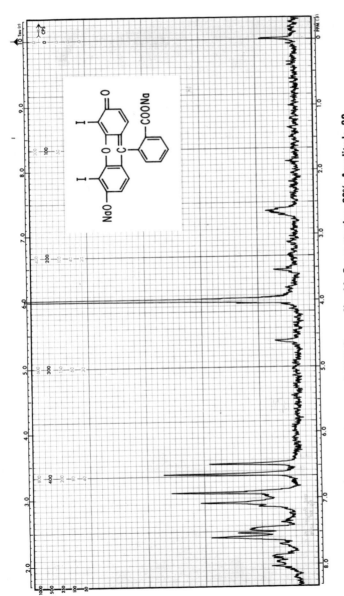

Figure 2d Spectrum of D&C Orange No. 11: Concentration, 20%; Amplitude, 80

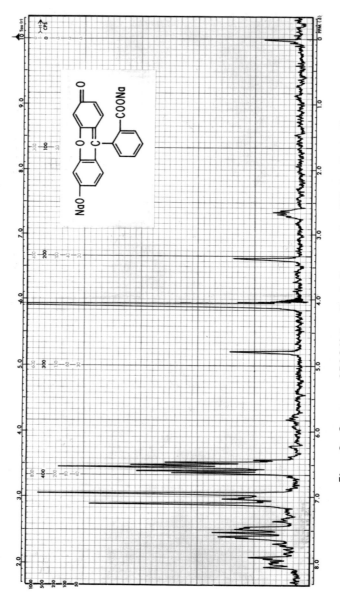

Figure 2e Spectrum of D&C Yellow No. 8: Concentration, 20%; Amplitude, 63

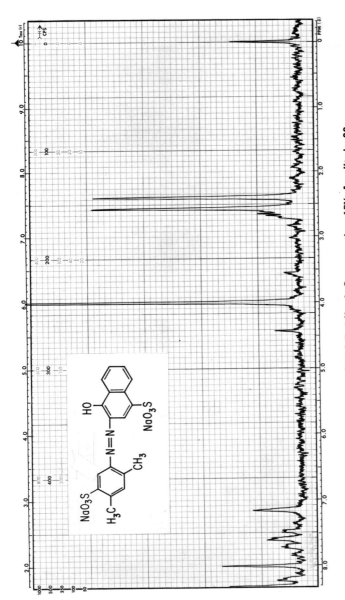

Figure 2f   Spectrum of FD&C Red No. 4: Concentration, 15%; Amplitude, 80

141

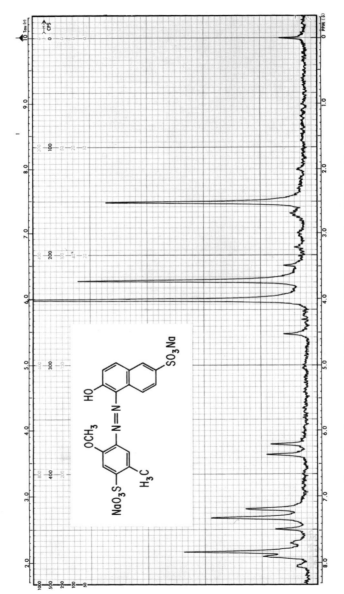

Figure 2g Spectrum of FD&C Red No. 40: Concentration, 20%; Amplitude, 40

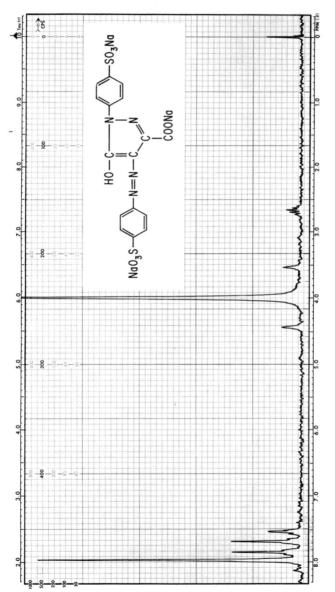

Figure 2h Spectrum of FD&C Yellow No. 5: Concentration, 20%; Amplitude, 25

143

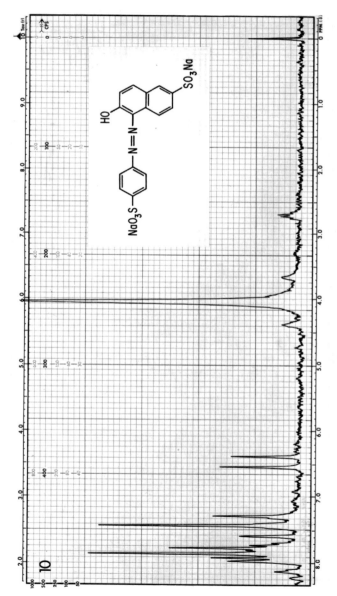

Figure 2*i* Spectrum of FD&C Yellow No. 6: Concentration, 20%; Amplitude, 50

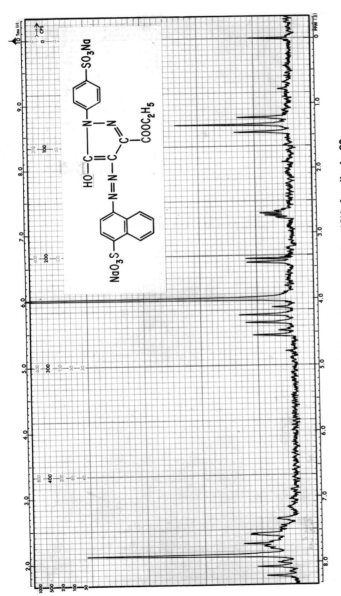

Figure 2*j* Spectrum of Orange B: Concentration, 10%; Amplitude, 63

145

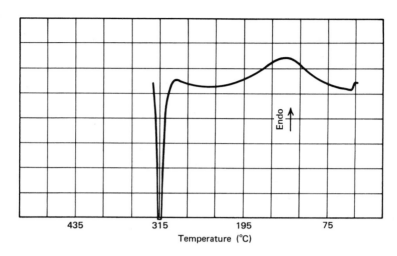

Figure 3a Thermogram of FD&C Blue No. 1

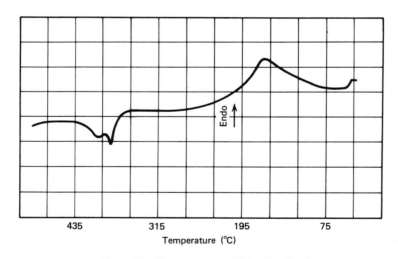

Figure 3b Thermogram of FD&C Red No. 3

Figure 3 Thermograms Obtained Using Perkin-Elmer DSC-1B Calorimeter and the Following Conditions: Sample Weight, 9.8-10 mg (3.5 mg for D&C Red No. 17); Scan Rate, 40°/min; Range, ×32; Atmosphere, N_2

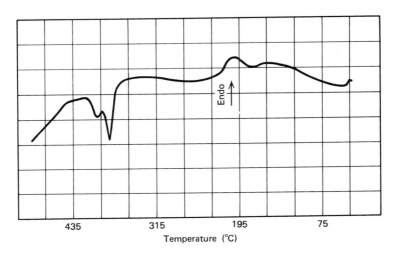

Temperature (°C)

Figure 3c Thermogram of FD&C Red No. 4

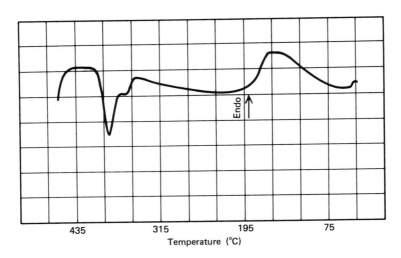

Temperature (°C)

Figure 3d Thermogram of FD&C Yellow No. 5

147

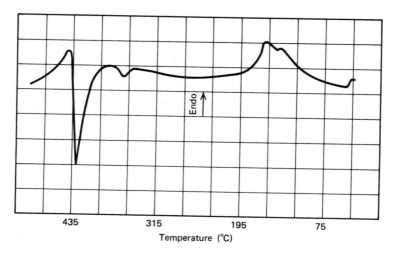

Figure 3*e* Thermogram of FD&C Yellow No. 6

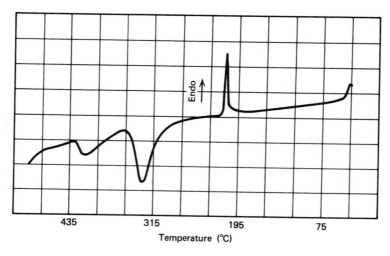

Figure 3*f* Thermogram of D&C Red No. 17

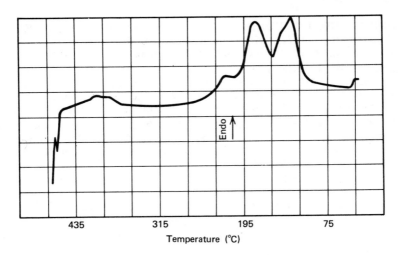

Figure 3*g* Thermogram of D&C Yellow No. 8

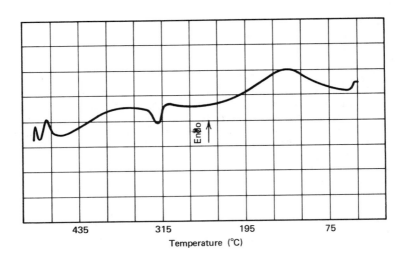

Figure 3*h* Thermogram of D&C Yellow No. 10

BIBLIOGRAPHY

BAUERNFEIND, J. C., BUNNELL, R. H. Food Technol. *16*, 76-82 (1962). β-Apo-8′-Carotenal—A New Food Color. Includes visible spectrum of β-apo-8′-carotenal in cyclohexane and IR spectrum as a KBr disk.

BROWN, C. W., LYNCH, P. F. J. Food Sci. *41*, 1231-1232 (1976). Identification of FD&C Dyes by Resonance Raman Spectroscopy. FD&C Red No. 40, FD&C Red No. 4, and Amaranth were studied by Raman spectroscopy using a 4880-Å laser line with about 300-mW power at the sample.

BROWN, J. C. JSDC *85*, 137-146 (1969). The Chromatography and Identification of Dyes. Describes techniques suitable for the identification of dyes, including paper and thin-layer chromatography, UV and IR spectrophotometry, and electrophoresis.

DE GORI, R., GRANDI, F. Boll. Lab. Chem. Provinciali *12*, 60-80 (1961). Spectrophotometric Identification of Colors for Food Use. Spectrophotometric curves are presented in neutral, alkaline, and acid media for 13 colorants, including FD&C Red No. 3 and FD&C Yellow No. 5.

PLÁ-DELFINA, J. M. J. Soc. Cosmet. Chemists *13*, 214-244 (1962). Systematic Identification of Food, Drug and Cosmetic Azo Dyes. Presents a simple, systematic, paper-chromatographic method for separating and identifying water-soluble azo colors.

DOLINSKY, M., JONES, J. H. JAOAC *37*, 197-209 (1954). The Infrared Spectra of Some Unsulfonated Monoazo Dyes. The IR technique as applied to monoazo dyes.

DOLINSKY, M. JAOAC *37*, 805-808 (1954). Report on Subsidiary Dyes in FD&C Colors. I. Higher Sulfonated Dyes in FD&C Yellow No. 6. Shows visible spectra of FD&C Yellow No. 6 in water, $0.1N$ HCl, and $0.1N$ NaOH.

DOLINSKY, M., STEIN, C. Anal. Chem. *34*, 127-129 (1962). Solubilization of Sulfonic Acids for Infrared Studies. Describes the use of Amberlite LA-2 (a liquid anion exchange resin) for solubilizing sulfonated dyestuffs for IR analysis. Extraction procedure: dissolve 100-200 mg of sample in 50 ml 2% (v/v) aqueous HCl. Extract with two 10-ml then one 5-ml portion of 5% Amberlite LA-2 in carbon disulfide. Concentrate as needed and dry. Ethanolic procedure: heat 100-200 mg of sample on a steam bath with 25 ml of 5% Amberlite LA-2 in 95% ethanol until the odor of alcohol can no longer be detected. Dissolve the residue in warm carbon disulfide, make to volume, and dry.

EVANS, W. H., MAC NAB, J. A., WARDLEWORTH, D. F. J. Sci. Food Agric. *21*, 207-210 (1970). Infrared Identification of Synthetic Food Colors. Describes the preparation of Nujol mulls and alkali halide disks of 55 colorants. Includes no spectra.

FOPPEN, F. Chromatog. Rev. *14*, 133-298 (1971). Tables for the Identification of Carotenoid Pigments. Includes data on paper, thin-layer, and column chromatography; visible, UV, IR, and mass spectrometry; melting points; partition coefficients.

FRANC, F., STRÁNSKÝ, Z. Collection Czechoslov. Chem. Commun. *24*, 3611-3623 (1959). Chromatography of Organic Compounds. III. Identification of Organic Compounds by Means of Chromatographic Spectra. Compounds are chromatographed in a series of 12 systems of stationary and mobile phases. The R_f values obtained are plotted in a fixed sequence on graph paper to obtain a "chromatographic spectrum" characteristic for each compound.

FREEMAN, J. F. Can. Textile J. (February) 83-89 (1970). An Introduction to Modern Methods of Dye Identification—Chromatography and Spectrophotometry. Presents a general survey of the classical and modern methods for dye identification.

FUJII, S., KAMIKURA, M., HOSOGAI, Y. Eisei Shikenjo Hôkoku *75*, 29-31 (1957). Paper Chromatography of the Reduction Products of Monoazo Dyes. Dissolve 0.1 g of dye in water or ethanol and reduce at room temperature to a colorless solution by the dropwise addition of fresh 10% sodium hydrosulfite. Resolve 0.01 ml of the colorless solution on Toyo filter paper, No. 50, 8 cm X 40 cm. The most useful solvents reported are BuOH:EtOH:0.5N NH$_4$OH (6:2:3) and BuOH:EtOH:0.5N AcOH (6:2:3).

GRAICHEN, C., MOLITOR, J. C. JAOAC *42*, 149-160 (1959). Studies on Coal-Tar Colors. XXII. 4,5-Dibromofluorescein and Related Bromofluoresceins. Visible spectrum of D&C Orange No. 5 in 0.5% NH$_4$OH.

CHR. HANSEN'S LABORATORY, INC., Milwaukee, Wisconsin. Annatto Food Colors. Shows spectrum of oil-soluble annatto in chloroform.

HARROW, L. S., JONES, J. H. JAOAC *36*, 914-923 (1953). The Identification of Azo Dyes by Spectrophotometric Identification of Their Reduction Products. Describes the reduction of colorants with sodium hydrosulfite or titanium trichloride and the resolution of the reduction products by extraction and steam distillation and their identification by UV spectrometry.

HOODLESS, R. A., PITMAN, K. G., STEWART, T. E., THOMSON, J., ARNOLD, J. E. J. Chromatog. *54*, 393-404 (1971). Separation and Identification of Food Colours. I. Identification of Synthetic Water-Soluble Food Colours Using Thin-layer Chromatography. A scheme is described that identifies unknown dyes by observing their TLC behavior relative to Orange G (CI Acid Orange 10) and Amaranth (CI Food Red 9) in several solvent systems.

JONES, J. H., CLARK, G. R., HARROW, L. S. JAOAC *34*, 135-148 (1951). A Variable Reference Technique For Analysis By Absorption Spectrophotometry. Describes a procedure whereby a sample is run spectrophotometrically versus a known reference solution that is continually varied until it compensates for all the sample absorption. The method can be used to prove the identity of a sample or to analyze mixtures.

JONES, J. H., HARROW, L. S. JAOAC *34*, 831-842 (1951). The Identification of Azo Dyes By Spectrophotometric Identification of Their Reduction Products. Water-soluble colors are reduced in water with sodium hydrosulfite, and oil-soluble colors are reduced in alcohol with titanium trichloride. The reduction products are resolved by distillation, extraction, and/or chromatography and identified spectrophotometrically.

KAMIYA, I., IWAKI, R. Bull. Chem. Soc. Jap. 39, 264-269 (1966). Studies of the Chemiluminescence of Several Xanthene Dyes. II. The Chemiluminescence Emission Spectra of Uranine and Eosine. The emission spectra of uranine (D&C Yellow No. 8) and eosine (D&C Red No. 22) are described.

KARASZ, A. B., DE COCCO, F., BOKUS, L. JAOAC 56, 626-628 (1973). Detection of Turmeric in Foods by Rapid Fluorometric Method and by Improved Spot Test. Contains excitation and emission fluoresence spectra of turmeric in butanol from 650 nm to 250 nm.

KITAHARA, S., MIYAZAKI, S., HIYAMA, H. Kôgyô Kagaku Zasshi 61, 189-193 (1958). Detection of Reduction Products of Basic and Acid Azo Dyes by Paper Chromatography. Colorants are reduced by heating with Sn-HCl or Zn-HCl, chromatographed on paper at $20^{\circ}C$ using BuOH:HCl (4:1) or 2% HCl, and then the spots are detected with 1-naphthol-4-sulfonic acid or aqueous $FeCl_3$ and identified by comparison with knowns. Color additives discussed include FD&C Yellow No. 5 and FD&C Yellow No. 6.

MARMION, D. M. JAOAC 53, 244-249 (1970). Evaluation of Color Additives Using a Differential Scanning Calorimeter. Includes thermograms of FD&C Red No. 4, FD&C Yellow No. 5, FD&C Yellow No. 6, FD&C Blue No. 1, FD&C Red No. 3, D&C Red No. 17, D&C Yellow No. 8 and D&C Yellow No. 10 and describes the use of DSC as a tool for identification.

MARMION, D. M. JAOAC 54, 131-136 (1971). Analysis of Allura Red AC (A Potential New Color Additive). Contains the visible spectrum of FD&C Red No. 40 in distilled water.

MARMION, D. M. JAOAC 57, 495-507 (1974). Applications of Nuclear Magnetic Resonance to the Analysis of Certified Food Colors. Includes the NMR spectra of all the certified food colors permitted in the United States except FD&C Blue No. 2.

PRZYBYLSKI, W., MC KEOWN, G. G. JAOAC 43, 800-804 (1960). Absorption Spectra of 1-Arylazo-2-Naphthol Food Colors. Contains spectra of Citrus Red No. 2 in hexane, absolute ethanol, aqueous ethanol, and chloroform.

PUCHE, R. C. T. Publ. Inst. Invest. Microquim., Univ. Nacl. Litoral 25, Nos. 23-24, 58-62 (1959-1960). Absorption maxima and minima for 13 food dyes are listed.

REITH, J. F., GIELEN, J. W. J. Food Sci. 36, 861-864 (1971). Properties of Bixin and Norbixin and the Composition of Annatto Extracts. Includes IR spectra (KBr disk) of α-bixin, β-bixin, α-norbixin, and β-norbixin.

SADTLER STANDARD SPECTRA, Sadtler Research Laboratories, 3316 Spring Garden St., Philadelphia, Pa. 19104. A collection of visible, UV, and IR spectra. The color additives included in this collection are listed as follows by page number.

	Page in Sadtler Collection	
Color Additive	IR	Visible/UV
FD&C Blue No. 1	X2082	U394
FD&C Blue No. 2	X2081	U393
FD&C Green No. 3	X2090	U396
FD&C Red No. 3	X2068	U391
FD&C Red No. 4	X2055	U388
FD&C Yellow No. 5	X2042	U387
FD&C Yellow No. 6	X2041	U386
D&C Brown No. 1	X21	U2
D&C Blue No. 4	X10	
D&C Blue No. 6	X12	
D&C Blue No. 9	X15	
D&C Green No. 5	X31	
D&C Green No. 6	X32	
D&C Orange No. 4	X42	U3
D&C Orange No. 5	X43	
D&C Orange No. 17	X55	
D&C Red No. 6	X66	
D&C Red No. 7	X67	
D&C Red No. 8	X68	
D&C Red No. 9	X69	
D&C Red No. 17	X77	
D&C Red No. 19	X79	U4
D&C Red No. 21	X81	
D&C Red No. 22	X82	
D&C Red No. 27	X87	
D&C Red No. 28	X88	U5
D&C Red No. 31	X91	
D&C Red No. 33	X93	U6
D&C Red No. 34	X94	
D&C Red No. 36	X96	
D&C Red No. 37	X97	
D&C Red No. 39	X99	
D&C Violet No. 2	X106	
D&C Yellow No. 7	X118	
D&C Yellow No. 8	X119	
D&C Yellow No. 10	X121	U7
D&C Yellow No. 11	X122	
Ext. D&C Violet No. 2	X172	
Ext. D&C Yellow No. 7	X184	U9 and U10

SHELTON, J. H., GILL, J. M. T. J. Assoc. Public Analysts, *1* 88-91 (1963). Paper Chromatographic Identification of Food Dyes. The chromatographic method of Yanuka is used to identify food colors permitted in the United Kingdom.

SUZUKI, M., NAKAMURA, E., NAGASE, Y. Analytical Studies for Dyes by Using Infrared Spectrum. I. Yellow Dyes for Food, Auramine and Butter Yellow. Yakugaku Zasshi *79*, 1116-19 (1959); II. Green Dyes for Food and Malachite Green. Ibid. *79*, 1209-1211 (1959); III. Red Dyes for Food and Rhodamine B. Ibid. *80*, 916-919 (1960). Describes the KBr method for the identification of color additives. The discussions include Ext. D&C Yellow No. 7, FD&C Yellow Nos. 5 and 6, FD&C Green No. 3, FD&C Red Nos. 3 and 4, and D&C Red Nos. 19, 22, and 28.

SZOKOLAY, A., PAGACOVA, A. Prumysl Potravin *12*, 656-658 (1961). Identification of Food Colorants by Light-Absorption Measurement. Includes visible and ultraviolet spectra of various colorants.

TONET, N. Mitt Geb. Lebens. Hyg., *60*, 201-205 (1969). Use of High-Voltage Electrophoresis as a Supplementary Tecnhique for the Identification of Water-Soluble Dyes. Seventy-five synthetic dyes and several natural colorants were studied by electrophoresis at 4500 V with 20% acetic acid or 0.1 M-aqueous NH_3:3.3 mM acetic-acid buffer adjusted to pH = 10.3.

VILLANÚA, L., CARBALLIDO, A., MUÑIZ, J. An. Bromatol. *20*, 113-136 (1968). Artificial Food Colours. XI. Ultra-Violet and Visible Spectrometry of Some Water-Soluble Colours. Spectrophotometric parameters in the range 350-700 nm are reported for 46 water-soluble dyes permitted in Spain.

YANUKA, Y., SHALON, Y., WEISSENBERG, E., NIR-GROSFELD, I. Analyst *87*, 791-796 (1962). A Paper-Chromatographic Method for the Identification of Food Dyes. Dyes are run in BuOH:EtOH:H_2O(1:1:1) adjusted to eight different pH values; R_f values are then plotted to form a characteristic curve.

Chapter 7 Determination of Strength

Color additives, like most commercial products, are rarely 100% pure. Although the impurities present are usually little more than inorganic salts and water, there is a constant need to know a colorant's strength. The Food and Drug Administration uses "pure dye" as a means of monitoring an additive's overall purity and of ensuring a consistency in its batch-to-batch manufacture. The economic importance of a dyestuff's strength is obvious.

To a dye chemist terms such as "strength," "pure dye," and "coal-tar dye content" relate to the absolute measure of the dyestuff's principal active ingredient and usually serve as the basis of sale and a criterion for purity. This is in contrast to the relative or effective strength used by the colorist to compare batchs of colorant of equal chemical composition but different application properties. The latter is more of a physical phenomenon dependent on particle size and crystal structure and is most important when using colorants in the solid state, such as pigments, lakes, and plating types. Absolute dye strength is typically measured by titrimetry, gravimetry, elemental analysis, or spectrophotometry or some other procedure based on chemistry or chemical physics, whereas effective strength is generally determined visually after plating the colorant on sugar or making filter-paper pourouts of colorants in solution or draw-downs of pigments pasted in oil. When determining effective or relative strength a control or "type" is run along with the sample tested. Absolute strength and effective strength are not necessarily directly related.

TITRATION WITH TITANOUS CHLORIDE

Reduction with titanous chloride is currently the "wet" method most widely used for determining pure dye and is the titration procedure favored by both the FDA and the color manufacturers. In principle, the method is applicable to any colorant that is readily and quantitatively reduced to a colorless compound. The types of compounds that can be analyzed by this technique include azo, nitro, and nitroso.

Reduction of azo dyes: $R-N=N-R' + 4H^+ + 4Ti^{+3} \rightarrow R-NH_2 + R'-NH_2 + 4Ti^{+4}$

Reduction of nitro dyes: $R-NO_2 + 6H^+ + 6Ti^{+3} \rightarrow R-NH_2 + 2H_2O + 6Ti^{+4}$

Reduction of nitroso dyes: $R-NO + 4H^+ + 4Ti^{+3} \rightarrow R-NH_2 + H_2O + 4Ti^{+4}$

Several variations of the $TiCl_3$-reduction method are now in use, and the proper technique is dictated by the color analyzed. The methods to use as well as the milliequivalent weights and the volumes of $0.1N$ titanous chloride needed per gram of color titrated are shown in Table 19 for a number of colorants.

Table 19 $TiCl_3$ Titration Factors

Color	Milliequivalent Weight	Milliliters of 0.1N $TiCl_3$/g of Color	Appropriate Method
FD&C Blue No. 1	0.3964	25.23	3
FD&C Blue No. 2	0.2332	42.89	3
FD&C Green No. 3	0.4044	24.73	3
FD&C Red No. 4	0.1201	83.26	3
FD&C Red No. 40	0.1241	80.58	3
FD&C Yellow No. 5	0.1336	74.86	3
FD&C Yellow No. 6	0.1131	88.42	1
Citrus Red No. 2	0.07708	129.7	8
Orange B	0.1476	67.74	3
D&C Blue No. 4	0.3915	25.54	3
D&C Blue No. 6	0.1311	76.26	7
D&C Brown No. 1	0.05605	178.4	3
D&C Green No. 5	0.3113	32.12	2
D&C Green No. 6	0.2092	47.79	
D&C Orange No. 4	0.08758	114.2	3
D&C Red No. 6	0.1076	92.95	3
D&C Red No. 7	0.1061	94.24	6
D&C Red No. 8	0.09970	100.3	6
D&C Red No. 9	0.1111	89.99	6
D&C Red No. 17	0.04405	227.0	5
D&C Red No. 19	0.2395	41.75	
D&C Red No. 30	0.1967	50.85	7
D&C Red No. 31	0.07783	128.5	6
D&C Red No. 33	0.1168	85.58	3
D&C Red No. 34	0.1151	86.87	6
D&C Red No. 36	0.08193	122.1	
D&C Red No. 37	0.3635	27.51	
D&C Red No. 39	0.08233	121.4	
D&C Yellow No. 7	0.1662	60.18	5
D&C Yellow No. 8	0.1881	53.15	5
Ext. D&C Violet No. 2	0.2157	46.36	
Ext. D&C Yellow No. 7	0.02985	335.0	3

In general, titanous chloride titrations require a certain degree of skill if one is to obtain consistent results. Care must be taken to exclude all oxygen from the reduction flask during the titration by purging the system with carbon dioxide or nitrogen. In addition, close attention must be given to the volumes of reagents used, the reagent blank, and the end-point timing and color.

In some cases the titration end point is indicated by a sharp decoloration of the sample. Often, though, the change in color is so gradual that it is better to add a slight excess of titrant then back titrate the excess with a standard solution of a suitable dye such as methylene blue (CI No. 52015). At other times the addition of an internal indicator that is reduced after the sample has reacted with $TiCl_3$ works best. Light Green SF Yellowish (CI No. 42095, formerly FD&C Green No. 2) is good for this purpose.

Standard titanous chloride is usually prepared from commercial $TiCl_3$ solution (Lamotte Chemical, Chestertown, Maryland 21620, or similar). A procedure is given in the text that follows.

Preparation of 0.1N TiCl$_3$

Add 500 ml of HCl and 500 ml of 20% $TiCl_3$ solution to a 7-liter plastic bottle containing about 5 liters of distilled water. Dilute to 7 liters with water and mix well. Pass nitrogen through the solution for 1 hr, stopper, and let the solution stand for 2 days.

Weigh 3.0 g of ferrous ammonium sulfate ($FeSO_4(NH_4)_2SO_4\cdot6H_2O$) into the titration flask shown in Fig. 4. Purge the system with nitrogen or carbon dioxide and add 50 ml of recently boiled water and 25 ml of 40% (w/w) H_2SO_4. Then, without interrupting the flow of gas, add 40 ml of 0.1N potassium dichromate (4.9037 g of dry NBS $K_2Cr_2O_7$/liter of solution). Slowly add about 90% of the $TiCl_3$ solution calculated as required and then quickly add 5 g of ammonium thiocyanate (NH_4CNS) and complete the titration. Determine a reagent blank and correct for it.

$$\text{Normality} = \frac{\text{ml } K_2Cr_2O_7 \times \text{normality } K_2Cr_2O_7}{\text{net ml } TiCl_3}$$

Titanous chloride solutions should be stored under an inert gas and restandardized at least weekly.

Titration Procedures

Method 1: Prepare a 1% aqueous sample solution. Pipette an amount equivalent to about 20 ml of 0.1N titanous chloride into a 500-ml wide-mouthed Erlenmeyer flask. Add 15 g of sodium citrate and adjust the volume to 150-200 ml with water. Using the apparatus shown in Fig. 4, blanket the sample with nitrogen or carbon dioxide and, while boiling, titrate with 0.1N $TiCl_3$ to the disappearance of color.

Method 2: Prepare a 0.5% ethanolic sample solution. Proceed as in Method 1, substituting 50% ethanol for water.

Method 3: Proceed as in Method 1, substituting 15 g of sodium acid tartrate

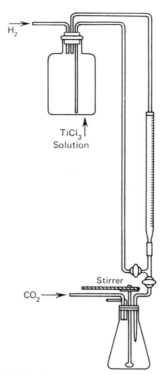

Figure 4   Titanous Chloride Titration Apparatus (Reprinted with the permission of the Association of Official Analytical Chemist)

for sodium citrate.

Method 4: Proceed as in Method 3, using about 10 mg of Light Green SF Yellowish as an indicator. Determine a reagent blank and correct accordingly.

Method 5: Prepare a 0.5% ethanolic sample solution and proceed as in Method 4, substituting 50% ethanol for water.

Method 6: Dissolve 0.200 g of sample in 5 ml of sulfuric acid in a 500-ml wide-mouthed Erlenmeyer flask. Add 100 ml of ethanol and heat and stir the sample until in solution. Dissolve 20 g of sodium acid tartrate in 100 ml of boiling water and then add 20 ml of 30% sodium hydroxide solution. Stirring rapidly, add this solution to the sample. Attach the flask to the apparatus shown in Fig. 4, blanket the sample with nitrogen or carbon dioxide, and titrate with $0.1N$ titanous chloride.

Method 7: Place a sample equivalent to about 20 ml of $0.1N$ titanous chloride in a 50-ml beaker. Pour 2 ml of fuming sulfuric acid (20% free SO_3) down the side of the beaker. Mix and place on a steam bath for 30 min. Pour the solution into a 500-ml wide-mouthed Erlenmeyer flask containing 100 g of ice. Add ice to the beaker and wash any remaining color into the flask. Add 50 ml of ethanol and 20 g of sodium acid tartrate to the flask. Attach the flask to the apparatus in Fig. 4 and titrate the sample as in Method 1.

Method 8: Add 0.150 g of sample and 125 ml of acetone to a 500-ml Erlenmeyer flask; heat cautiously to dissolve. Dissolve 15g sodium acid tartrate in 75 ml of distilled water and add this to the flask. Connect the flask to the apparatus in Fig. 4, blanket with nitrogen or carbon dioxide, and titrate with $0.1N$ titanous chloride.

GRAVIMETRIC DETERMINATIONS

Because of their insolubility in dilute acid, xanthene and fluoran colorants can be assayed gravimetrically. Those that are water-soluble sodium salts are precipitated from water solution and then weighed and corrected for molecular weight differences using Method 1 (below). Colorants that are free acids and hence water insoluble are first dissolved in dilute alkali and then precipitated with acid (see Method 2).

Gravimetric procedures are slow in comparison to titrimetric and spectrophotometric methods but require little, if any, sophisticated equipment and are generally more precise and accurate.

Method 1: Transfer 0.500 g of sample to a 400-ml beaker; add 100 ml of distilled water and then heat to boiling. Add 25 ml of dilute HCl (1:49) and bring to a boil. Wash down the sides of the beaker with distilled water and then cover and keep on a steam bath for several hours or overnight. Cool to room temperature and then quantitatively transfer the precipitate to a tared Gooch crucible with dilute HCl (1:99). Wash the precipitate with two 15-ml portions of distilled water and then dry the crucible for 3 hr at 135°C. Cool in a desiccator and weigh.

$$\text{Percent pure dye} = \frac{\text{Weight of precipitate} \times \text{conversion factor} \times 100}{0.500}$$

Color	Gravimetric Conversion Factors
FD&C Red No. 3	1.074 to disodium salt, monohydrate
D&C Orange No. 11	1.075 to disodium salt
D&C Red No. 22	1.068 to disodium salt
D&C Red No. 28	1.056 to disodium salt

Method 2: Transfer 0.500 g of sample to a 400-ml beaker. Add 50 ml of $0.1N$ NaOH and swirl to dissolve. Add 50 ml of distilled water and then proceed as in Method 1, beginning with "heat to boiling."

$$\text{Percent pure dye} = \frac{\text{Weight of precipitate} \times 100}{0.500}$$

No conversion factors are needed here since the colors for which this method is useful are free acids. These include D&C Orange Nos. 5 and 10 and D&C Red Nos. 21 and 27.

SPECTROPHOTOMETRIC DETERMINATIONS

Many of today's additives can be assayed spectrophotometrically. The procedures used are, for the most part, simpler, more expedient and to some extent more specific than wet methods and have the added advantage of being able to provide qualitative information through evaluation of the spectra generated. As with most benefits, these are costly, and the price here is usually paid in the form of more expensive equipment and less precise results.

The majority of "spectro" methods in use necessitate little more than dissolution of the sample in a suitable solvent and measurement of its absorbance versus that of a standard. These are treated as a group, and the calculations given in this general procedure serve as a guide for the remaining methods. A few procedures require more care and are described in greater detail.

The absorption coefficients or absorptivities used to calibrate spectro methods are normally obtained in one of two ways—by measuring the absorption of a highly purified sample of the compound of interest that has been shown by chemical analysis to be essentially 100% pure, or by establishing a relationship between the absorbance of a large number of typical commercial samples and their strengths determined by some chemical means, usually $TiCl_3$ reduction. In most instances the numbers are the same. Published absorptivities, including those shown here, should be considered as only close approximations of the true values. For best results each laboratory should obtain its own factors using its standards and its spectrometer.

Spectrophotometric Procedures

General Procedure: Dissolve a portion of well-mixed sample in a spectrograde quality of the solvent shown in Table 20 and dilute this solution to obtain an absorbance at the absorption maximum within the range recommended as the most accurate for the spectrophotometer used. Correct this absorbance for solvent blank and apply the following formula:

$$\text{Percent pure dye} = \frac{A \times 100}{a \times b \times c}$$

where: A is blank-corrected sample absorbance, a is the color's absorptivity (in liters/g-cm) (see Table 20), b is the absorption cell's path length (in cm), c is the concentration (in g/liter) of the solution presented to the spectrophotometer, and 100 is the factor for conversion to percent.

D&C Blue No. 6: Transfer 0.5 g of sample to a 50-ml beaker. Add 5 ml of 100% H_2SO_4 and blend the mixture into a smooth paste with a stirring rod. Place the beaker and rod in an oven at $100 \pm 5°C$ for 30 ± 5 min. Cool the sample to room temperature and then drown it in about 400 ml of distilled water. Transfer the solution to a 1000-ml volumetric flask and then dilute to volume with water and mix. Pipette 10 ml of this solution into a 500-ml volu-

TABLE 20 ABSORPTION COEFFICIENTS

	Solvent System[a]	Wavelength of Maximum Absorbance (in nm)	Absorptivity (in liters/g-cm)
FD&C Blue No. 1	Water	630	164
FD&C Blue No. 2	Water	610	47.8
FD&C Green No. 3	Water	625	156
FD&C Red No. 3	Water	527	110
FD&C Red No. 4	Water	502	54.0
FD&C Red No. 40	Water	502	55.6
FD&C Yellow No. 5	Water	428	53
FD&C Yellow No. 6	Water	484	55
Citrus Red No. 2	$CHCl_3$	515	70
Orange B	Water	437	35.5
D&C Blue No. 4	Water	630	170
D&C Blue No. 6	$CHCl_3$	603	79.0
D&C Blue No. 9	96% H_2SO_4	458	18.9
D&C Green No. 5	Water	610	21.3
D&C Green No. 6	$CHCl_3$	648	39.2
D&C Green No. 8	Water (OH)⁻	454	51
D&C Orange No. 4	Water	484	65.3
D&C Orange No. 5	Water (OH)⁻	503	163
D&C Orange No. 10	Water (OH)⁻	510	122
D&C Orange No. 17	$CHCl_3$	480	77
D&C Red No. 6	Alcohol & water (1 + 1)	511	61.5
D&C Red No. 7	Alcohol & water (1 + 1)	516	157.2
D&C Red No. 8	Alcohol & water (1 + 1)	486	52.2
D&C Red No. 9	Alcohol & water (1 + 1)	486	47
D&C Red No. 17	$CHCl_3$	514	94
D&C Red No. 19	Water	554	236
D&C Red No. 21	Water (OH)⁻	518	150
D&C Red No. 22	Water	518	140
D&C Red No. 27	Water (OH)⁻	537	129
D&C Red No. 28	Water	537	122
D&C Red No. 30	$CHCl_3$	537	44
D&C Red No. 30	Xylene	496[b]	24.8
D&C Red No. 31	Alcohol & Water (1 + 1)	518	80
D&C Red No. 33	Water	530	66.2
D&C Red No. 34	Alcohol & Water (1 + 1)	526	65.5
D&C Red No. 36	$CHCl_3$	490	84
D&C Red No. 37	Water	545	154
D&C Violet No. 2	$CHCl_3$	588	36.3
D&C Violet No. 2	Toluene	580	35.7
D&C Yellow No. 7	Water (OH)⁻	489	247
D&C Yellow No. 8	Water	489	228
D&C Yellow No. 10	Water	413	92
D&C Yellow No. 11	$CHCl_3$	420	135
Ext. D&C Yellow No. 7	Water	430	49

[a]All neutral water systems are buffered with 0.01N ammonium acetate.

[b]The wavelength of maximum absorbance of D&C Red No. 30 in xylene is actually 537 nm. The isosbestic point of the *cis* and *trans* forms of D&C Red No. 30 present in xylene is 496 nm [Hobin, N. K., JAOAC *54*, 215 (1971)].

metric flask and dilute to volume with water. Determine the sample's absorbance in a 1-cm cell versus distilled water at the absorption maximum near 608 nm. Correct for solvent blank.

$$\text{Percent pure dye} = \frac{A \times 100}{a \times b \times c} = \frac{A \times 100}{81.1 \times 1 \times 0.01}$$

[Phthalocyaninato (2-)] Copper: Prepare a dispersing solution by transferring 50 ml of 2-ethoxyethanol ("cellosolve", Eastman Kodak No. 1698) and 20 ml of Dispersol VL (Ahcowet VL, Arnold Hoffman Co.) to a 1000-ml volumetric flask and diluting to volume with distilled water. Place 150 ml of this solution into a 250-ml beaker and stir magnetically at a rate to give a vortex about 2/3 the depth of the solution. Pipette 50 ml of sample solution (0.04 g in 250 ml of 96% H_2SO_4) into this vortex and mix well. Transfer to a 250-ml volumetric flask, dilute to volume with dispersing solution, and measure the solution's absorbance in a 5-cm cell at the absorption maximum near 600 nm (A_1) and the absorption minimum near 470 nm (A_2). Correct for solvent blank.

$$\text{Percent pure dye} = \frac{A \times 100}{a \times b \times c} = \frac{(A_1 - A_2) \times 100}{39.6 \times 5 \times 0.032}$$

Results for duplicate dispersions should agree within 3% relative.

Annatto, oil soluble: Dissolve 0.1 g of sample in chloroform and dilute with chloroform to 100 ml in a volumetric flask. Using a 1-cm cell, determine the sample's absorption spectrum between 600 nm and 400 nm. Measure the absorbance at the absorption maximum near 503 nm and at 404 nm and correct for solvent blank.

$$\text{Percent pure dye (as Bixin)} = \frac{[A_{503} + A_{404} - 0.256 \, (A_{503})] \, (100)}{(282.6) \, (1) \, (1)}$$

where A represents absorbances of the sample solution at the indicated wavelengths, 1 is cell path length (in cm), 1 is sample concentration (in g/liter), 0.256 is the factor relating the absorbances of bixin in chloroform at 404 nm and 503 nm, and 282.6 is the absorptivity of bixin at 503 nm (in liters/g-cm).

Determine the blank-corrected absorbance of a chloroform solution of sample at the maximum near 467 nm. Calculate percent pure dye as bixin using 320 liters/g-cm as the absorptivity.

Annatto, Water Soluble: Dissolve the sample in 2-5% aqueous KOH and measure its absorbance at the maximum near 480 nm. Calculate percent pure dye as bixin using 287 liters/g-cm as the absorptivity.

Annatto, Emulsions: Dissolve the sample in chloroform:methanol; 1:1 v/v. Make just acid with a few drops of glacial acetic acid and measure its absorbance against the same solvent at the absorption maximum near 500 nm. Calculate percent pure dye as bixin using 287 liters/g-cm as the absorptivity.

Canthaxanthin, All-*trans* Crystals: Measure the absorbance of a cyclohexane solution of sample at the absorption maximum near 470 nm. Calculate percent pure dye as *trans* canthaxanthin, using 220 liters/g-cm as the absorptivity.

Canthaxanthin, Water-dispersible Beadlets: Dissolve 0.5 g of reagent-grade iodine in 50 ml of 3A alcohol. Dilute a portion of this solution 1000-fold with cyclohexane.

Carry out the following work in subdued light. Transfer 0.1 g of beadlets into a 200-ml volumetric flask. Add about 100 ml of distilled water and warm on a steam bath until well dispersed. Cool to room temperature, make to volume with distilled water, and mix well. Transfer 10 ml of this solution into a 50-ml glass-stoppered centrifuge tube. Add 2 ml of 1N-HCl and 25-ml of chloroform and shake for 5 min. Break any emulsion that forms with two or three drops of 3A alcohol and then centrifuge at 2500 rpm for about 3 min. If the aqueous phase is not colorless, add about 1 g of sodium chloride, shake for an additional 5 min, and then recentrifuge. Transfer about 15 ml of the chloroform extract into a second glass-stoppered centrifuge tube containing about 2 g of anhydrous sodium sulfate. Mix well then centrifuge for about 3 min. Pipette 4 ml of the clear chloroform extract into a 50-ml volumetric flask. Place the flask in a 40°C water bath and evaporate to near dryness under a stream of nitrogen. Dissolve the residue in about 30 ml of cyclohexane, add 2.5 ml of iodine solution (prepared as described in the preceding paragraph) and dilute to volume with cyclohexane. Mix and then store in the dark for 2 hr at room temperature. Measure the absorbance of this solution in a 1-cm cell versus cyclohexane at the absorption maximum near 470 nm.

$$\text{percent pure dye (as } cis\text{-}trans \text{ equilibrium mixture)} = \frac{A \times 100}{a \times b \times c} = \frac{A \times 100}{197 \times 1 \times 0.016}$$

β-Carotene, All-*cis* Crystals: Prepare the sample in subdued light and use low-actinic glassware. Transfer 0.065 g of sample to a 100-ml volumetric flask. Dissolve the sample with 10 ml of acid-free chloroform and then dilute to volume with cyclohexane and mix well. Pipette 5 ml of this solution into a 50-ml volumetric flask, dilute to volume with cyclohexane and mix. Pipette 5 ml of this second solution into a 50-ml volumetric flask. Dilute to volume with cyclohexane and mix. Using cyclohexane as the reference, determine the blank-corrected sample absorbance in a 1-cm cell at the absorption maximum near 340 nm.

$$\text{Percent pure dye (as } cis \text{ β-carotene)} = \frac{A \times 100}{a \times b \times c} = \frac{A \times 100}{101 \times 1 \times 0.0065}$$

β-Carotene, All-*trans* Crystals: Prepare the sample solutions in subdued light and use low-actinic glassware. Transfer 0.05 g of sample to a 100-ml volumetric flask, dissolve it with 10 ml of acidfree chloroform, and dilute to volume with cyclohexane. Mix well. Pipette 5 ml of this solution into a 100-ml volumetric flask and dilute to volume with cyclohexane. Pipette 5 ml of this second solution into a 50-ml volumetric flask and dilute to volume with cyclohexane. Using a 1-cm cell, determine the blank-corrected absorbance of the final solution at the absorption maximum near 455 nm.

$$\text{Percent pure dye (as } \textit{trans } \beta\text{-carotene)} = \frac{A \times 100}{a \times b \times c} = \frac{A \times 100}{250 \times 1 \times 0.025}$$

β-Carotene, Water-dispersible Beadlets: Transfer 0.25 g of sample into a 250-ml low-actinic glass separatory funnel containing 50 ml of distilled water. Swirl to disperse the beadlets. Add 50 ml of 3A denatured alcohol and extract with 50-ml portions of petroleum ether, shaking for 3 min each time until the aqueous layer is colorless (ca. three extractions). Combine the extracts in a 250-ml low-actinic glass volumetric flask and dilute to volume with petroleum ether. Add 2-3 g of granular anhydrous sodium sulfate and shake for approximately 3 min. Allow the sulfate to settle, then pipette 5 ml of the clear solution into a 50-ml low-actinic glass volumetric flask, and evaporate this to dryness with a stream of nitrogen. Use no heat. Dissolve the residue and dilute it to volume with cyclohexane. Measure the absorbance of this solution in a 1-cm cell at the maximum near 452 nm, using cyclohexane as the reference.

$$\text{Percent pure dye (as } \textit{cis-trans} \text{ equilibrium mixture)} = \frac{A \times 100}{a \times b \times c} =$$

$$\frac{A \times 100}{223 \times 1 \times 0.1}$$

β-Carotene, Vegetable-oil Suspension: Weigh 0.1 g of sample and assay as *trans* β-carotene using the procedure given under All-*trans* Crystals. (Make proper adjustment in calculation for differences in sample weight.)

β-Carotene, Emulsions: Weigh 0.25 g of sample and assay as *cis-trans* isomers, using the procedure reported under Water-dispersible Beadlets. (Make proper adjustment in calculations for differences in sample weight.) If an emulsion forms during the extractions with ether, add about 5 g of sodium sulfate or sodium chloride to the separator.

β-Apo-8′-Carotenal, All-*trans* Crystals: Weigh 0.04 g of sample and analyze, using the procedure reported under β-Carotene, All-*trans* Crystals. Determine the net absorbance in a 1-cm cell at the absorption maximum near 461 nm.

$$\text{Percent pure dye} = \frac{A \times 100}{a \times b \times c} = \frac{A \times 100}{264 \times 1 \times 0.002}$$

β-Apo-8′-Carotenal, Water-dispersible Beadlets: Prepare the sample in subdued light and use low-actinic glassware throughout the analysis. Transfer 0.15 g of sample to a 200-ml volumetric flask containing 100 ml of distilled water. Warm the sample on a steam bath and swirl to effect complete solution. If an emulsion forms, add two or three drops of 3A alcohol. Cool to room temperature and then make to volume with distilled water and mix. Pipette 5 ml of this solution into a 50-ml glass-stoppered centrifuge tube and add 1 ml of 1N-HCl and 20 ml of chloroform. Shake vigorously for 5 min and then centrifuge at 2500 rpm for 3 min. If the water layer is not colorless, shake again for 5 min and recentrifuge. Transfer 15 ml of the chloroform extract to a 50-ml glass-stoppered centrifuge tube containing 5 g of anhydrous sodium sulfate, shake well, then centrifuge for about 3 min. Pipette 5 ml of this dried solution into a 50-ml volumetric flask and evaporate to dryness on a water bath (40°C) under a stream of nitrogen. Dissolve the residue in cyclohexane, make to volume with cyclohexane, and mix well. Determine the blank-corrected absorbance of the solution in a 1-cm cell at the absorption maximum near 457 nm.

$$\text{Percent pure dye (as } cis\text{-}trans \text{ equilibrium mixture)} = \frac{A \times 100}{a \times b \times c} =$$

$$\frac{A \times 100}{240 \times 1 \times 0.01875}$$

Carmine (Carminic Acid): Dissolve 0.1 g of sample in 30 ml of boiling 2N HCl. Cool and dilute to 1 liter with water. Using a 1-cm cell, determine the blank-corrected absorbance of this solution at the absorption maximum near 494 nm.

$$\text{Percent pure dye (as carminic acid)} = \frac{A \times 100}{a \times b \times c} = \frac{A \times 100}{13.9 \times 1 \times 0.1}$$

Cochineal Extract: Transfer 1 g of sample to a 500-ml volumetric flask containing 30 ml of boiling 2N HCl. Cool and make to volume with deionized water. Filter a portion of the sample through Whatman No. 1 filter paper and then determine the blank-corrected absorbance of this solution at the absorption maximum near 494 nm.

$$\text{Percent pure dye (as carminic acid)} = \frac{A \times 100}{a \times b \times c} = \frac{A \times 100}{13.9 \times 1 \times 2}$$

Riboflavin: Protect the sample solution from direct sunlight throughout the entire procedure. Dry the sample for 2 hr at 105°C, weigh 0.5 g, and transfer it to a 1-liter volumetric flask with water. Add 5 ml of glacial acetic acid and dilute to about 800 ml with water. Heat on a steam bath until dissolved. Cool to about 25°C and dilute to 1 liter. Dilute a 10-ml aliquot with water to 1 liter and measure the fluorescence of this solution in a fluorometer at the maximum near 460 nm. Immediately add 0.01 g of sodium hydrosulfite to the sample solution, mix to dissolve, and again measure the fluorescence at the maximum near 460 nm. Similarly prepare and measure the fluorescence of a USP riboflavin reference standard.

$$\text{Percent riboflavin} = \frac{A - B}{A_s - B_s}(100)$$

where A is fluorescence of the sample before sodium hydrosulfite addition, B is fluorescence of the sample after sodium hydrosulfite addition, A_s is fluorescence of the standard before sodium hydrosulfite addition, and B_s is fluorescence of the standard after sodium hydrosulfite addition.

Turmeric (Curcumin): Weigh 0.1 g of powdered sample into a 100-ml volumetric flask. Add 60 ml of glacial acetic acid and place the mixture on a water bath (90°C) for 1 hr. Add 2 g of boric acid and 2 g of oxalic acid to the flask and place it on the bath for an additional 10 min. Cool to room temperature and dilute to volume with glacial acetic acid. Pipette 5 ml of this solution into a 50-ml volumetric flask and dilute to volume with acid. Determine the absorbance in a 1-cm cell at 540 nm against a similarly prepared standard.

ASSAY BY ELEMENTAL ANALYSIS

In principle, colorants can be assayed by determining the amount of any of its elements and relating the percentage found according to theory. In comparison to spectroscopy, gravimetry, and titrimetry, elemental analysis is usually more time consuming, requires greater skill on the part of the analyst, and often produces less satisfying results. The advent of automated equipment for the determination of carbon, nitrogen, hydrogen, sulfur, and other elements has minimized some of the problems associated with elemental analysis, but these tools are costly and, unless one is interested in using them for structural determinations, it would very likely be wiser to spend the money on a different method of assay.

General methods are included here for the determination of sulfur and nitrogen since these are probably the most useful elements to analyze for (particularly when dealing with certified colors) and because they represent two different procedures for the decomposition of colorants, both of which are useful when determining other elements. In using either method it is important to realize that neither is capable of identifying the source of the element measured, and thus it is necessary to know the nature of the sample, particularly if it is suspected to contain such materials as sodium sulfate or sodium chloride. Some specific methods are also included.

Assay Through Organic Sulfur Content

Weigh 0.2 g of sample into a Parr peroxide bomb. Mix with about 14 g of sodium peroxide, 1 g of sugar, and 0.1 g of potassium chlorate, and ignite the bomb. Place the opened bomb in a beaker, cover with water, heat at 50-60°C until the reaction ceases, and then rinse the bomb into the beaker with water. Cool the solution and neutralize it with concentrated hydrochloric acid; then add 5 ml excess. Filter the solution into a second beaker, washing the filter paper with water. Heat the filtrate to boiling and while boiling, add 25 ml of 10% barium chloride solution. Digest on a steam bath for 1-2 hr or allow the sample to stand in a warm place overnight. Filter through a tared Gooch crucible and wash the precipitate with water until it is free of chloride. Dry at 100°C for 20 min, and then at 700°C for 30 min.

$$\text{Percent total sulfur} = \frac{W(0.1373)(100)}{w}$$

$$\text{Percent organic sulfur} = \% \text{ total sulfur} - \% \text{ inorganic sulfur}$$

$$\text{Percent pure dye} = \frac{(\% \text{ organic sulfur})(100)}{\text{theoretical } \% \text{ sulfur}}$$

where W is weight of barium sulfate precipitate in g and w is sample weight, in g.

Assay Through Organic Nitrogen Content

For Colors Requiring Reduction Prior to Digestion: Weigh a quantity of sample containing 2 mg of nitrogen and transfer it to a 20-ml Kjeldahl flask. Add 0.5 ml of a 9:1 mixture of 50% (w/v) hydriodic acid and 50% (w/v) hypophosphorous acid. Reflux 5-10 min, turning the flask occasionally to ensure solution and reduction. Remove most of the liquid by distillation. Remove the flask from the heater, cool, add 5 ml of 1:1 sulfuric acid, and evaporate to fumes. If considerable iodine remains, add 1-2 ml of water and again evaporate to fumes. Remove the flask from the heater, cool, and wash down the sides with 0.5 ml of water. Add 0.6 g of anhydrous sodium sulfate and 0.5 ml of 20% mercuric acetate solution. Place on a heater and digest until the solution clears and then heat for an additional hour. If the dye contains ring nitrogen, heat for 2.5 hr after clearing and add more sulfuric acid if needed. Remove the flask, cool, and add 10-12 ml of water to dissolve the salts.

Transfer the sample to a micro-Kjeldahl distilling apparatus using about 10 ml of water. Set the electric controller of the steam generator to distill 20 ml in about 10 min. Add 5-6 ml of 50% sodium hydroxide solution and 3 ml of 21% sodium thiosulfate solution. Prepare an indicator solution by dissolving first 0.3 g of methyl red in 60 ml of ethanol and diluting with water to 100 ml, then 0.2 g of methylene blue in 100 ml of 50% ethanol, and mixing the two solutions. Add three drops of the indicator solution to a 50-ml Erlenmeyer flask containing 5 ml of 2% boric acid solution. Position the flask so that the outlet from the condenser dips below the level of the liquid and steam distill for 5 min. Lower the receiving flask so that the condenser outlet is above the liquid in the flask and distill for 1-2 min to flush the condenser tube.

Titrate the solution in the receiving flask with standardized 0.02N hydrochloric acid. Make a blank determination.

$$\text{Percent nitrogen} = \frac{(A - B)N(0.014)(100)}{w}$$

where A is volume of titrant used for sample (in ml), B is volume of titrant used for blank (in ml), N is normality of titrant, and w is sample weight (in g).

$$\text{Percent pure dye} = \frac{\% \text{ nitrogen } (100)}{\text{theoretical } \% \text{ nitrogen}}$$

For Colors that Do Not Require Reduction Prior to Digestion: Weigh and transfer a sample as described in the preceding paragraphs. Add 5 ml of 1:1 sulfuric acid and heat until thoroughly charred. Cool, and then add 0.6 g of sodium sulfate and 0.5 ml of 20% mercuric acetate solution. Wash down the side of the flask with a minimum amount of water and proceed as described in the preceding paragraphs, starting with "Place on a heater and digest. . ."

Organically Combined Iodine in FD&C Red No. 3

Wash 0.5 g of sample into a 100-ml volumetric flask with about 50 ml of hot distilled water. Swirl the flask to dissolve the sample and then add 6 ml of 10% aqueous NaOH. Cool the solution, make it to volume with distilled water, and mix well.

Pipette 25 ml of this solution into a 500-ml Erlenmeyer flask and add 100 ml of distilled water, 25 ml of 7% aqueous $KMnO_4$, and a few glass beads. Boil the mixture for 5 min and then remove from the hot plate. When the boiling ceases, cautiously add 10 ml of HNO_3 and boil for an additional 5 min. Remove the flask from the heat and wash down the sides of it with distilled water. While swirling, quickly add 4 ml of 12% aqueous $NaNO_2$ to the flask, then add more $NaNO_2$ dropwise until the suspension begins to clear. Continue the dropwise addition of $NaNO_2$, allowing each drop to react before the next one is added. Continue this addition until only a small amount of undissolved MnO_2 remains. Do not attempt to destroy or dissolve the last traces of MnO_2, but instead immediately add 1% $KMnO_4$ solution in 1-ml portions until the solution turns pink. (If more than 2 ml of 1% $KMnO_4$ is required or if a brown color appears, add 10 ml of $KMnO_4$ to the solution and again heat it to boiling. Repeat the dropwise addition of 12% $NaNO_2$ and 1% $KMnO_4$ to again obtain a pink color.)

Using suction, rapidly filter this solution through a Gooch crucible fitted with glass-fiber filter paper. Wash the flask and filter thoroughly with distilled water. (The filtrate must be pink at this point.) Add 12% $NaNO_2$ solution dropwise to the filtrate with shaking until 1 drop has been added in excess of that needed to decolorize the solution.

Add 10 ml of 10% aqueous sulfamic acid, wash down the sides of the flask, and then swirl the contents. Cool the solution to room temperature, add 2-3 g of solid KI, and titrate the liberated iodine with $0.1N$ $Na_2S_2O_3$, adding starch iodide indicator when the solution becomes lemon-yellow. Continue the titration until the blue color just disappears. Similarly determine a reagent blank.

$$\text{Percent total iodine} = \frac{(A - B)(N)(0.02115)(100)}{0.125}$$

where A is ml of titrant used for sample, B is ml of titrant used for blank, N is normality of titrant, 0.02115 is milliequivalent weight of I, 100 is factor for conversion to percent, and 0.125 is sample weight (in g).

Organically Combined Bromine and Chlorine in D&C Red No. 28

Weigh 0.2 g of sample into a Parr oxygen bomb containing 15 ml of distilled water. Assemble and ignite the bomb. Using distilled water, quantitatively transfer the contents of the bomb to a 400-ml beaker. Adjust the sample volume to about 200 ml with distilled water, add 15 ml of HNO_3, and stir well. Add 25 ml of $0.1N$ $AgNO_3$ to the sample and place it on a magnetic stirrer. While stirring vigorously, add 5 ml of nitrobenzene. Add 5 ml of ferric alum indicator, stir for about 3 min, and then titrate with 0.1N KSCN to a faint persistent red-brown color.

$$\text{Percent total halogen as chlorine} = \frac{(AN_1 - BN_2)(0.03546)(100)}{0.200}$$

where A is ml of $AgNO_3$ solution = 25, N_1 is normality of $AgNO_3$, B is ml of KSCN solution, N_2 is normality of KSCN, 0.03546 is milliequivalent weight of chlorine, 100 is factor for conversion to percent, and 0.200 is sample weight (in g).

Bomb a second 0.2 g sample as above and quantitatively wash the combustion products into a 500-ml iodination flask with distilled water. Adjust the sample volume to about 125 ml with distilled water.

While working in a fume hood, add 12 ml of 85% phosphoric acid, 3 ml of 5% potassium cyanide, and 5 ml of saturated potassium permanganate to the flask wetting the sides of the flask as each reagent is added. Stopper the flask and mix the contents by gentle swirling, wetting the entire inside surface. Allow the sample to stand for at least 7 min. Add about 2 g of solid ferrous ammonium sulfate hexahydrate and then wash down the sides of the flask with distilled water and swirl the sample to mix it. (A clear, nearly colorless solution should result. If the solution is still colored, add more ferrous ammonium sulfate; a 2-g excess does no harm.)

Add 2 g of potassium iodide and immediately titrate the liberated iodine with 0.05N sodium thiosulfate to a pale yellow color. Add 2-3 ml of starch iodide indicator solution and continue titrating with 0.05N sodium thiosulfate to the disappearance of the blue starch iodide color. Similarly determine a reagent blank.

$$\text{Percent bromine} = \frac{(A - B)(N)(0.03996)(100)}{0.200}$$

where A is ml of $Na_2S_2O_3$ required to titrate sample, B is ml of $Na_2S_2O_3$ required to titrate blank, 0.03996 is milliequivalent weight of bromine, 100 is factor for conversion to percent, and 0.200 is sample weight (in g).

$$\text{Percent chlorine} = (\% \text{ total halogen as chlorine}) - (\% \text{ bromine})\frac{35.457}{79.916}$$

where 35.457 is atomic weight of chlorine and 79.916 is atomic weight of bromine.

Organically Combined Bromine in D&C Red No. 22

Determine bromine using the method described for D&C Red No. 28.

MISCELLANEOUS PROCEDURES

Calcium Carbonate: Weigh 1 g of calcium carbonate, previously dried at

$200°C$ for 4 hr, and transfer to a 250-ml. beaker. Moisten thoroughly with a few milliliters of water and then add, dropwise, sufficient diluted hydrochloric acid to effect complete solution. Transfer the solution to a 250-ml volumetric flask, dilute to volume, and mix. Pipette 50 ml of the solution into a suitable container, add 100 ml of water, 15 ml of 4% (w/v) sodium hydroxide, and 300 mg of hydroxy naphthol blue indicator, and titrate with 0.05 M disodium ethylenediaminetetraacetate until the solution is a distinct blue in color. One milliliter of 0.05 M disodium ethylenediaminetetraacetate is equal to 5.004 mg of $CaCO_3$.

Ferrous Gluconate: Prepare an *o*-phenanthroline indicator solution as follows: Dissolve 1.48 g of ferrous sulfate, $FeSO_4 \cdot 7H_2O$, in 100 ml of water. Immediately dissolve 0.15 g of *o*-phenanthroline in 10 ml of this solution. Add 1.5 g of sample, 75 ml of water, and 15 ml of 10% (w/v) sulfuric acid to a 300-ml Erlenmeyer flask. Stir to dissolve. Add 0.25 g of zinc dust. Close the flask with a stopper containing a Bunsen valve and allow it to stand for 20 min at room temperature. Filter the sample through a Gooch crucible fitted with an asbestos mat coated with a thin layer of zinc dust. Wash the crucible with 10 ml of 10% (w/v) sulfuric acid and then 10 ml of water. Immediately titrate the filtrate to an *o*-phenanthroline end point with 0.1N ceric sulfate. Similarly determine a blank. One milliliter of 0.1N ceric sulfate is equal to 0.04462 g of anhydrous ferrous gluconate ($C_{12}H_{22}FeO_{14}$).

Titanium Dioxide: Prepare zinc amalgamate by dissolving 10 g of granulated zinc in about 20 ml of mercury, heating and stirring at $150°C$. A liquid amalgam is formed on cooling.

Mix 0.3 g of sample with 3 g of potassium hydrogen sulfate in a platinum dish and fuse. Cool the fusion and dissolve it in 150 ml of about 2N sulfuric acid. Activate a freshly charged zinc amalgam reductor by passing 100 ml of 2N sulfuric acid through the column, followed by 100 ml of water. Pass 200 ml of N sulfuric acid through the column, followed by 100 ml of water. Collect the effluent in a receiver containing 50 ml of 15% ferric ammonium sulfate in 0.5N sulfuric acid. Titrate the solution with 0.1N potassium permanganate; this is the blank. Reactivate the column with 100 ml of N sulfuric acid. Pass the sample solution through the column followed by 100 ml of N sulfuric acid and 100 ml of water. Titrate the effluent and subtract the blank titration. One milliliter of 0.1N potassium permanganate is equal to 0.00799 g of titanium dioxide.

Insoluble Ca, Ba, and Sr salts and Lakes of Some Colors:

Reagents

10% EDTA solution—Dissolve 25g of ethylenediaminetetraacetic acid in 165 ml of 10% aqueous NaOH. Dilute to 250 ml with distilled water.

Dilute EtOH—Mix equal volumes of 95% EtOH and distilled water.

Weigh 0.1 g of sample into a 250-ml beaker. Add 7 ml of 10% EDTA solution, 3 ml of 10% aqueous NaOH, and 15 ml of distilled water. Cover the beaker and bring to a boil. Stir the sample until wetted and boil an additional 2 min longer. Remove the beaker from the hot plate and stir for 30-60 sec. Add 25 ml of 95% EtOH and 125 ml of dilute EtOH; mix. Cover the beaker and digest the sample just below the boiling point until all the color is in solution. Cool to room temperature and quantitatively transfer to a 250-ml volumetric flask with dilute EtOH. Dilute to volume with same. (If talc or TiO_2 insolubles are present, filter the sample through a fine-porosity sintered-glass filter before transferring to the volumetric flask.) Determine the pure dye content spectrophotometrically against a standard similarly prepared.

Relative Methods: Because of their indefinite composition, certain colorants are analyzed on a relative basis in comparison with a house standard (a "type") or a chemically nonrelated color standard such as colored glass or solutions of inorganic salts. Such measurements are made either visually or with the aid of a spectrometer. The methods are usually empirical in nature and are more a measure of color value or tinctorial strength than of purity or assay. The procedures are rarely standardized within an industry, and the numbers obtained with them are often quoted in terms that are meaningless out of context, so extreme care must be exercised in interpreting them. Colorants frequently analyzed this way include paprika, turmeric, and caramel. Paprika and its oleoresins are rated visually versus Lovibond glasses, colorimetrically versus standard solutions composed of potassium dichromate and cobaltous chloride, and spectrophotometrically by measuring the absorbance of sample solutions in acetone at 458 nm. Turmeric oleoresins also are often measured spectrophotometrically in acetone at 422-425 nm, whereas caramel is measured colorimetrically using either a Klett-Summerson color comparator equipped with No. 52 and No. 64 glasses, or a spectrophotometer operated at 640 nm and 510 nm. Typical examples of such methods are given in the paragraphs that follow.

Colorimetric Color Value (CV) of Paprika Oleoresin: Weigh 1 g of sample into a 100-ml volumetric flask. Make to volume with acetone; mix well. Dilute the sample with acetone as needed (see table that follows).

Pipette the required amount of dilute solution (see table that follows) into a 100-ml Nessler tube. Bring the volume almost to the mark with acetone and compare it through the length of the tube against a blank. Make small additions until sample and blank match. Calculation is as follows:

$$CV = \frac{100 - (A \times B)}{A \times B} \times 100$$

where A is ml of dilute solution used and B is percent of dilute solution.

Blank: Into a 100-ml Nessler tube pipette 10 ml of 0.1N potassium dichromate solution (4.904 g $K_2Cr_2O_7$ per liter) and 1 ml of 0.5N cobaltous chloride solution (5.948 g $CoCl_2 \cdot 6H_2O$ per 100 ml) and make to 100 ml with distilled water.

Color Value	Percent Dilution	Milliliters of Solution 1 used for dilution	Milliliters Dilute Solution to be Used
100,000	0.01	1 ml/100 ml of acetone	8-10
50,000	0.02	1 ml/50 ml of acetone	8-10
40,000	0.02	1 ml/50 ml of acetone	10-13
30,000	0.02	1 ml/50 ml of acetone	15-20
20,000	0.04	2 ml/50 ml of acetone	10-13
10,000	0.10	5 ml/50 ml of acetone	8-10

Spectrophotometric Color Value of Turmeric Oleoresin: Weigh 100-200 mg of a well-mixed sample into 100-ml volumetric flask. Add approximately 75 ml of acetone and shake until the sample is completely in solution. Bring to volume and mix thoroughly. Pipette a 1-ml aliquot into a 50-ml volumetric flask and bring to volume with acetone; mix thoroughly. Using a Beckman Model B (or similar) spectrometer, a 1-cm cell, and a tungsten light source, obtain the sample absorbance at 422-425 nm, using acetone as a blank. If the absorbance is not between 0.2 and 0.4 absorbance units, adjust the concentration accordingly by varying the size of the aliquot taken from the first solution.

$$CV = \frac{Absorbance}{Sample\ wt.\ @\ 1/5000}$$

$$Percent\ Curcumin = \frac{CV\ (1/5000)}{33.00}$$

where 33.00 is CV of *Curcumin* (EK No. 1179) at 1/5000 in acetone, 422-425 nm.

BIBLIOGRAPHY

CERMA, E. Rass. chim. per chim. e ind. *12*, 13-20 (1960). Polarographic Assays of Permitted Dyes in Coloring Foods. Discusses the polarography of 13 food colors permitted in Italy, including FD&C Red No. 2, FD&C Red No. 3, FD&C Yellow No. 5, FD&C Yellow No. 6, and FD&C Blue No. 2. The samples were run under nitrogen at $25° \pm 1°$C using dropping Hg and standard HgCl electrodes. The sample solutions were buffered with 0.2N Me$_4$NOH or 0.2N NaOH, and 1% gelatin was added to eliminate interfering maxima.

DENDY, D. A. V. J. Sci. Food Agric. *17*, 75-76 (1966). The Assay of Annatto Preparations by Thin-Layer Chromatography. Thin-layer chromatography is used to separate bixin, which is then assayed spectrophotometrically.

ETTLESTEIN, N. JAOAC *34*, 792-794 (1951). EDTA as an Aid in the Analysis of Certain Coal-Tar Color Lakes.

GRAICHEN, C., HEINE, K. S., Jr., JAOAC *37*, 905-912 (1954). Studies on Coal-Tar Colors. XVI. FD&C Red No. 4. Describes the preparation of a pure sample of colorant and spectrophotometric and chemical methods of assay.

HOBIN, N. K. JAOAC *54*, 215 (1971). Determination of Pure Color in Commercial Samples of D&C Red No. 30. Describes the preparation of a pure sample of colorant and a spectrophotometric procedure for assay.

JEKABSONS, E. JAOAC *52*, 110-112 (1969). Fluorimetric Analysis for Fluorescein Sodium in Ophthalmic Solutions. A method is calibrated as follows. Heat 100 mg of fluorescein diacetate with 10 ml of ethanol and 2 ml of 10% aqueous NaOH on a steam bath for 20 min. Cool the solution and dilute it to 100 ml with water. Dilute the sample to contain about 1 μg of sodium fluorescein per ml, then mix 3 ml of this solution with 20 ml of pH=9 borate buffer and make to 100 ml with water. Measure the fluorescence at 515 nm with excitation at 460 nm.

JONES, J. H., HARROW, L. S., HEINE, K. S., Jr., JAOAC 38, 949-977 (1955). Studies on Coal-Tar Colors. XX. FD&C Blue No. 2. Describes the preparation of a sample of pure color and spectrophotometric and chemical methods of assay.

MOSTER, J. B., PRATER, A. N. Food Technol. 6, 459-463 (1952). Color of Capsicum Spices. I. Measurement of Extractable Color. A method is described for measuring the color of alcohol extracts of capsicum spices in terms of Gentry units. The procedure involves spectrophotometric measurement at two wavelengths; either 569 nm and 663 nm or 577.5 nm and 663 nm, depending on the color of the extract.

MOSTER, J. B., PRATER, A. N. Food Technol. 11, 146-148 (1957). Color of Capsicum Spices. II. Extraction of Color. Extracts are prepared by shaking 0.1 g of sample with 50 ml of 99% isopropanol at 70° ± 1°C for 3 hr in the dark. The color is then measured by the Gentry method.

MOSTER, J. B., PRATER, A. N. Food Technol. 11, 226-229 (1957). Color of Capsicum Spices. IV. Oleoresins Paprika. The color of paprika oleoresins is determined spectrophotometrically at 470 nm in acetone. The method is compared to the potassium dichromate-cobaltous chloride procedure.

NEY, M. Deut. Lebensm. Rundschall 63, 167-170 (1967). Carmine Dyes and Archil. For assay, weigh 0.1 g of dye into a 100-ml flask. Add 20 ml of 1:1 HCl and reflux for 30 min. Transfer the mixture to a 250-ml Erlenmeyer flask with 100 ml of water, add 30-40 ml of $0.1N$ Chloramine T, and stopper for 10 min. Add 1 g of KI and titrate the liberated iodine with $0.1N$ $Na_2S_2O_3$. In the case of an ammoniacal solution of carmine, neutralize 5 g of sample with dilute HCl, add 100 ml of water, 20 ml of 1:1 HCl, and heat on a water bath for about 30 min until a clear orangered solution is obtained. Transfer to an Erlenmeyer flask and proceed as described above.

RAO, G. G., RAO, N. V. Talanta 8, 539-546 (1961). Titrimetric Determination of Indigosulfonate with Potassium Iodate. Conditions are given for the room-temperature titration of indigo carmine with KIO_3. The titration is carried out in a medium that is $6\text{-}8N$ with respect to HCl (at the end of the titration). The blue color of the indigo disappears sharply at the end point, giving a clear yellow solution.

RAO, T. S. S., SASTRY, L. V. L., SIDDAPPA, G. S. Sci Cult. 31, 27-29 (1965). Estimation of Food Colors Using Stannous Chloride. Watersoluble colors in food are determined by measuring the volume of standard $SnCl_2$ needed to decolorize it. As little as 0.25 mg of dye can be estimated.

RAYMOND, P., DAGNEAUX, E. L. K. Chem. Weekblad 53, 134-136 (1957). Dye Strength Determinations. A discussion of the methods used to assay food colors in Holland. Includes titrimetric ($TiCl_3$), gravimetric, and elemental (N) methods, as well as paper chromatographic procedures applicable to color blends.

ROSEBROOK, D. D. JAOAC *54*, 37-38 (1971). Collaborative Study of a Method for Extracting Color in Paprika and Paprika Oleoresin. The color in paprika or other capsicum spices is extracted with acetone and the absorbance is measured at 460 nm.

SCHOLZ, F., REPPEL, L., STARK, A., WAGLER, M. Zentbl. Pharm., Pharmakother. Lab.-diagnostik *110*, 967-968 (1971). Indigo Carmine (CI Food Blue 1) (Proposals for DAB VII). The product is described and methods of analysis are given. A procedure is included for assay at 607 nm in phosphate buffer.

STEIN, C., FREEMAN, K. A. JAOAC *35*, 491-495 (1952). Studies in Coal-Tar Colors, XI: D&C Red No. 30. Describes the preparation of a pure sample as well as spectrophotometric and chemical methods of assay.

SUZUKI, M., NAKAMURA, E., KANAYA, Y., NAGASE, Y. Tokyo Yakka Daigaku Kenkyu Nempo *11*, 120-123 (1961). Indigo Carmine Determination. Indigo carmine is dried at $105°C$ for 2 hr. Then 0.2 g is placed in an iodine flask and dissolved in 50 ml of H_2O, 20 ml of 10% H_2SO_4 and 50 ml of 0.05N $K_2Cr_2O_7$ are added, and the flask is stoppered and placed in the dark. The sample is occasionally shaken until the blue color has completely disappeared and then 3 g of KI is added. The flask is again stoppered and allowed to stand 10 min and then the liberated iodine is titrated with 0.05N $Na_2S_2O_3$.

TRUHAUT, R., CASTAGNOU, R., LARCEBAU, S., LASSALLAE-SAINT-JEAN, V. Bull. Soc. Pharm. Bordeaux *100*, 145-158 (1961). Photocolorimetric Method for the Measurement of the Coloring Power of Caramels and Their Acid Resistance. The coloring power of caramels was measured photocolorimetrically at 430 nm and the results compared with those obtained by the Lovibond method.

Chapter 8 Insoluble Matter

The amount of insolubles in a dyestuff is an indication of purity and is used as such by the colorant manufacturer, the consumer, and the Food and Drug Administration alike. In addition, insolubles represent practical problems for the user since their presence can lead to cloudy soda pop, gritty toothpaste, plugged plant filters, and so on. As a consequence, government, to protect the public health, has established limits on the insolubles content of many colorants, whereas industry has striven to produce products of even higher quality to meet the needs of their customers.

The methods used to determine insolubles are generally simple gravimetric procedures. Those developed and tested by the Association of Official Analytical Chemists are given here.

Preparation of Gooch Crucibles

Digest a good grade of retentive asbestos with HCl (1 + 3), wash it free of acid, and decant the supernatant liquid to remove any fine particles. Prepare a well-packed asbestos mat of suitable thickness in a Gooch crucible, wash it with hot water, dry, ignite, rewash, dry at 135°C, cool in a desiccator, and weigh. Repeat the washing, heating, and drying until the weight is constant.

Water-insoluble Matter

Dissolve a 2-g sample in 200 ml of hot water, then let the solution cool to room temperature. Filter through a tared Gooch crucible, wash with cold water until the washings are colorless, dry for 3 hr at 135°C, cool in a desiccator, and weigh. Calculate any increase in weight as water-insoluble matter.

Alkaline-insoluble Matter

Proceed as described for water-insoluble matter, substituting 1% sodium hydroxide solution or 1:14 ammonium hydroxide for water.

Carbon Tetrachloride Insoluble

Method A: Mix 0.2-0.5 g of sample with 100 ml of CCl_4 in a 250-ml beaker, stir, and heat to the boiling point. Filter hot through a tared Gooch crucible, transfer any residue in the beaker to the filter, and then wash with 10-ml portions of CCl_4 until the washings are colorless. Dry for 3 hr at 100-105°C, weigh, and calculate any weight increase as CCl_4-insoluble matter.

Carbon Tetrachloride Insoluble

Method B: Fit a Gooch crucible with an asbestos mat and a cotton pad, then wash, dry, and tare it. Weigh 1 g of the sample into the crucible, placing the pad on top of the sample. Support the crucible in a Soxhlet extraction apparatus so that the bottom of the crucible is slightly above the top of the siphon tube. Extract with carbon tetrachloride until no more color is removed. Dry to a constant weight at 100-105°C. Cool in a desiccator and weigh.

Toluene, Benzene, Acetone, Alcohol, and Xylene Insolubles

Using the appropriate solvent, proceed as described under Carbon Tetrachloride Insoluble.

Chapter 9 Inorganic Salt Content

The inorganic salts present in colors are there as a result of a step in the manufacturing process—neutralization, isolation, iodination, and so on—or because of the deliberate addition by the manufacturer to meet the strength demands of the customer. The salts most often found are sodium sulfate and sodium chloride. Others less frequently present include sodium acetate, sodium iodide, and sodium carbonate.

The procedures that follow were developed chiefly for the analysis of certifiable colors since they most often contain inorganic salts and generally have specifications controlling their salt content. The methods should, however, be applicable to many of the colorants exempt from certification, either as is or after minor modification.

Most of the techniques used in the following procedures are classical in nature. Some have been instrumented, but their principles remain the same.

A powerful new procedure almost certain to be used extensively in the future is ion chromatography. A preliminary report on the use of this tool for analyzing fluoride, chloride, nitrite, phosphate, bromide, nitrate, sulfate, and iodide in water-soluble color additives was given at the 91st Annual Meeting of the Association of Official Analytical Chemists (Washington, D.C., D. D. Fratz, Paper No. 196). The technique was also used to analyze water extracts of lakes and water-insoluble colorants.

SODIUM CHLORIDE

Volhard Method for Acid Dyes (4): Dissolve 2 g of dye in 100 ml of water and add 10 g of activated carbon (Norit SG No. 2, American Norit Co., Jacksonville, Fla., or equivalent) that is free of chloride and sulfate. Boil gently for 2-3 min. Cool to room temperature, add 1 ml of 6N nitric acid, and stir. Dilute with water to 200 ml in a volumetric flask, then filter through dry paper. Repeat the treatment with 2-g portions of carbon until no apparent color rises when filter paper is dipped into the filtrate.

Transfer 50 ml of filtrate to a 250-ml flask. Add 2 ml of 6N nitric acid, 5 ml of nitrobenzene, and 10 ml or more of standardized 0.1N silver nitrate solution, depending on the chloride content. Shake the flask until the silver chloride coagulates. Prepare a saturated solution of ferric ammonium sulfate and add just enough concentrated nitric acid to discharge the red color; add 1 ml of this solution to serve as the indicator. Titrate with $0.1N$ ammonium thiocyanate solution that has been standardized against the silver nitrate solution until the color persists after shaking for 1 min.

$$\text{Percent sodium chloride} = \frac{(A)(N)(0.05844)(195)(100)}{(w)(50)}$$

where A is net volume of silver nitrate solution required (in ml), N is normality of silver nitrate solution, and w is sample weight (in g).

Note: Calculation is based on a 195-ml volume since 10 g of carbon occupies 5 ml.

Volhard Method for Basic Dyes (4): Dissolve a 2-g sample in water and dilute to 200 ml in a volumetric flask. Add 10 g of carbon, stir for 1 min, and test for complete adsorption as described previously. Repeat the carbon treatment if necessary. Filter through dry paper.

Evaporate a 50-ml aliquot to dryness and then heat to volatilize ammonium chloride. Transfer the residue to a 250-ml flask and determine sodium chloride as described previously.

$$\text{Percent sodium chloride} = \frac{(A)(N)(0.05844)(200)(100)}{(w)(50)}$$

Potentiometric Procedure for Water-Soluble Certifiable Colors, Except D&C Red No. 19 (2): Use a Fisher Titralyzer ® Model 740 equipped with Fisher Nos. 9-313-216 (Ag/AgCl) and 13-639-122 (Ag) electrodes. Immediately prior to use, polish the Ag electrode with a water paste of levigated alumina and then rinse with distilled water. Install the electrode, making sure that the buret delivery tip remains close to and pointing toward the electrode.

If the samples to be analyzed are sodium salts of fluorescein colors, dissolve 2.5 g in water, dilute to about 200 ml, add 15 ml of $1.5N$ nitric acid, and mix well. Then dilute to 250 ml, mix, and filter through fluted paper. Collect 50 ml of the filtrate for analysis and dilute to about 100 ml.

If the samples are sulfonated water-soluble colors, dissolve 0.5 g in about 100 ml of water. Set the time delay on the Titralyzer at 15 sec and for manual operation, in the millivolt mode. Set millivolt dial at 580 (short of end-point setting). Pipette 5-ml aliquots of a 0.58% NaCl solution into three beakers, add about 100 ml of water, and place in positions 1-3. Start pump to deliver silver nitrate. When pump stops, change MV setting in small steps, causing addition of small increments of silver nitrate. Changes must be made in 15-sec intervals to avoid ending cycle. As this is done, watch milliliter dial, and when it reaches calculated value, lock in position. Allow next two beakers to titrate automatically. If titers differ from calculated value by >0.03 ml, make small adjustment in millivolt setting and titrate two more standards in automatic mode.

Place samples in vacant positions and allow instrument to titrate. Place 100 ml of water in last position of turntable. This serves as a blank, which should be nil (i.e., ca. 0.02 ml) and is not subtracted from titers. If higher, contamination is indicated.

Percent NaCl = $N \times T \times 11.69$

where N is normality of silver nitrate solution and T is titer (in ml).

Report results to nearest 0.1%. In halogenated fluoresceins, determine any iodide or bromide as NaCl.

Sodium Chloride in FD&C Yellow No. 5, and FD&C Yellow No. 6 Using a a Specific Ion Electrode: Use a pH meter with an expanded scale (Orion Model 801 or similar) and equipped with an Orion Model 90-01 single-junction sleeve-type reference electrode and an Orion 94-17 chloride electrode.

Using the colorant of interest, prepare 5% (w/v) solutions containing 0%, 3%, 6%, and 9% NaCl. Transfer 20 ml of these solutions to 50-ml beakers and, while stirring, determine the chloride activity of each sample in millivolts. Prepare a calibration curve on semilogarithmic paper of percent NaCl (logarithmic axis) versus millivolts (linear axis). Similarly prepare and measure the potential of the sample solutions and determine their NaCl content from the calibration curve.

SODIUM SULFATE

Titration with Barium Chloride, Acid Dyes (4): Dissolve a 2-g sample and treat it with carbon as described under Sodium Chloride, Volhard Method for Acid Dyes. Place 25 ml of the filtrate obtained in a 125-ml Erlenmeyer flask, and add one drop of a solution of 0.5% phenolphthalein in 50% ethanol. Make alkaline with 0.05N sodium hydroxide, and then add 0.002N hydrochloric acid until the indicator is decolorized. Add 25 ml of ethanol and about 0.2 g of tetrahydroxyquinone indicator (THQ prepared sulfate indicator, Betz Laboratories, Inc., Trevase, Pa.). Titrate with 0.03N barium chloride solution to a red end point. Make a blank determination.

$$\text{Percent sodium sulfate} = \frac{(A - B)(N)(0.07102)(195)(100)}{(w)(25)}$$

where A is volume of barium chloride solution required for sample (in ml), B is volume of barium chloride solution required for blank (in ml), N is normality of barium chloride solution, and w is sample weight (in g).

Titration with Barium Chloride, Basic Dyes (4): Dissolve a 2-g sample and treat it with carbon as described under Sodium Chloride, Volhard Method for Basic Dyes. Place 25 ml of the filtrate obtained in a 125-ml Erlenmeyer flask and continue as directed previously for sodium sulfate in acid dyes.

$$\text{Percent sodium sulfate} = \frac{(A - B)(N)(0.07102)(200)(100)}{(w)(25)}$$

General Gravimetric Method (7): Weigh 5 g of sample, dissolve it in 100 ml of warm water, and transfer to a 250-ml volumetric flask. Add 35 g of sulfate-free sodium chloride, stopper the flask, and let it stand for 1 hr. Swirl at frequent intervals. Dilute to volume with saturated sodium chloride solution and filter through dry filter paper. Precipitate the sulfate with barium chloride; filter, dry, and weigh the precipitate.

Turbidimetric Determination in Water-soluble Certifiable Color Additives (3): Weigh 2 g of sulfate-free color into each of eight 400-ml beakers. Add 0 ml, 2 ml, 5 ml, 10 ml, 20 ml, 30 ml, 40 ml, and 50 ml of 0.1% w/v aqueous Na_2SO_4 stock solution to the eight beakers consecutively. Add 10 g of activated carbon and 200 ml of water to each beaker and decolorize as described under Sodium Chloride, Volhard Method for Acid Dyes.

Transfer the mixtures to 250-ml volumetric flasks, dilute to volume with water, mix thoroughly, and filter through dry paper. Collect filtrates when clear, and taking two 50-ml aliquots of each, add 10 ml of conditioning solution to one and 25 ml of conditioning solution to the other. Add 0.2 g of finely ground barium chloride to each sample, stir vigorously, and then immediately obtain spectra of the samples from 400-460 nm, washing the absorption cell with ethanol and water between samples. Prepare a plot on linear graph paper of sample absorbance at 440 nm against percent Na_2SO_4.

Similarly prepare and determine the absorbance of the test sample. If the sample absorbance at 440 nm is $\geqslant 0.16$ when a 50-ml aliquot of the clear filtrate is treated with 10 ml of conditioning solution and 0.2 g of barium chloride, repeat the reading on a second 50-ml aliquot of sample to which 25 ml of conditioning solution has been added.

The conditioning solution is prepared by dissolving 120 g of NaCl in 400 ml of water, adding 10 ml of concentrated HCl and 500 ml of glycerol, and diluting the mixture to one liter with distilled water.

Potentiometric Titration of Sodium Sulfate in Certifiable Water-soluble Sulfonated Colors (1) (Note: This Procedure Not Applicable to Fluorescein-type Colors): Use a Hiranuma recording autotitrator Model RAT-11 equipped with a WB-11 double-action buret and a C-11 autocycle attachment (Rainin Instrument Co., Inc., 555 Main St., Fort Lee, N. J. 07024) and a Coleman 3-571 silver-billet reference electrode and a 3-551 platinum cap indicating electrode.

Prior to use, clean the reference electrode with steel wool and then soak it for at least 4 hr in a solution of 252 mg $K_4Fe(CN)_6 \cdot 13H_2O$ and 164 mg of $K_3Fe(CN)_6$ in 200 ml of water. Rinse the coated electrode with distilled water just prior to use. Use a freshly coated electrode each day prepared from coating solution no more than 1 week old. The indicating electrode should be cleaned after 2 days of use by electrolyzing at + 22.5 V in HCl-water (1+3).

Pipette 10 ml of a 1-2% sample solution (depending on color and Na_2SO_4 content) to a 200-ml tall-form Berzelius beaker. Add 10 ml of water and 100 ml of ethanol solution (360 ml of H_2O + 5640 ml of 95% EtOH) to the sample and, to ensure an end point, add 5 ml of 0.005 M Na_2SO_4. Similarly prepare standards using 10 ml of water in place of color solution.

Using the following control settings, titrate the sample to an end point with 0.01M $Pb(NO_3)_2$: range, 10 MV/cm; dwell time, 90 sec; interval, 10 sec; delivery, about 0.1 ml; sensitivity, 8; stirrer, 5.

$$\text{Percent } Na_2SO_4 = \frac{\text{ml } Pb(NO_3)_2 \times 0.01 \times 0.142 \times 100}{\text{Sample weight}}$$

SODIUM ACETATE (6)

Prepare silver toluenesulfonate by dissolving silver oxide or carbonate in a solution containing 10% excess p-toluenesulfonic acid, evaporating to dryness, and drying at 135°C for 8 hr.

Pretreat p-toluenesulfonic acid by drying the monohydrate at 110°C overnight, and then cooling and powdering the material.

To a 500-ml Erlenmeyer flask add 100 ml of water, one drop of m-cresol purple indicator, and sufficient 0.1N sodium hydroxide or 0.1N hydrochloric acid to make the solution just yellow. Assemble the apparatus shown in Fig. 5 and place the Erlenmeyer flask under the condenser. Place 30 ml of absolute ethanol in the distillation flask, then add through a powder funnel 5 g of sample, 1 g of silver toluenesulfonate, and 5 g of p-toluenesulfonic acid. Wash the funnel and neck of the flask with 25 ml of absolute ehtanol; add three or four boiling stones, shake the flask to mix the contents, and attach it to the condenser.

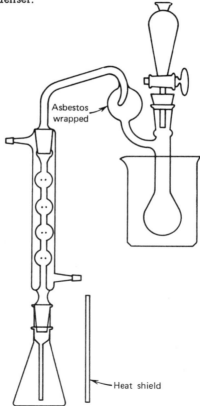

Figure 5   Apparatus Used in Sodium Acetate Determination (Reprinted with the permission of the Association of Official Analytical Chemists)

Immerse the distillation flask in a beaker of hot water and boil the water. Collect about 25 ml of distillate, remove the heat, and slowly add 25 ml of absolute ethanol. Replace the heat source and collect an additional 25 ml of distillate. Make a third addition of absolute ethanol and distill as before. Finally, boil until the distillation rate is slow (ca. 30 min total distillation time from the beginning of the first distillation).

Wash down the condenser into the receiver with 50 ml of water. Add 50 ml of standardized $0.1N$ sodium hydroxide and three or four boiling stones. Connect to a reflux condenser fitted with an absorption tube containing Ascarite; reflux for 10 min. Cool to room temperature, add a few drops of m-cresol purple indicator, and titrate with $0.1N$ hydrochloric acid to a yellow-green end point.

Determine a blank by duplicating the same procedure without a sample.

Calculate the NaOAc content from the net volume of standard NaOH required. One milliliter of $0.1N$ NaOH is equal to 0.0082 g of $C_2H_3O_2Na$.

SODIUM HALIDES

In Fluorescein Colors (4): Place 5 g of sample in a 400-ml beaker and add about 150 ml of distilled water. If the sample is a color acid, add sufficient 10% NaOH to effect solution. Heat the mixture nearly to boiling and then add 5 ml of H_3PO_4. Digest the solution until the precipitate formed is well coagulated. Cool to room temperature, transfer to a 250-ml volumetric flask, dilute to volume with distilled water, and mix well. Filter through dry fluted paper.

Transfer a 100-ml aliquot of the filtrate to a 500-ml tall-form beaker, add 2.5 ml of 30% NaOH, and then determine iodide as described on p. 168 beginning with "25 ml of 7% aqueous $KMnO_4$. . . ."

Take a second 100-ml aliquot and determine bromide as described on p. 169 beginning with "3 ml of 5% potassium cyanide. . . ."

Place a third 100-ml portion of filtrate in a 400-ml beaker, heat to boiling, and add enough 10% $AgNO_3$ solution to precipitate halides. Digest the solution until the precipitate is well coagulated, cool, and transfer to a weighed Gooch crucible. Wash with water and alcohol, dry at $135°C$, cool, and weigh as NaCl. Correct the weight for any NaBr or NaI present in the sample.

Sodium Iodide in FD&C Red No. 3(5): Prepare an eluant solution by dissolving 500 g of ammonium sulfate in water and diluting to 2 liters, and then mixing the solution with 200 ml of SDH No. 30 alcohol.

Weigh 0.5 g of sample, transfer it to a 150-ml beaker, and dissolve it in 15 ml of water and 10 ml of alcohol. Wash 10 g of Whatman Column Chromedia CF11 with the eluant solution. (Some lots of adsorbent contain an impurity with a spectrum similar to that of sodium iodide, which must be removed.) Add the washed adsorbent and 15 ml of the eluant to the sample and mix. Then add 50 g of ammonium sulfate and mix well. Using 25 ml of eluant, transfer the sample to a chromatographic column prepared and eluted as described on p. 202. Record the UV spectra of the fractions as eluted. Usually sodium iodide cannot be detected in the first two fractions and is completely eluted in the first eight fractions. It is characterized by an absorption maximum near 222 nm.

BIBLIOGRAPHY

1. BAILEY, J. E., GRAICHEN, C. JAOAC 57, 353-355 (1974).
2. GRAICHEN, C., BAILEY, J. E. JAOAC 57, 356-357 (1974).
3. HOBIN, N. K. JAOAC 53, 242-243 (1970).
4. *Official Methods of Analysis,* 12th ed. Association of Official Analytical Chemists, Washington, D. C., 1975, p. 644.
5. Private Communication, Food & Drug Administration, Division of Color Technology, Washington, D.C.
6. SCHIFFERLI, J., SCHRAMM, A. T. JAOAC 32, 614-617 (1949).
7. *Specifications for Identity and Purity of Food Additives,* Vol. 2, *Food Colors.* Food and Agriculture Organization of the United Nations, Rome, 1963.

Chapter 10 Metals

For the most part, the trace metals present in color additives are a result of the equipment used in the processes, the starting materials, or impurities in them. Their amounts are monitored in an effort to ensure product uniformity and safety. In recent years, the list of metals tested for has grown with the increased awareness of, and the concern over, the affects certain of them have on man's physical well being. There has also been a trend toward analyses for specific items rather than groups of them such as "heavy metals."

The battery of methods now in use for the determination of metals ranges from the classical to the modern and includes such procedures as the Gutzeit method for arsenic as well as the determination of chromium by atomic absorption spectroscopy (AAS). In general, increased emphasis has been placed on the use of modern, automated instrumental techniques.

Many of the newer instrumental methods have proven quite useful. Unfortunately, some, such as flame AAS, have not been the panacea they were predicted to be since matrix interferences frequently necessitate the wet ashing of samples prior to their analysis, a step AAS was once thought capable of obviating. Flameless AAS methods such as that for mercury, methods dependent on the generation of volatile hydrides—including those for arsenic and selenium—and graphite-furnace techniques for metals such as tin and lead should prove useful for the analysis of color additives. However, few, if any, applications of these procedures have been published in this area as yet. The growing need, though, for an ever increasing number of fast, accurate, and precise analyses for increasingly more metals in color additives leaves little doubt that these tools or similar ones will eventually replace most, if not all, of the existing classical procedures.

ARSENIC

Reaction with Silver Diethyldithiocarbamate (6)

Reagents:

(a) Standard arsenic solution—Dissolve 0.132 g of dry, powdered As_2O_3 in 5 ml of NaOH solution (1:5). Neutralize the solution with 10% (w/v) H_2SO_4, add 10 ml of excess, and dilute to 1000 ml with recently boiled H_2O. Pipette 10 ml of this solution into a 100-ml volumetric flask, add 10 ml of 10% (w/v) H_2SO_4, and dilute to volume with recently boiled H_2O. Use within 3 days.

(b) Silver diethyldithiocarbamate solution—Dissolve 1 g of $(C_2H_5)_2NCSSAg$ in 200 ml of reagent-grade pyridine. Store in a light-resistant container and use within 1 month.

(c) Stannous chloride solution—Dissolve 40 g of reagent-grade $SnCl_2 \cdot 2H_2O$ in 100 ml of HCl. Store in glass and use within 3 months.

(d) Lead-acetate-impregnated cotton—Soak cotton in a saturated solution of reagent-grade lead acetate, squeeze out the excess solution, and dry in a vacuum at room temperature.

CAUTION—Some substances may react unexpectedly with explosive violence when digested with H_2O_2. Appropriate safety precautions must be employed at all times. If halogen-containing compounds are present, use a lower temperature while heating the sample with H_2SO_4, do not boil the mixture, and add the peroxide, with caution before charring begins, to prevent loss of trivalent arsenic.

Transfer 1 g of sample into the generator flask, add 5 ml of H_2SO_4 and a few glass beads, and digest on a hot plate until charring begins. (Additional H_2SO_4 may be necessary to completely wet some samples, but the total volume added should not exceed 10 ml.) After the sample has been initially decomposed by the acid, cautiously add 30% H_2O_2 dropwise, allowing the reaction to subside and reheating between drops. The first few drops must be added very slowly with sufficient mixing to prevent a rapid reaction, and heating should be discontinued if foaming becomes excessive. Maintain oxidizing conditions at all times during the digestion by adding small quantities of peroxide whenever the mixture turns brown or darkens. Continue digestion until the organic matter is destroyed, fumes of H_2SO_4 are evolved, and the solution becomes colorless. Cool, cautiously add 10 ml of H_2O, evaporate to strong fuming, and cool. Add 10 ml of H_2O, wash the sides of the flask with a few ml of H_2O, and dilute to 35 ml $\pm$ 2 ml.

Add 20 ml of dilute H_2SO_4 (1:5), 2 ml of potassium iodide solution (15:100), and 0.5 ml of stannous chloride solution; mix. Allow the mixture to stand for 30 min at room temperature. Pack the scrubber tube (Fig. 6c) with two plugs of lead acetate-impregnated cotton, leaving a small air space between the plugs, lubricate joints (b) and (d) with stopcock grease, and connect the scrubber unit with the absorber tube (e). Transfer 3 ml of silver diethyldithiocarbamate solution to the absorber tube, add 3 g of granular Zn (20-mesh) to the mixture in the flask, and immediately insert the standard-taper joint in the flask. Allow the evolution of hydrogen and color development to proceed at room temperature ($\geqslant 25°$) for 45 min, swirling the flask gently at 10-min intervals. Transfer the diethyldithiocarbamate solution to a 1-cm absorption cell and determine its absorbance at 525 nm, using silver diethyldithiocarbamate as the blank. The absorbance due to any red color from the sample solution should not exceed that produced by 3 ml of standard arsenic solution (3 μg of As) when treated in the same manner and under the same conditions as the sample. The room temperature during the generation of arsine from the standard should be held to within $\pm 2°$ of that observed during the determination of the sample.

Interferences. Metals or salts of metals such as chromium, cobalt, copper, mercury, molybdenum, nickel, palladium, and silver are said to interfere with the evolution of arsine. Antimony, which forms stibine, is the only metal

Figure 6   Apparatus for Arsenic Test (Courtesy Fisher Scientific Co.)

likely to produce a positive interference in the color development with the silver diethyldithiocarbamate. Stibine forms a red color that has a maximum absorbance at 510 nm, but at 525 nm the absorbance of the antimony complex is so diminished that the results of the determination would not be altered significantly.

Iodimetric Procedure (8)

Reagents:

(a) Potassium iodide solution—Dissolve and dilute 15 g of recrystallized KI to 100 ml with H_2O.

(b) Stannous chloride—Dissolve 40 g of $SnCl_2 \cdot 2H_2O$ in 100 ml of concentrated HCl.

(c) Absorbing solution—Dissolve and dilute 3.2 g of $HgCl_2$ (recrystallized from H_2O) and 0.1 g of powdered USP gum arabic to 200 ml with H_2O.

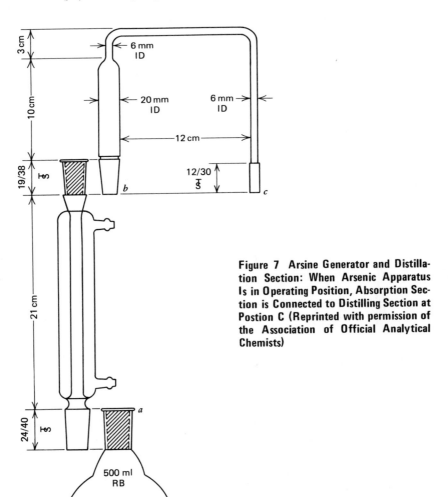

Figure 7 Arsine Generator and Distillation Section: When Arsenic Apparatus Is in Operating Position, Absorption Section is Connected to Distilling Section at Postion C (Reprinted with permission of the Association of Official Analytical Chemists)

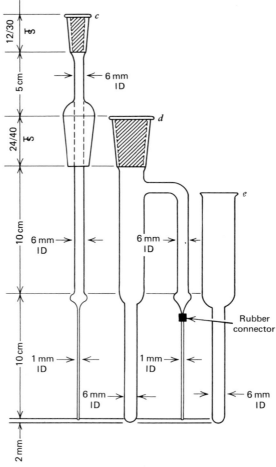

Figure 8   Arsine Absorber and Titrating Tubes (Reprinted with the permission of the Association of Official Analytical Chemists)

(d) Lead acetate solution—Dissolve 10 g of $Pb(C_2H_3O_2)_2$ in 80 ml of H_2O. Make just acid to litmus paper with acetic acid. Dilute to 100 ml with H_2O.

(e) Standard iodine solution—Dissolve 6.35 g of I_2 and 12.7 g of KI in a little H_2O; filter; dilute the filtrate to 1 liter. Dilute 100:1000, 20:1000 and 10:1000 with H_2O. Add 25 g of KI to each liter of dilute I_2 solution.

(f) Standard arsenic solutions—Dissolve 1 g of As_2O_3 in 25 ml of 20% aqueous NaOH. Saturate the solution with CO_2 and dilute to 1 liter with freshly boiled H_2O. Dilute 100:1000, 20:1000, and 10:1000 with H_2O.

(g) Starch indicator—Make 1 g of soluble starch into a thin paste with cold

H_2O. Pour into 200 ml of hot H_2O, and while still hot add two or three small crystals of HgI_2.

(h) Buffer solution—Dissolve and dilute 10 g of $Na_2HPO_4 \cdot 12H_2O$ to 100 ml with H_2O.

Apparatus (see Fig. 7):

Soak glass wool in 10% $Pb(Ac)_2$ solution, thoroughly dry, and insert loosely in the trap above the condenser. All connections between the condenser and the absorber should be dry. All glassware should be washed with dichromate cleaning solution and rinsed with distilled H_2O prior to use.

CAUTION—Perchloric acid is a strong oxidant. Contact with organic material may cause fire or explosion. See MCA Chemical Safety Sheet SD-11.

Weigh 10 g of sample into a 800-ml Kjeldahl digestion flask, add 10 ml of concentrated H_2SO_4 and 10 ml of concentrated HNO_3. Digest over a low flame until the mass begins to clear. Add 15 ml of H_2O and digest until clear. Add successive 5-ml portions of concentrated HNO_3 until all the organic matter is in solution. Slowly add 10 ml of a (1 + 1) mixture of concentrated HNO_3 and 60-70% $HClO_4$. Digest until the mixture in the flask is only lightly colored. Add an additional 5 ml of the $HClO_4$ mixture. Heat until the initial vigorous reaction subsides. Continue the addition of 5-ml portions of $HClO_4$ mix until the digest is water-white. Heat to SO_3 fumes. Cool. Slowly add 20 ml of saturated ammonium oxalate solution and heat to SO_3 fumes. Cool and add 50 ml of H_2O. Swirl and transfer to the arsine generator (Fig. 7a) using three 20-ml portions of H_2O. Add 5 ml of solution (a) (preceding list), 1 ml of soluiton (b), and dilute to 90-100 ml with H_2O. Place one ml of solution (c) in each receiving tube and place in the operating position. Add 5 g of 20-30 mesh As-free Zn to the generator. Wash down the flask neck with a few milliliters of H_2O and connect at point (a) (Fig. 7).

Heat the generator to reflux. Lower the flame and continue heating for 12-15 min, then disconnect the first receiver and the delivery tube at (c) (Fig. 7). Raise the delivery tube at (d) (Fig. 8) and add 2 ml of buffer solution (h) through the tube. Wash down the outside with 3-5 ml of H_2O. Place the transfer tube in position at (d) and suck the contents of the second receiver into the first receiver. Wash the outside of the second delivery tube with two 1-ml portions of H_2O and suck the washings into the first receiver. Remove the second receiver (e) and the transfer tube.

In an empty receiver tube place 2 ml of solution (h) and 2 ml of H_2O. Add 3 ml of solution (e), five drops of solution (g), and titrate to a colorless end point with solution (f). Agitate the soln during the titration by alternately sucking and blowing through the stirring tube.

$$\frac{\text{mmg As}_2O_3}{\text{ml I}_2} = \frac{\text{mmg As}_2O_3/\text{ml} \times \text{ml As}_2O_3 \text{ soln.}}{\text{ml I}_2 \text{ soln.} \times 4}$$

Place the stirring tube in the first receiver containing the sample. Add solution (e) slowly until the orange precipitate that initially forms just disappears. Add 5 drops of solution (g) and titrate with solution (f) to the disappearance of the blue color. Run a blank on all reagents.

$$\text{Percent As}_2O_3 =$$

$$\frac{\text{ml I}_2 \text{ soln.} \times \text{I}_2 \text{ factor} - \dfrac{[\text{ml } \dfrac{\text{As}_2O_3}{4} \times \text{mmg As}_2O_3/\text{ml}]}{\text{sample weight}}}{} - \text{blank}$$

BARIUM (SOLUBLE) IN D&C RED NO. 9

Rub 10 g of color into a paste with a little water and four to eight drops of alcohol. Add 150 ml of water and 40 ml of 1 + 249 HCl. (On a separate aliquot test the pH of the suspension with meta cresol purple indicator. It should be pH 2 ± 0.1). Warm on a water bath for 30 min, stirring frequently. Cool, transfer to a 250-ml volumetric flask, make to volume, mix, and filter on a dry filter. To 200 ml of filtrate, add 20 ml of concentrated HCl and heat to boiling. Add 5 ml of (1 + 3) H_2SO_4, again heat to boiling, and allow to settle for 30 min on a water bath. Filter, wash the precipitate with hot (1 + 199) H_2SO_4, ignite, and weigh as $BaSO_4$. Report as $BaCl_2$.

$$BaSO_4 \times 0.8923 = BaCl_2$$

If strontium salts are present, Ba and Sr may be separated as follows. Prepare a paste, dilute with HCl, treat on a water bath, make to volume, and filter as described above. Make the filtrate alkaline to methyl red with NH_4OH and acidify with acetic acid. Add 5 ml of a 5% solution of $K_2Cr_2O_7$. Yellow $BaCrO_4$ precipitates, which may be filtered, washed with dilute acetic acid and water, redissolved in HCl, and reprecipitated as $BaSO_4$.

CHROMIUM IN FD&C BLUE NO. 1

Weigh 5 g of sample into a platinum dish. Mix in 8 g of a fusion mix consisting of equal portions of sodium carbonate and potassium carbonate. Slowly fuse the sample over a flame until all organic material is destroyed. Place the sample in a 800-850°C muffle furnace until it is completely ashed (ca. 2 hr). Remove the dish from the furnace, allow it to cool, and wash the contents into a 150-ml beaker with 50 ml of water. Add 25 ml of 6% hydrogen peroxide solution, stir, and boil for 10 min. Cool and then filter through Whatman No. 1 filter paper into a 100-ml volumetric flask. Rinse the beaker and filter paper into the flask and then dilute to volume. Using a 5-cm cell, determine the absorption spectrum from 450 nm to 310 nm by plotting against water. Construct a baseline connecting the absorption minimum near 320 nm and 450 nm. Draw a line that is parallel to the absorbance axis and passes through the spectrum and the baseline at 370 nm. The absorbance at the intersection of this line with the sample spectrum equals A_1. The absorbance at the intersection of this line with the baseline equals A_2:

$$\text{Parts per million chromium} = (A_1 - A_2)(50.4)$$

where 50.4 is the reciprocal of the slope of a least-squares fit of the regression of A at 370 nm on ppm of chromium determined by this method.

HEAVY METALS (16)

Reagents:

(a) Standard Pb(NO$_3$)$_2$ solution—Dissolve 0.1598 g of Pb(NO$_3$)$_2$ in 1% (v/v) aqueous HNO$_3$. Dilute with H$_2$O to 1000 ml. Dilute 10 ml to 100 ml with H$_2$O. (Dilute solution must be freshly prepared.)

Weigh 1 g of sample into a porcelain crucible and ignite at low temperature until thoroughly charred. Add 2 ml of HNO$_3$ and five drops of H$_2$SO$_4$ to the crucible and carefully heat to SO$_3$ fumes. Ignite at 500-600°C to remove carbon. Cool, add 2 ml of HCl, and evaporate to dryness on a steam bath. Moisten the residue with one drop of HCl, add 10 ml of hot H$_2$O, and digest for 2 min. Add dilute NH$_4$OH dropwise until the solution is just alkaline to litmus paper. Add dilute acetic acid until the solution is slightly acid to litmus paper, and then add 2 ml excess. Filter (if necessary) into a 50-ml Nessler Tube. Wash the filter and crucible into the tube with 10 ml of H$_2$O and dilute to 25 ml with H$_2$O.

Pipette 2 ml of dilute acetic acid into a second 50-ml Nessler tube. Add a volume of solution (a) containing the amount of Pb to be tested for. Add H$_2$O to 25 ml.

Add 10 ml of saturated H$_2$S solution to each tube. Mix and allow to stand 10 min. Compare the sample and standard visually. Report Pb as more than or less than standard.

LEAD

In Colors That Do Not Contain Calcium, Strontium, or Barium (5): Prepare a 10% hydroxylamine hydrochloride solution by dissolving 10 g of NH$_2$OH· HCl in 20 ml of water and slightly alkalizing with ammonium hydroxide. Extract any lead with dithizone (diphenylthiocarbazone). Remove excess dithizone with chloroform and boil off any chloroform remaining in the aqueous phase. Acidify with hydrochloric acid and dilute to 100 ml.

Prepare a stripping reagent by adding 10 ml of glacial acetic acid to 20 ml of saturated sodium acetate solution and diluting to 100 ml.

Prepare a 0.5% starch solution by pasting 1 g of soluble starch with several ml of cold water, pouring into 200 ml of hot water, and while still hot, adding two or three small crystals of mercuric iodide as a preservative.

Prepare a standardized sodium thiosulfate solution by adding, to a 0.1N sodium thiosulfate solution, 5 ml of isoamyl alcohol per liter as a preservative. On the day the analysis is performed, dilute 1:100 or 1:20 with carbon dioxide-free water (depending on the lead concentration). Standardize against a lead solution prepared by dissolving 0.3197 g of lead nitrate in 100 ml of 1% nitric acid.

Transfer 5 g of sample to a 500-ml Kjeldahl flask. Add 10 ml of concentrated sulfuric acid and 10 ml of concentrated nitric acid and heat to sulfur trioxide fumes. Add 5 ml of concentrated nitric acid and again heat to sulfur trioxide fumes. Repeat the addition of nitric acid each time sulfur trioxide fumes appear, until the dye is in solution and the digest is yellow. Then add 10 ml of a 1:1 mixture of concentrated nitric acid and 60-70% perchloric acid, and continue heating until the digest is colorless or pale yellow and the bulk of the sulfuric acid is evaporated.

Cool the flask and neutralize with ammonium hydroxide. Add 20 ml of 50% citric acid solution and adjust to pH = 8.5-9 with ammonium hydroxide using thymol blue indicator. Add 5 ml of 10% potassium cyanide solution and transfer the sample to a separatory funnel. Extract the lead with 20 ml of a 0.002% solution of dithizone in chloroform. (If there is enough iron present to cause excessive oxidation of the dithizone, as indicated by a yellow color in the chloroform layer, add 10 ml of the 10% hydroxylamine hydrochloride solution to reduce the iron.) Let the chloroform layer settle; drain it into a second funnel. Repeat the extraction until the red lead dithizonate is completely removed.

Wash the combined chloroform extracts with 25 ml of water containing one drop of ammonium hydroxide. Drain the washed chloroform layer into a third funnel. Add 110 ml of 1:99 nitric acid and shake for 1 min. Drain and discard the chloroform layer and about 1 ml of the acid layer. Drain the acid layer through cotton, discarding the first 3 ml. Electrolyze a 100-ml aliquot by heating a platinum gauze cylindrical anode to red heat in the oxidizing flame of a burner. Cool the anode and place it in the sample solution. Start the electrode rotating, heat the sample solution to 60-70°C, and add 0.1 g of potassium dichromate. Electrolyze at 70-80°C, using 100 mA. Remove the heat and siphon the sample from the beaker while playing a stream of water directly on the anode; keep the level of liquid above the deposit at all times. The acid is entirely removed when the current falls to zero.

Dissolve the deposit from the anode with 4-5 ml of stripping reagent and 1 ml of freshly prepared 2% potassium iodide solution, contained in a flat-bottom vessel of such size that the solution just covers the anode. Add a few drops of the starch solution and titrate the liberated iodine with $0.001N$ sodium thiosulfate solution using the anode as a stirrer. Use a $0.005N$ thiosulfate solution if the lead content is 1-5 mg. Similarly determine a blank.

In Aluminum Lakes (9): Weigh 2 g of sample into a 500-ml Kjeldahl digestion flask, add 10 ml of concentrated H_2SO_4 and 10 ml of concentrated HNO_3, and digest on a low flame until SO_3 fumes appear. Add successive 5-ml portions of concentrated HNO_3 (waiting until SO_3 fumes appear before adding each succeeding portion) until all organic matter is in solution. Slowly add 5-10 ml of a (1:1) mixture of concentrated HNO_3 and 60-70% $HClO_4$, and continue the digestion until the white precipitate formed shows the first signs of spattering. Allow the flask to cool and cautiously add 5 ml of H_2O and then a few drops of concentrated NH_4OH. Swirl the flask vigorously and cool under running water. Add 20 ml of 50% (w/v) citric acid solution and adjust the pH to 3-3.4 (bromophenol blue) with concentrated NH_4OH. Add 1 ml of $CuSO_4$ solution containing 1 mg of Cu per ml and transfer the solution to the precipitation tube (b) of the sulfiding apparatus (Fig. 9). Bubble H_2S thru the solution at an approximate rate of two bubbles per second for 3-5 min and filter the resulting suspension through (c) at an approximate rate of one drop per second. When filtration is complete remove the receiver containing the filtrate and attach a suction test tube. Add 3 ml of hot concentrated HNO_3 through the separatory funnel (a) and draw through

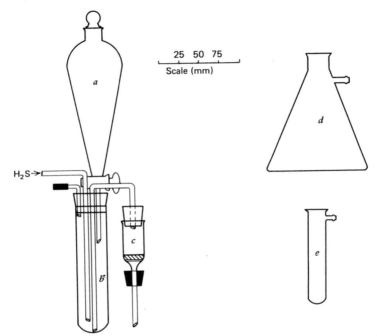

25 50 75
Scale (mm)

Figure 9 Sulfiding Apparatus (Reprinted with the permission of the Association of Official Analytical Chemists)

the filter, followed with 2 ml of hot water. Detach the filter and pass an additional 3 ml of hot concentrated HNO_3 through the filter, wetting all sides. Again follow with 2 ml of hot water. If the filter is still colored with PbS, wash again with hot concentrated HNO_3 and water. Wash the dissolved sulfides into the precipitation tube (b), wetting all sides to take up any residual lead sulfide and then into a 50-100-ml glass-stoppered conical flask. Stopper and shake for a few seconds and then remove the stopper and boil until the solution clears to remove the last traces of H_2S and to coagulate any free sulfur present.

Transfer the solution to a 250-ml separatory funnel. Wash the flask with two 5-ml portions of distilled water and add the washings to the main solution. Add 10 ml of 50% (w/v) citric acid solution, 5 ml of 10% sodium cyanide solution, and a few drops of hydroxylamine hydrochloride solution to prevent oxidation of the dithizone; adjust the pH to 8.5-9.5 (thymol blue) with concentrated NH_4OH.

Extract with dithizone and electrolyze as described above.

In Barium, Calcium, and Strontium Lakes (9): Place 2 g of lake, 4 g of Na_2CO_3, 6 g of K_2CO_3, and 0.5 g of $NaNO_2$ in a platinum crucible of suitable size. Mix thoroughly. Heat carefully until the color is carbonized, then heat to about 850°C, and hold at that temperature for 15 min. If a controlled muffle furnace is available, it is only necessary to place the fusion mixture in the cold furnace and raise the temperature gradually to 850°C over a 2-h period. Usually 15-30-min heating at 850°C is sufficient to complete the fusion.

When fusion is complete, allow the crucible and contents to cool below 100°C and then add 2-3 ml of water and heat over a low flame, using care to prevent spattering, until the contents can be separated from the crucible. Transfer the fused mixture to a 150-ml beaker with the aid of about 25 ml of hot water. Boil until the caked material is completely disintegrated, and then filter through a retentive filter paper. Wash the residue on the filter with two 15-ml portions of hot 5% Na_2CO_3 solution. Lead will be in both filtrate and residue. Transfer the filtrate to a separatory funnel and extract the lead from the filtrate as directed under aluminum lakes. Dissolve the residue on the filter in 10-20 ml of the hydrochloric acid solution, wash the filter with water, and add washings to the solution. Boil the solution to expel carbon dioxide and then transfer to a separatory funnel and extract the lead as directed above. Combine with the chloroform extracts from the soluble portion of the fusion products and determine the total lead by the electrolytic method.

MERCURY

Ion Exchange Paper—X-ray Emission Method (11): Place a 1.5-in. disk of ion-exchange paper (Reeve Angel, Grade SB-2 Amberlite ion resin-loaded papers, anion exchanger, strong base-type, containing Amberlite IRA-400 resin, Cl⁻ form) in the joint of a two-piece, 0.75-in.-ID chromatographic column, joined by a threaded aluminum coupling and having a Teflon stopcock with fine adjustment control; tighten the join. Using light suction, draw water up through the paper. Wash the column and paper by passing 50 ml of $0.5N$ hydrochloric acid through the column at 1 ml/min. Leave several ml of solution in the column.

Dissolve the sample in dilute hydrochloric acid and adjust the acid concentration to $0.5N$ hydrochloric acid. Filter through fine filter paper previously washed with $0.5N$ hydrochloric acid and rinsed with water. Dilute to 200 ml or more with $0.5N$ hydrochloric acid. Divide the sample in two; add a known amount (1-5 μg) of mercury to one portion. Carry a reagent blank and a standard, containing a known amount of mercury in 100 ml of $0.5N$ hydrochloric acid, through the rest of the procedure.

Pass each solution through a column such as that described above at 1 ml/min. Follow the sample with 25 ml of $0.5N$ hydrochloric acid, also at 1 ml/min. Drain the solution from below the resin paper, remove the paper, and dry at room temperature.

Using a standard solution containing 1.354 μg/ml of mercuric chloride, set up an X-ray emission spectrograph, equipped with a molybdenum tube, lithium fluoride crystal, scintillation counter, and pulse-height analyzer, on the mercury Lα line $2\theta = 36°$. Mount the resin-loaded disc in the instrument holder. Using at least 16,000 counts, determine the counts per second at $2\theta = 35°$, $2\theta = 36°$, and $2\theta = 37.1°$. For best results take an average of four readings, rotating the sample $90°$ between readings. Calculate as follows:

$$R_{36°/35°} = \frac{\text{counts/sec at } 2\theta = 36°}{\text{counts/sec at } 2\theta = 35°}$$

$$R_{37.1°/35°} = \frac{\text{counts/sec at } 2\theta = 37.1°}{\text{counts/sec at } 2\theta = 35°}$$

For blank, sample, and standards, $R_{37.1°/35°}$ should be equal if tungsten is absent. If tungsten is absent, calculate μg of mercury in the sample aliquot from $R_{36°/35°}$ as follows:

$$\mu g \text{ mercury} = \frac{(R_s - R_b)Y}{R_{s+y} - R_s}$$

where R_s is calculated $36°/35°$ value for the sample aliquot, R_b is calculated $36°/35°$ value for the blank, R_{s+y} is calculated $36°/35°$ value for the sample aliquot to which Y μg of mercury has been added, and Y is μg of mercury added to the aliquot.

Photometric Mercury Vapor Method (17, 18): Use the apparatus shown in Fig. 10. Preheat the furnace to $650°C$ and adjust the nitrogen flow to 1 liter/min. Standardize the mercury vapor meter following the manufacturer's instructions. Adjust the attenator so that the recorder scale is 200 mV.

Calibrate the meter by placing aliquots of mercuric chloride solution containing 0.01-0.03 μg of mercury on separate pieces of ignited asbestos in individual combustion boats. Cover the asbestos with 1-2 g of anhydrous sodium carbonate. Place the boats one at a time in the tube furnace and close the inlet. After 1 min start the nitrogen flow. Prepare a plot of response vs amount of mercury.

Treat 0.025 g of organic sample or 0.25 g of inorganic sample in the same way. Iodine interferes with the determination.

Colorimetric Method (14): Use NF-grade chloroform throughout this procedure. Prepare a 20% hydroxylamine hydrochloride solution. Remove heavy metals by shaking with 50-ml portions of 100 mg/liter dithizone in chloroform. Wash with several portions of chloroform to remove excess dithizone.

Prepare a 40% potassium bromide solution and remove heavy metals as described in the preceding paragraph. Make alkaline with one or two drops of 10% sodium hydroxide before storing.

Prepare a 50% ammonium acetate solution and remove heavy metals as described above.

Prepare a mercury standard solution, using mercuric oxide, and containing 5 mg/liter of mercury.

Prepare a dithizone solution containing 6 mg/liter of dithizone in chloroform, and use this throughout the portion of the procedure that follows.

Weigh a 1-g sample and transfer it to a two-neck, 500-ml digestion flask fitted with a Freidrichs condenser and a 50-ml dropping funnel. (If the sample is a triphenylmethane or oil-soluble dye, use a 0.5-g sample.) Add 10 ml of 1:1:1 sulfuric acid-nitric acid-water and allow to stand for about 5 min. Heat

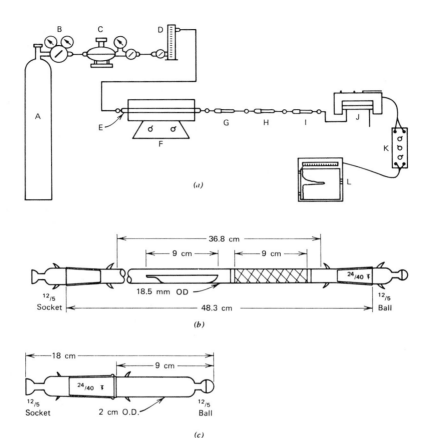

Figure 10 (a) Schematic Diagram of Apparatus for Photometric Mercury Vapor Method:

A. Tank of Nitrogen

B. Two-stage Pressure Regulator

C. Low-pressure Regulator

D. Flowmeter

E. Combustion Tube

F. Combustion-tube Furnace

G. Dehydrite Trap

H. Ascarite Trap

I. Aluminum Oxide Trap

J. Mercury Vapor Meter

K. Attenuator

L. Recorder

(b) Quartz Combustion Tube with Boat and Copper Oxide Packing; (c) Schematic Diagram of Trap Used to Contain Ascarite, Dehydrite, and Aluminum Oxide (Reprinted with the permission of the Association of Official Analytical Chemists) gently for about 5 min, add 2 ml of 70% perchloric acid, and reflux for 2 hr (3 hr for triphenylmethane or oil-soluble dyes; 1 hr for lakes having a low dye content). Concurrently run two blanks containing all the reagents.

Allow the solution to cool. Wash the condenser and the funnel, using sufficient water to bring the volume to about 100 ml. Add 10 ml of the 20% $NH_2OH \cdot HCl$ solution and reflux for 10 min. Cool. Wash the condenser and funnel with 30 ml of water and transfer the sample to a 250-ml separatory funnel, filtering if necessary. Dilute to 200 ml. Add 20 ml of the mercury standard solution to one blank; this is the standard. The second blank is the reagent blank.

Add 10 ml of the dithizone solution and shake vigorously for 1 min. Allow the chloroform layer to settle and transfer it to a second 250-ml funnel containing 50 ml of 0.25N hydrochloric acid. Pass 5 ml of chloroform through the first funnel and add it to the second one. Repeat the extraction with 10-ml portions of dithizone solution until the green color of the dithizone remains unchanged. Wash the contents of the first funnel with 10 ml of chloroform and add the wash to the dithizone extracts.

Shake the second funnel vigorously for 1 min. Allow the layers to settle and transfer the chloroform layer to a third 250-ml funnel containing 50 ml of 0.25N hydrochloric acid and 5 ml of the potassium bromide. Wash the contents of the second funnel with 10 ml of chloroform and add the wash to the third funnel.

Shake the third funnel vigorously for 1 min. Drain and discard the chloroform phase. Wash the aqueous phase with 10-ml portions of chloroform until the chloroform and aqueous phases are colorless. Discard the chloroform layer. Add 10 ml of chloroform and 20 ml of 50% ammonium acetate solution. Shake for 10 sec. Remove the funnel stopper and allow the chloroform film on the surface to evaporate. Drain the chloroform layer.

Add 10 ml of the dithizone solution; shake for 1 min. Drain the chloroform layer through absorbent cotton, discarding the first milliliter. Within 1 hr, determine the absorbances of the filtered chloroform solutions of the sample, the blank, and the standard at 490 nm.

SELENIUM (4)

Decompose 2-5 g of sample with a mixture of hot H_2SO_4, HNO_3 and $HClO_4$, cool the solution, dilute with water, heat to remove oxides of nitrogen, cool, and then make to volume with water. Treat an aliquot of this solution with hydroxylammonium chloride, formic acid, and citric acid, adjust to pH 2 with dilute aqueous NH_3, then treat with 0.5% aqueous 3,3'-diaminobenzidine hydrochloride[a] at 43°C (in diffuse light). Adjust the solution to pH 7, extract with $CHCl_3$, then measure the extract's absorbance at the absorption maximum near 420 nm versus a standard. For identification purposes, confirm that the extract has an absorption minimum near 372 nm and a second maximum near 340 nm.

[a]CAUTION—This compound is a suspected carcinogen and should be handled with care.

THALLIUM (10)

Dissolve 5 g of sample in 70 ml of 3 M HBr and then extract the solution with three 15-ml portions of ethyl ether. (If the sample is water-insoluble, first digest it with a mixture H_2SO_4 and HNO_3.) Combine the ether extracts, evaporate the composite to dryness in a current of warm air, dissolve the residue in 2 ml of aqueous Br, and then boil the solution to evaporate excess Br. Dilute the solution with 20 ml of water, add 0.5 ml of a 0.01% solution of methyl violet, and then extract the $TlBr_4$-methyl violet complex into 5 ml of isoamyl acetate. Filter the organic phase into a 1-cm absorption cell and determine its absorption at 580 nm versus a reagent blank.

BIBLIOGRAPHY

1. BERVENMARK, H. Acta pharm. suec. 6, 579-588 (1968). Homogeneity Variations of Talc and Their Consequences for Quality Control. Includes a discussion of the determination of Ca, Cu, and Fe in talc by atomic absorption spectrophotometry.

2. CHRISTENSEN, R.E., BECKMAN, R. M., BIRDSALL, J. J. JAOAC 51, 1003-1010 (1968). Some Mineral Elements of Commerical Spices and Herbs as Determined by Direct Reading Emission Spectroscopy. Fourteen elements were determined in 33 spices using a direct-reading emission spectroscopic method.

3. CLARK, G. R. Proc. Sci. Sect. Toilet Goods Assoc. 34, 49-52 (1960), Some Analytical Applications of X-Ray Fluorescence Spectrometry.

4. DOMENECH, R. Chim. Analyt. 51, 440-443 (1969). Detection and Determination of Traces of Selenium in Dyes for Use in Foodstuffs.

5. ETTELSTEIN, N. JAOAC 30, 552-555 (1947). The Application of the Dithizone Method to the Determination of Lead in Coal-Tar Colors.

6. Food Chemicals Codex, 2nd ed. National Academy of Sciences, National Research Council, Washington, D. C., 1972, pp. 865-868.

7. FORD, A., YOUNG, B., MELOAN, C. J. Agric. Food Chem. 22, 1034-1036 (1974). Determination of Lead in Organic Food Coloring Dyes by Atomic-Absorption Spectroscopy.

8. HARROW, L. S. JAOAC 34, 396-404 (1951). Arsenic and Antimony in Coal-Tar Colors.

9. HARROW, L. S. JAOAC 31, 677-683 (1948). Determination of Lead in Lakes of Coal Tar Colors.

10. KROELLER, E. Deut. Lebensm. Rundschau 71, 73-74 (1975). Sensitive Method for the Determination of Thallium in Food Dyes.

11. LINK, W. B., HEINE, K. S. Jr., JONES, J. H. WATTLINGTON, P. JAOAC 47, 391-394 (1964). Ion Exchange Paper-X-ray Emission Procedure for Determination of Microgram Quantities of Mercury.

12. MOTEN, L. JAOAC 53, 916-922 (1970). Quantitative Determination of Chromium in Triphenylmethane Color Additives by Atomic Absorption Spectroscopy.

13. MOTEN, J. JAOAC *55*, 1145-1149 (1972). Quantitative Determination of Cadmium in Water-Soluble Color Additives by Atomic Absorption Spectroscopy.

14. STEIN, C. JAOAC *33*, 409-412 (1950). Report on Heavy Metals in Coal-Tar Colors—Mercury.

15. SULSER, H. Mitt. Gebiete Lebensm. Hyg. *57*, 66-97 (1966). Paper Chromatographic Detection and Approximate Determination of Trace Metals in Food Dyes.

16. *The United States Pharmacopeia*, 17th ed. (XVII). Mack Publishing Co., Easton, Pa., 1965, p. 695.

17. WENNINGER, J. A., JONES, J. H. JAOAC *46*, 1018-1021 (1963). Determination of Submicrogram Amounts of Mercury in Inorganic Pigments by the Photometric Mercury Vapor Procedure.

18. WENNINGER, J. A. JAOAC *48*, 826-832 (1965). Direct Microdetermination of Mercury in Color Additives by the Photometric-Mercury Vapor Procedure.

Chapter 11 Organic Impurities

Several types of organic impurities can be present in color additives. If the colorants are factory made, they often contain traces of the reagents or "intermediates" from which they were synthesized, impurities originally present in the starting materials that survive the process unchanged, products formed by side reactions, isomeric colors, and subsidiary colors. Natural colors, of course, can contain analogous impurities, depending on the colorant, its origin, method of isolation, and so on. In general, the organic impurities found in color additives can be divided into two groups—colorless and colored.

Because of the problems associated with large-scale manufacture it's virtually impossible to obtain perfect stoichiometric balance in the ingredients used in a synthesis, complete mixing of a batch, and flawless washing of an isolated product. As a consequence, most colorants contain small amounts of at least one of the intermediates used in its preparation. Also, since some of the intermediates employed, such as 2-naphthol-6-sulfonic acid (Schaeffer's salt) can be and frequently are contaminated with traces of isomers like 2-naphthol-8-sulfonic acid and 1-naphthol-5-sulfonic acid as well as higher and lower sulfonated homologs, including 2-naphthol and 2-naphthol-3,6-disulfonic acid, the colorants prepared from these intermediates often contain small amounts of these impurities and the respective isomeric and subsidiary dyes formed from them. Such impurities are logical and anticipated.

Less expected are those impurities that stem from side reactions in the processes and unanticipated contaminants in the starting materials. Impurities of this type include 6,6'-oxybis (2-naphthalenesulfonic acid) found in FD&C Red No. 40 and FD&C Yellow No. 6, and 4,4'-(diazoamino)-dibenzenesulfonic acid also present in FD&C Yellow No. 6. The former is a colorless, noncoupling impurity only recently discovered in commercial Schaeffer's salt which survives both the Red 40 and Yellow 6 processes intact. The latter is a pale yellow side product formed during the manufacture of Yellow 6 when insufficient acid or nitrite is present during diazotization of the sulfanilic acid used.

The level of organic impurities found depends, of course, on the contaminants and colorant in question. For the most part, however, colorless impurities range from 0-0.3% each, whereas colored contaminants vary from 0-1% each.

The majority of the impurities are analyzed for by one form or another of chromatography. Thin-layer and paper chromatography are used for semi-quantitative and "go-no-go" tests, whereas column procedures are preferred when more accurate results are needed. High-pressure liquid chromatography is currently proving to be the most effective method of analysis from the standpoint of speed, accuracy, sensitivity, selectivity, and resolving power.

Methods for the determination of uncombined intermediates and other low-molecular weight, colorless impurities can be found in Chapter 12. Procedures for determining colored impurities including homologous and isomeric colorants can be found in Chapter 13.

Chapter 12 Uncombined Intermediates and Other Low-molecular-weight Impurites

GENERAL COLUMN-CHROMATOGRAPHIC SCREENING PROCEDURE (7, 20)

With this method low-molecular-weight impurities usually separate from the colorants in question, but not always from each other. The analyst may have to depend on taking smaller cuts, the use of simultaneous equations, spectral shifts with changes in pH, juggling of chromatographic conditions, or some combination of these changes to make good, quantitative determinations. The modifications in chromatographic conditions usually most conducive to improved resolution include reduction of sample size, increase in the column length: width ratio, and changes in eluant strength, including the use of full or step gradients.

Procedure

Affix a short length of clean rubber tubing to the tip of the glass-chromatographic column shown in Fig. 11. Attach a pinchcock and place a glass-wool plug in the constriction above the column tip. Slurry 60 g of Whatman Column Chromedia CF11 in 500 ml of the eluant given in Table 21. With the pinchcock open, pour the slurry into the column. Wash the column with 200 ml of eluant and let it drain until the liquid level is 2-3 ml above the level of the packed cellulose.

Place 0.5 g of sample in a 150-ml beaker and add the solvent indicated in Table 21. (The solvent indicated does not necessarily dissolve the dye sample; however, it usually extracts the impurities.) Add 10 g of Chromedia that has been previously washed with the appropriate eluant and mix. Add 50 g of ammonium sulfate powder to salt out the dye and mix. Using 50 ml of

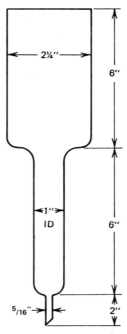

Figure 11   Glass-chromatographic Column

eluant, transfer the mixture to the column. Let the column drain to the surface of the cellulose. Add eluant to the column at a rate equivalent to the rate of flow through the column. Collect as many 100 ml ± 1 ml fractions as can be obtained before the dye itself begins to emerge from the column. Similarly prepare and elute a blank column to which no sample has been added.

Treat the eluate fractions as directed in Table 21 and then record the UV spectra of the fractions versus blank (eluant plus reagents) and compare them with spectra of known compounds similarly prepared. For the ideal case in which there is no interference;

$$\text{Percent intermediate} = \frac{(\Sigma A)(f + x)(100)}{(a)(b)(w)(1000)}$$

where ΣA is sum of the absorbances (corrected for column blank) of the sample fractions containing the intermediate, a is absorptivity in liters/g-cm of the intermediate at the wavelength at which A is measured, b is absorption cell length (in cm), w is sample weight (in g), f is volume of the fraction collected (in ml), and x is volume of reagents added (in ml).

TABLE 21. SOLVENTS, ELUANTS, AND TREATMENT OF ELUATE FRACTIONS FOR INDIVIDUAL COLORS

Color	Solvent[a]	Eluant (w/v) solution[b]	Treatment of Fractions
FD&C Blue No. 1	Water, 25 ml	Ammonium sulfate, 35%	Divide each fraction equally; add 0.5 ml of concentrated hydrochloric acid to one portion and 0.5 ml of conc. ammonium hydroxide to the other
FD&C Blue No. 2	Water, 25 ml	Ammonium sulfate, 25%	Run as eluted
FD&C Green No. 3	Water, 25 ml	Ammonium sulfate, 35%	Same as FD&C Blue No. 1
FD&C Red No. 3	Ethanol, 5 ml; water, 25 ml	Ammonium sulfate, 25%, containing 10% ethanol (v/v)	Divide each fraction equally; add 0.5 ml of concentrated hydrochloric acid to one portion; run the other as is
FD&C Red No. 4	Water, 25 ml	Ammonium sulfate, 25%	Add 1 ml of concentrated ammonium hydroxide to each fraction
FD&C Yellow No. 5	Two drops of Conc. hydrochloric acid in 25 ml of water	Ammonium sulfate, 25%, containing 0.5% hydrazine sulfate	Same as FD&C Blue No. 1
FD&C Yellow No. 6	Water, 25 ml	Ammonium sulfate, 35%,	Run as eluted
Citrus Red No. 2	Ethanol, 10 ml; water, 25 ml	Ammonium sulfate, 5%, containing 5% ethanol	Run as eluted
Orange B	Two drops of Conc. hydrochloric acid in 25 ml of water	Ammonium sulfate, 50%, containing 1% hydrochloric acid (v/v) for 600 ml, then 25% ammonium sulfate containing 1% hydrochloric acid (v/v) for remainder	Same as FD&C Blue No. 1

D&C Blue No. 4	Water, 25 ml	Ammonium sulfate, 35%	Same as FD&C Blue No. 1
D&C Blue No. 6	Slurry with 10 ml of ethanol; then mix in 10 ml of eluant	Ammonium sulfate, 10%	Same as FD&C Red No. 4
D&C Blue No. 9	Water, 25 ml	Ammonium sulfate, 10%	Run as eluted
D&C Brown No. 1	Water, 25 ml	Ammonium sulfate, 25%	Same as FD&C Red No. 4
D&C Green No. 5	Water, 25 ml	Ammonium sulfate, 25%	Same as FD&C Blue No. 1
D&C Green No. 6	Slurry with 10 ml of ethanol, then mix in 10 ml of eluant	Ammonium sulfate, 10%	Same as FD&C Red No. 4
D&C Green No. 8	Water, 25 ml	Ammonium sulfate, 40%	Run as eluted
D&C Orange No. 4	Water, 25 ml	Ammonium sulfate, 25%	Same as FD&C Blue No. 1
D&C Orange No. 5	Dissolve in a minimum of concentrated ammonium hydroxide	Ammonium sulfate, 25%	Same as FD&C Blue No. 1
D&C Orange No. 10	Dissolve in a minimum of concentrated ammonium hydroxide	Ammonium sulfate, 25%	Same as FD&C Red No. 3
D&C Orange No. 11	Water, 25 ml	Ammonium sulfate, 25%	Same as FD&C Red No. 3
D&C Orange No. 17	Slurry with 10 ml of ethanol	Ammonium sulfate, 10%, containing 10% ethanol (v/v)	Run as eluted
D&C Red No. 6	Slurry with 10 ml of ethanol, then mix in 10 ml of eluant	Ammonium sulfate, 10%	Divide each fraction equally; add 0.5 ml of concentrated hydrochloric acid to one portion; run the other as is
D&C Red No. 7			
D&C Red No. 8	Slurry with 10 ml of ethanol, then mix in 10 ml of eluant	Ammonium sulfate, 10%	Same as D&C Red No. 6
D&C Red No. 9			

TABLE 21 Continued

Color	Solvent[a]	Eluant (w/v) solution[b]	Treatment of Fractions
D&C Red No. 17	Slurry with 10 ml of ethanol, then mix in 10 ml of eluant	Ammonium sulfate, 10%	Same as D&C Red No. 6
D&C Red No. 19	Slurry with 5 ml of ethanol	Ammonium sulfate, 35%	Same as FD&C Blue No. 1
D&C Red No. 21	Dissolve in a minimum of concentrated ammonium hydroxide, then add 5 ml of ethanol	Ammonium sulfate, 25%	Same as FD&C Blue No. 1
D&C Red No. 22	Slurry with 5 ml of ethanol	Ammonium sulfate, 25%	Same as FD&C Blue No. 1
D&C Red No. 27	Dissolve in a minimum of concentrated ammonium hydroxide, then add 5 ml of ethanol	Ammonium sulfate, 30%, containing 4% ammonium hydroxide (v/v)	Same as FD&C Blue No. 1
D&C Red No. 28	Slurry with 5 ml of ethanol	Ammonium sulfate, 30%, containing 4% ammonium hydroxide (v/v)	Same as FD&C Blue No. 1
D&C Red No. 30	Water, 25 ml	Ammonium sulfate, 10%	Run as eluted

Dye	Preparation	Eluant	Standard
D&C Red No. 31	Slurry with 10 ml of ethanol and 10 ml of eluant	Ammonium sulfate, 10%	Same as FD&C Blue No. 1
D&C Red No. 33	Water, 25 ml	Ammonium sulfate, 25%	Same as FD&C Blue No. 1
D&C Red No. 34	Slurry with 10 ml of ethanol, then mix in 10 ml of eluant	25% ammonium sulfate, 1200 ml, followed by 10% ammonium sulfate	Run as eluted
D&C Red No. 36	Slurry with 10 ml of ethanol	Ammonium sulfate, 10%	Run as eluted
D&C Red No. 37	Slurry with 5 ml of ethanol	Ammonium sulfate, 35%	Same as FD&C Blue No. 1
D&C Red No. 39	Slurry with 10 ml of ethanol	Ammonium sulfate, 35%	Same as FD&C Blue No. 1
D&C Violet No. 2	Slurry with 10 ml of ethanol	Ammonium sulfate, 10%	Same as FD&C Red No. 4
D&C Yellow No. 7	Water, 25 ml	Ammonium sulfate, 35%	Same as FD&C Blue No. 1
D&C Yellow No. 8	Water, 25 ml	Ammonium sulfate, 35%	Same as FD&C Blue No. 1
D&C Yellow No. 10	Water, 25 ml	Ammonium sulfate, 40%	Same as FD&C Blue No. 1
D&C Yellow No. 11	Slurry with 5 ml of ethanol	Ammonium sulfate, 10%	Same as FD&C Blue No. 1
Ext. D&C Violet No. 2	Water, 25 ml	Ammonium sulfate, 25%	Same as FD&C Blue No. 1
Ext. D&C Yellow No. 7	Water, 25 ml	Ammonium sulfate, 35%	Run as eluted

[a] The solvent indicated may not dissolve the dye but only leach it free of impurities.
[b] The eluant should be essentially free of iron and other UV-absorbing impurities.

HIGH-PRESSURE LIQUID-CHROMATOGRAPHIC PROCEDURE (7, 34-36)

Intermediates and other impurities can be determined in water-soluble certifiable food colors by high-pressure liquid chromatography (HPLC). Although most of the work reported to date has been done by ion exchange chromatography, some progress has been made using ion-pairing systems with reverse phase columns. The procedure given here is the most widely used and is applicable to a variety of colorants. The conditions quoted are only approximate and will generally require modification depending on the condition of the particular column, the chromatograph used, and other variables.

Reagents and Apparatus:

Primary eluant—0.01 M aqueous $Na_2B_4O_7$.

Secondary eluant—see specific colorant.

Liquid chromatograph—DuPont Model 830 with a gradient elution accessory.

Column—Dupont strong anion exchange (SAX), 1 m $\times$ 2.1 mm-ID (DuPont No. 830950405). Conditon column by heating at 50°C for 50 hr with primary eluant flowing. The conditioning needed may vary from column to column.

Operating Conditions: To equilibrate the system, run a gradient of 0-100% secondary eluant at 10%/min and then pump primary eluant through the column for 10 min. Inject the sample and chromatograph as indicated in the list that follows.

FD&C Blue No. 1:

Secondary eluant—0.25 M $NaClO_4$ in aqueous 0.01 M $Na_2B_4O_7$.

Sample size—5 μl of a 1% solution.

Flow rate—0.25 ml/min.

Gradient—linear, 0-100% secondary at 4%/min.

Order of elution—(1) m-sulfobenzaldehyde, (2) o-sulfobenzaldehyde, (3) N-ethyl-N-(3-sulfobenzyl)-sulfanilic acid, (4) FD&C Blue No. 1, (5) ethylbenzylaniline sulfonic acid.

FD&C Blue No. 2:

Secondary eluant—0.50 M $NaClO_4$ in aqueous 0.01 M $Na_2B_4O_7$.

Sample size—5 μl of a 1% solution.

Flow rate—1.00 ml/min.

Gradient—slow start exponential 3, 0-100% secondary at 1%/min.

Order of elution—(1) isatin, (2) isatin 5-sulfonic acid, (3) FD&C Blue No. 2, (4) 5,7'-disulfonated indigo, (5) monosulfonated indigo.

FD&C Red No. 40:

Secondary eluant—0.5 M $NaClO_4$ in aqueous 0.01 M $Na_2B_4O_7$.

Sample size—5 μl of a 1% solution.

Flow rate—0.5 ml/min.

Gradient—linear, 0-25% secondary at 2%/min.

Order of elution—(1) cresidinesulfonic acid, (2) Schaeffer's salt, (3) 4,4′-diazoaminobis (5-methoxy-2-methylbenzenesulfonic acid), (4) FD&C Red No. 40, (5) 6,6′-oxybis (2-naphthalenesulfonic acid).

FD&C Yellow No. 5:

Secondary eluant—0.20 M $NaClO_4$ in aqueous 0.01 M $Na_2B_4O_7$.

Sample size—5 μl of a 2% solution.

Flow rate—1 ml/min.

Gradient—slow start exponential 5, 0-90% secondary at 3%/min.

Order of elution—(1) phenylhydrazine-p-sulfonic acid, (2) sulfanilic acid, (3) 3-ethylcarboxy-1-(4-sulfophenyl)-5-pyrazolone, (4) 3-carboxy-1-(4-sulfophenyl)-5-pyrazolone, (5) 4,4′-(diazoamino)-dibenzenesulfonic acid, (6) FD&C Yellow No. 5.

FD&C Yellow No. 6:

Secondary eluant—0.25 M $NaClO_4$ in aqueous 0.01 M $Na_2B_4O_7$.

Sample size—5 μl of a 1% solution.

Flow rate—0.5 ml/min.

Gradient—slow start exponential 3, 0-80% secondary at 3%/min.

Order of elution—(1) sulfanilic acid, (2) Schaeffer's salt, (3) 4,4′-(diazoamino)-dibenzenesulfonic acid, (4) R-salt dye, (5) FD&C Yellow No. 6, (6) 6,6′-oxybis (2-naphthalenesulfonic acid).

DETERMINATION OF β-NAPHTHOL

β-Naphthol is extracted with an appropriate solvent, coupled, and then determined by titanous chloride titration or spectrophotometrically.

Titration Procedure

With Titanous Chloride (30): Prepare a 0.05N diazotized sulfanilic acid solution by dissolving 4.78 g of sulfanilic acid in 500 ml of 1:99 hydrochloric acid. Then dissolve 1.04 g of sodium nitrite in 300 ml of water. Place 40 ml of the sulfanilic acid solution in a 100-ml volumetric flask. Cool to 5°C, add 44 ml of the sodium nitrite solution, and allow it to diazotize. Test for excess nitrous acid with starch iodide paper; destroy any excess with a few milligrams of sulfamic acid. Dilute to volume with water. Store at 5°C.

Weight a 10-g sample into a 2.5 cm × 8-cm extraction thimble, place it in a Soxhlet apparatus, and extract for 8 hr with petroleum ether (boiling range 35-60°C).

Disconnect the extractor and add 150 ml of 1:10 hydrochloric acid to the flask. Gently boil off the ether on a hot plate and filter the solution through glass wool into a 250-ml volumetric flask. Rinse the extraction flask several times with small quantities of 1:10 hydrochloric acid, filtering the rinses through the glass wool and collecting them in the volumetric flask. Dilute to volume with water. Mix and then divide into two equal portions. Neutralize each portion with a dilute sodium hydroxide solution, using phenolphthalein as an indicator.

Add 10 g of sodium acetate trihydrate to one portion and cool to 5°C in an ice bath. Slowly add 25 ml of the diazotized sulfanilic acid solution, stir for 5 min, and test for excess reagent with alkaline β-naphthol solution on spot paper. If necessary, add additional reagent until a positive test is obtained. Let stand for 1 hr or longer.

Heat on a water bath for 30 min to decompose any excess reagent. Test with β-naphthol solution on spot paper to determine whether decomposition is complete. To both the coupled and uncoupled protions of solution add 10 g of sodium bitartrate dissolved in 50 ml of hot water. Titrate the uncoupled blank with $0.1N$ TiCl$_3$ to a colorless end point. Titrate the coupled portion until the dye reduces and becomes yellow. Add 1-2 ml of excess TiCl$_3$ solution and immediately back titrate with a standardized methylene-blue solution or another suitable dye.

Subtract the volume of titrant required for the blank from the volume required for the coupled portion, and calculate percent β-naphthol. One milliliter of $0.1 N$ TiCl$_3$ is equal to 0.0036 g of β-naphthol.

Spectrophotometric Methods

General Methods: Prewash all isopropyl ether once with $0.1N$ sodium hydroxide.

Prepare a p-nitrobenzenediazonium chloride solution by dissolving 0.02 g of p-nitroaniline in 2 ml of hydrochloric acid and diluting to 200 ml with water. Then add 100 g of crushed ice and stir until the solution temperature is 5-10°C. Add 2 ml of 10% sodium nitrite solution and stir for 10-15 min. Add small portions of 10% sulfamic acid solution until the reagent gives a negative test on starch iodide paper. Store at 5°C.

For colors soluble in water: Dissolve 2 g of sample in 250 ml of water. Acidify with 5 ml of $6N$ hydrochloric acid and extract with six 30-ml portions of prewashed isopropyl ether. Wash the combined ether extracts with 20 ml of $0.1N$ hydrochloric acid and discard the aqueous layer. Extract with six 30-ml portions of $0.1N$ sodium hydroxide.

Cool the combined sodium hydroxide extracts to 5-10°C with crushed ice and add p-nitrobenzenediazonium chloride solution slowly with constant stirring. Stir the mixture for 15 min. Heat to 90°C on a steam bath, remove from the bath, and cool to room temperature. Extract with 20-ml portions of chloroform until the extracts are colorless. Wash the combined extracts with 30 ml of $0.1N$ sodium hydroxide. Filter the chloroform layer through cotton into a 500-ml flask. Dilute to volume with chloroform. Measure the absorbance of the sample solution at 490 nm; determine the β-naphthol content by comparison with a standard solution prepared by dissolving 0.1 g of 1-(4-nitrophenylazo)-2-hydroxynaphthalene in 200 ml of chloroform and by diluting with chloroform to obtain a concentration of 5-10 mg/liter.

$$\text{Percent } \beta\text{-naphthol} = \frac{(A)(c)(144)(0.05)}{(B)(w)(293)}$$

where A is absorbance of the sample solution, B is absorbance of the standard solution, c is concentration of the standard solution (in mg/liter), and w is sample weight (in g).

For Colors Soluble in Isopropyl Ether: Weight a 2-g sample and add 250 ml of prewashed isopropyl ether; warm into solution on a steam bath. Extract with six 30-ml portions of 0.1N sodium hydroxide. Wash the combined extracts with 30 ml of prewashed isopropyl ether and proceed as described in the preceding paragraph.

For Colors Insoluble in Water or Isopropyl Ether: Extract a 10-g sample with prewashed isopropyl ether for 8-10 hr in a Soxhlet apparatus. Transfer the extract to a 1-liter separatory funnel. Rinse the Soxhlet extractor with two 20-ml portions of the isopropyl ether and add the rinses to the separatory funnel. Extract with six 30-ml portions of 0.1N sodium hydroxide and wash the combined extracts with 30 ml of the isopropyl ether. Dilute the sodium hydroxide extract to 500 ml with 0.1N sodium hydroxide, place a 100-ml aliquot in a beaker, and proceed as described above.

Method for D&C Red No. 36 (19): Prepare a p-nitrobenzenediazonium chloride solution by dissolving 0.040 g of p-nitroaniline in 2 ml of hydrochloric acid. Then chill with 35 g of ice and 15 ml of water. Diazotize with 0.04 g of sodium nitrite, stir, and let stand for 5 min. Store cold.

Stir 5 g of sample with 90 ml of ethanol and 10 ml of 10% sodium hydroxide for 30 min. Transfer the mixture to a 1-liter volumetric flask containing 600 ml of water, agitate, and dilute to volume.

Filter a 400-ml aliquot and buffer with 1 g of sodium carbonate. Cool the solution with 100 g of ice and add the p-nitrobenzenediazonium chloride solution. Let it stand for 10 min, stirring occasionally. Heat to 90°C, cool, and acidify with hydrochloric acid. Remove the coloring matter with 40-ml portions of chloroform continuing until the extracts are colorless. Wash the combined extracts with three 40-ml portions of water, and then with 30 ml of 0.1N sodium hydroxide. Filter through absorbent cotten into a 500-ml volumetric flask. Dilute to volume with chloroform. Prepare a standard solution as described above and determine its absorbance and the absorbance of the sample at 490 nm. Calculate as under Colors Soluble in Water.

DETERMINATION OF PHTHALIC ACID DERIVATIVES (10, 11, 30)

Method A—in FD&C Red No. 3; D&C Orange Nos. 5, 10, 11, and 17; D&C Red Nos. 21 and 22; D&C Yellow Nos. 7 and 8—Colors as Salts: Weigh a 2-g sample and transfer it to a 250-ml beaker with about 100 ml of water. Heat nearly to boiling and while stirring, slowly add 1:9 hydrochloric acid until precipitation seems complete. Add an additional 8.5 ml of 1:9 hydrochloric acid and dilute to about 150 ml. Digest on a steam bath for 1-2 hr.

Cool and dilute with water in a 200-ml volumetric flask. Filter through dry paper.

Pipette 50 ml of filtrate into a 125-ml separatory funnel (do not grease stopcocks) and extract with 30 ml of absolute ethyl acetate. Transfer the aqueous phase to a second funnel and extract with 25 ml of ethyl acetate. Transfer the aqueous phase to a third funnel and extract with 20 ml of ethyl acetate. Pass three successive 50-ml portions of water through the funnels in the same order that the extractions were made. Discard the ethyl acetate layers, combine the aqueous extracts, and evaporate to dryness on a steam bath.

Dissolve the residue in water and transfer to a 100-ml volumetric flask. Add 8.5 ml of 1:9 hydrochloric acid and dilute to volume. Filter through dry paper and measure the absorbance at 230 nm, 262 nm, and 276 nm against $0.1N$ hydrochloric acid as the blank. Also measure the absorbance of a standard solution prepared by dissolving 0.13-0.135 g of potassium acid phthalate in 500 ml of water, and diluting a 10-ml aliquot to 200 ml with $0.1N$ hydrochloric acid [phthalic acid concentration, mg/100 ml = (mg of $KHC_8H_4O_4$)(0.00813)].

Calculate Y for both sample and standards as follows:

$$Y = A_{230} - (A_{230} - 0.7 A_{276}) - A_{262}$$

where A is absorbance of solution at the wavelength indicated;

$$\text{Percent phthalic acid} = \frac{(Y \text{ sample}/Y \text{ standard})(c)(0.4)}{w}$$

where c is phthalic acid concentration of standard solution (in mg/100 ml) and w is sample weight (in g).

Colors as Acids: To a 2-g sample add 6 ml of 10% sodium hydroxide and a few milliliters of water. Mix to dissolve. Dilute to about 100 ml and proceed as described above beginning with the second sentence under Method A ("Heat nearly to boiling. . .").

Method B- in D&C Red No. 19: Dissolve a 0.5-g sample in 20 ml of hot water. Cool and transfer to a 125-ml separatory funnel. Add 80 ml of chloroform and 2 ml of 10% sodium hydroxide and shake vigorously for 1 min. Drain the chloroform layer and wash the aqueous phase with two 30-ml portions of chloroform, discarding the chloroform layers. Add 7 ml of 1:9 hydrochloric acid to the aqueous phase and then wash with two 30-ml portions of chloroform, discarding the chloroform. Transfer the aqueous solution to a beaker with water and evaporate to dryness on a steam bath. Proceed as in Method A, (beginning with "Dissolve the residue in water. . .").

Method C—in D&C Yellow No. 10: Dissolve a 1-g sample in water and transfer it to a continuous extractor. Add 1 ml of 1:100 hydrochloric acid and extract for 8 hr with 250 ml of ethyl ether. Transfer the ether extract to a separatory funnel. Rinse the flask twice with ethyl ether and add the washings to the main extract. Wash the extract with four 10-ml portions of 1:199 hydrochloric acid. Back extract the combined washings with 50 ml of ethyl ether and add the ether layer to the main ether extract. Extract the combined ether layers with four 10-ml portions of 1% sodium hydroxide. Evaporate the alkaline extracts to dryness. Transfer the residue to a 200-ml volumetric flask with water, add 2 ml of hydrochloric acid, and dilute to volume. Filter through dry paper. Measure the absorbance as described in Method A and calculate percent phthalic acid in the same manner.

Method D—in D&C Yellow No. 11: Transfer a 0.5-g sample to a 125-ml separatory funnel with 80 ml of chloroform. Add 20 ml of 1% sodium hydroxide and shake vigorously for 1 min. Drain the chloroform layer, wash the aqueous phase with two 30-ml portions of chloroform, and discard the chloroform layers. Add 7 ml of 1:9 hydrochloric acid and wash with two 30-ml portions of chloroform, discarding the chloroform washings. Transfer the aqueous phase to a beaker and evaporate to dryness on a steam bath. Proceed as described in Method A (beginning with "Dissolve the residue in water. . .").

Sulfobenzaldehydes and *N*-Ethyl-*N*-(3-Sulfobenzyl) Sulfanilic Acid in FD&C Blue No. 1 (16, 33)

Slurry 24 g of Whatman Column Chromedia CF11 in 140 ml of eluant containing 400 g of ammonium sulfate per liter of water. Pour the slurry into the glass chromatographic tube shown in Fig. 12. Let the eluant drain at a rate of 5 ml/min or less until the liquid is 1-2 mm above the level of the packed cellulose. Close the pinch clamp.

Dissolve a 0.2-g sample in 10 ml of water. Add 2 g of the Chromedia and mix. Add 7 g of powdered ammonium sulfate and mix. Using 5 ml of the eluant, transfer the sample to the column. Drain until the flow nearly ceases. Elute the column at a rate of 5 ml/min or less, collecting as many 10 ml ± 0.05 ml fractions as necessary (ca. 30) to remove the intermediates from the column. Record the UV spectra of the fractions versus eluant in a 1-cm cell. Compare these against standard spectra obtained in the eluant.

o-Chlorobenzoic acid and o-sulfobenzoic acid elute just ahead of the sulfobenzaldehydes (SB) and may not be separated from them. They are identified by small maxima near 270 nm, but are not estimated by this method. To calculate a fraction as SB, the ratio $A_{252}:A_{274}$ must be 2 or greater. *ortho-*, *meta-*, and *para*-Sulfobenzaldehydes elute together, generally in fractions 7-15. They are calculated as total SB at 252 nm using an absorptivity of 51.6. This value is based on a mixture containing 46% o-sulfobenzaldehyde (o-SB), 46% m-sulfobenzaldehyde (m-SB) and 8% p-sulfobenzaldehyde (p-SB); 252 nm is the isoabsorptive point of o-SB and m-SB.

N-Ethyl-*N*-(3-sulfobenzyl) sulfanilic acid generally elutes between fractions 15 and 30. It is calculated at the maximum near 274 nm, using an absorptivity of 62.

Leuco Base in FD&C Blue No. 1 and FD&C Green No. 3

Oxidation with Oxygen and Cupric Chloride (5): Dissolve 0.12 g of FD&C Blue No. 1 or 0.13 g of FD&C Green No. 3 in distilled water and dilute with water to 1000 ml in a volumetric flask. Pipette 10 ml of this solution into a 250-ml volumetric flask containing 100 ml of water, add 50 ml of *N,N*-dimethylformamide (DMF), swirl to mix, cool to room temperature, and dilute

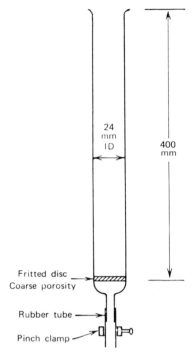

24
mm
ID

400
mm

Fritted disc
Coarse porosity

Rubber tube

Pinch clamp

Figure 12 Chromatographic Tube (Reprinted with permission of the Association of Official Analytical Chemists)

to volume with water; solution = A. Pipette a second 10 ml aliquot of sample solution into a 250-ml volumetric flask, add 50 ml of a solution containing 1 g of cupric chloride ($CuCl_2 \cdot 2H_2O$) in 100 ml of DMF, pass a rapid stream of air through the mixture for 30 min and then dilute to volume with water; solution = B. Pipette 50 ml of the 1% cupric chloride in DMF solution and 50 ml of DMF into separate 250-ml volumetric flasks containing 100 ml of water, swirl to mix, cool to room temperature, and then dilute to volume with water; solutions = C and D, respectively.

Using matched absorption cells, determine the spectra from 700 nm to 500 nm of: solution D versus solution D, solution A versus solution D, solution B versus solution C.

$$\text{Percent Leuco base} = \frac{(E - E_1)100}{a\,b\,c\,f}$$

where E is the blank corrected absorbance at the absorption maximum of solution B, E_1 is the blank corrected absorbance at the absorption maximum of solution A, b is length of cell path (in cm), c is effective sample concentration (in g/liter), f is molecular weight of the color divided by the molecular weight of the leuco base, and a is absorptivity of the color (in liters/g-cm).

The molecular weights of the colorants as disodium salts are: FD&C Blue No. 1 = 792.9, FD&C Green No. 3 = 808.9. The molecular weights of the leuco bases as the trisodium salts are: FD&C Blue No. 1 = 816.9, FD&C Green No. 3 = 832.9. Absorptivities (in liters/g-cm) are: FD&C Blue No. 1 = 164, FD&C Green No. 3 = 156.

Oxidation with Chloranil (17): Weigh a 0.5-g sample into a 250-ml volumetric flask and dilute to volume with water. Pipette a 10-ml aliquot of the solution into each of two 100-ml volumetric flasks. To the first flask, add 15 ml of a freshly prepared solution consisting of 0.04 g of chloranil in 100 ml of N,N-dimethylformamide. Place the flask in a boiling water bath for 60 min, cool, and dilute to volume with water. Dilute the aliquot in the second volumetric flask to volume with water. Pipette a 10-ml aliquot from each flask into separate 500-ml volumetric flasks and dilute to volume with water. Measure the absorbance of each solution in a 1-cm cell at the absorption maximum.

$$\text{Percent leuco base} = \frac{(A - A_1)(100)}{(a)(b)(c)(f)}$$

where A is absorbance of the sample solution treated with chloranil, A_1 is absorbance of the untreated sample solution, b is length of cell path (in cm), c is effective sample concentration (in g/liter) (=0.004), f is molecular weight of the color divided by the molecular weight of the leuco base, and a is absorptivity of the color (in liters/g-cm).

5-Amino-4-Methoxy-2-Toluenesulfonic Acid (CSA), Schaeffer's Salt, and 6,6'-Oxybis (2-Naphthalenesulfonic acid) (DONS) in FD&C Red No. 40 (21, 25)

Slurry 30 g of Whatman Column Chromedia CF11 in 200 ml of 40% aqueous $(NH_4)_2SO_4$ solution. Pour the slurry into a 24mm $\times$ 400 mm glass chromatographic column (Fig. 12) and let the eluant drain just to the surface of the cellulose. Wash the column with 200 ml of 40% $(NH_4)_2SO_4$, let it drain until the liquid is 1-2 mm above the level of the packed cellulose, and then close the pinch clamp.

Weigh 0.1 g of sample into a 50-ml beaker. Add 5 ml of water and stir to dissolve. Add 2 g of cellulose powder and mix. Add 7 g of $(NH_4)_2SO_4$ powder and mix. Using about 5 ml of 40% $(NH_4)_2SO_4$, transfer the mix to the column and let it drain just to the surface of the cellulose. Elute with 250 ml of 40% $(NH_4)_2SO_4$ and immediately begin collecting 50 ml $\pm$ 1 ml fractions. When the last of the eluant has just passed into the column elute with 20% $(NH_4)_2SO_4$ and continue collecting fractions until all the DONS elutes (ca. 1500 ml of 20% $(NH_4)_2SO_4$).

A blank column to which no sample has been added should be prepared and eluted as described above.

Record the UV spectra of the fractions in a 5-cm cell from 370 nm to 220 nm. Use 40% $(NH_4)_2SO_4$ as the reference for fractions 1-8 and 20% $(NH_4)_2SO_4$ as the reference for the remaining fractions. Compare these spectra against standards obtained in the appropriate eluant; CSA normally elutes in fractions 4-6, Schaeffer's salt in fractions 8-10, and DONS in fractions 16-23.

$$\text{Percent CSA} = 0.186 \; \Sigma \; (A_1 - A_2)_{252 \text{ nm}}$$

(as free acid, mw 217.25)

$$\text{Percent Schaeffer's salt} = 0.396 \; \Sigma \; (A_1 - A_2)_{282 \text{nm}}$$

(as monosodium salt, mw 246.22)

$$\text{Percent DONS} = 0.0585 \; \Sigma \; (A_1 - A_2)_{240 \text{ nm}}$$

(as disodium salt, mw 474.42)

where $(A_1 - A_2)_{x \text{ nm}}$ are the sums of the blank-corrected absorbances (also corrected for column blank where necessary) of the sample fractions containing the appropriate compound; $0.186 = 100/(53.8 \times 2 \times 5)$; $0.396 = 100/(25.3 \times 2 \times 5)$; $0.0585 = 100/(171 \times 2 \times 5)$; $100 =$ factor for conversion to percent; $2 =$ effective sample concentration (in g/liter); $5 =$ cell pathlength (in cm); 53.8, 25.3, and 171 = approximate absorptivities (in liters/g-cm of CSA, Schaeffer's salt, and DONS, respectively).

4,4'-Diazoaminobis (5-Methoxy-2-methylbenzenesulfonic Acid) (DMMA) in FD&C Red No. 40

High-Pressure Liquid Chromatographic Procedure (3, 27)

Apparatus and Reagents:

Liquid chromatograph—A DuPont Model 830 equipped with a 1 m $\times$ 2.1 mm-ID DuPont SAX column (strong anion exchange, No. 830950405), a gradient elution accessory, and a Model 835 multiwavelength detector fitted with a 365-nm filter (No. 835052-907).

Primary eluant—0.01 M aqueous $Na_2B_4O_7$.

Secondary eluant—0.5 M aqueous $NaClO_4$ in 0.01 M $Na_2B_4O_7$.

Instrument Parameters:

Eluant flow rate—1.5 ml/min; temperature: ambient; gradient: 0-60% secondary at 2%/min, slow start exponential function 5.

Sample Preparation and Resolution: Weigh 0.5 g of sample into a 5-ml volumetric flask. Dissolve in primary eluant and dilute to volume with same. Run a blank (0-60%) gradient, then pump primary eluant through the column for 14 min. Immediately inject 5 μl of sample solution into the chromatograph using a microliter syringe. Start the 0-60% gradient at once and elute until the chromatogram is complete (ca. 40 min). Pump primary eluant through the column for 14 min and then inject the next sample (DMMA elutes in ca. 11 min).

Gravity Column Procedure (9): Weigh 0.5 g of FD&C Red No. 40 into a 10-ml volumetric flask, add 7 ml of 0.01 M aqueous $Na_2B_4O_7$, swirl to dissolve, and dilute to volume with 0.01 M $Na_2B_4O_7$.

Slurry 3 g of Bio Rad Laboratories Cellex D (hydroxide form, standard-capacity DEAE cellulose) in 50 ml of 0.01 M $Na_2B_4O_7$ and pour into a 22 mm $\times$ 10-cm glass-chromatographic tube with a 200-ml resevoir top and a 22-mm fritted disk. Let the column drain to the surface of the packing and then wash with 50 ml of eluant (0.2 M aqueous $NaClO_4$ in 0.01 M $Na_2B_4O_7$) and then 100 ml of 0.01 M $Na_2B_4O_7$.

With the column clamped off, add 1 ml of the sample solution to the top of it. Let the sample enter the column and then wash the sides of the tube twice with 10 ml of 0.01 M $Na_2B_4O_7$ until all the sample has entered the packing. Carefully add 10 ml of eluant. Allow the column to drain just to the surface of the packing and then fill the reservoir with eluant and elute.

Discard the first 75 ml then collect the next 150 ml or so that contain the DMMA. Measure the exact volume of the DMMA fraction then determine its absorbance at 385 nm using a 5-cm cell or longer.

$$\text{Percent DMMA} = \frac{A \times V \times 100\%}{W \times L \times a}$$

where A = sample absorbance at 385 nm, V = volume of sample fraction (in ml), W = sample weight (in g), L = cell length (in cm), and a = absorptivity of standard at 385 nm (in liters/g-cm).

Sulfanilic Acid and 3-Carboxy-1-(4-Sulfophenyl)-5-Pyrazolone in FD&C Yellow No. 5 (38)

Transfer 1 g of sample into a 100-ml volumetric flask. Add 50 ml of water, swirl to dissolve, and then dilute to volume with methanol. Mix 1 ml of this solution with 1 ml each of m-chlorobenzoic acid (15 mg/ml) and 3-nitro-salicyclic acid (0.1 mg/ml), both prepared in methanol/water (50:50).

Inject 5 μl of this solution into a Waters Associates Model 202 liquid chromatograph equipped with a Model 6000 pump, a 280 nm detector, and a 30 cm $\times$ 4 mm-ID Micro Bondapack C_{18} column (Waters Associates). Elute at 0.9 ml/min using water-methanol-formic acid (400:400:1) containing 3 $\times$ 10^{-3} M tetrabutylammonium hydroxide and 0.6 $\times$ 10^{-4} M tridecylamine as the mobile phase.

Determine the amount of sulfanilic acid, 3-carboxy-1-(4-sulfophenyl)-5-pyrazolone, and FD&C Yellow No. 5 present using the m-chlorobenzoic acid and the 3-nitro-salicyclic acid as internal standards.

Sulfanilic Acid, Schaeffer's Salt, and 6,6'-Oxybis(2-Naphthalenesulfonic Acid) (DONS) in FD&C Yellow No. 6 (2, 22-24)

Slurry 30 g of Whatman Column Chromedia CF11 in 200 ml of 40% aqueous $(NH_4)_2SO_4$ solution. Pour the slurry into a 24 mm $\times$ 400 mm glass chromatographic column (Fig. 12) and let the eluant drain just to the surface of the cellulose. Wash the column with 200 ml of 40% $(NH_4)_2SO_4$, let it drain until the liquid is 1-2 mm above the level of the packed cellulose, and then close the pinchcock.

Weigh 0.5 g of sample into a 100-ml volumetric flask. Dissolve in distilled water and make to volume with same; mix. Pipette 5 ml of this solution into a 50-ml beaker. Add 2 g of cellulose powder and mix. Add 7 g of $(NH_4)_2SO_4$

powder and mix. Using about 5 ml of 40% $(NH_4)_2SO_4$, transfer the mixture to the column and let it drain just to the surface of the cellulose. Open the pinchcock and elute with 250 ml of 40% $(NH_4)_2SO_4$ and immediately begin collecting 50 ml $\pm$ 1-ml fractions. When the last of the eluant has just passed into the column, elute with 20% $(NH_4)_2SO_4$ and collect 50-ml fractions until color elutes.

A blank column to which no sample has been added should be prepared and eluted as described above.

Record the UV spectra of the fractions in a 5-cm cell from 370 nm to 210 nm. Use 40% $(NH_4)_2SO_4$ as the reference for fractions 1-8 and 20% $(NH_4)_2$-SO_4 as the reference for the remaining cuts. Compare these spectra versus standards obtained in the appropriate eluant. Sulfanilic acid normally elutes in fractions 3-6, Schaeffer's salt in fractions 8-10, and DONS in fractions 16-23.

$$\text{Percent sulfanilic acid} = 0.496 \ \Sigma \ (A_1 - A_2)_{250 \text{ nm}}$$
$$(\text{as sodium salt, mw } 195.2)$$

$$\text{Percent Schaeffer's salt} = 0.131 \ \Sigma \ (A_1 - A_2)_{232 \text{ nm}}$$
$$(\text{as sodium salt, mw } 246.2)$$

$$\text{Percent DONS} = 0.234 \ \Sigma \ (A_1 - A_2)_{240 \text{ nm}}$$
$$(\text{as disodium salt, mw } 474.4)$$

where $(A_1 - A_2)_{x \text{ nm}}$ are the sums of the blank-corrected absorbances (also corrected for column blank were necessary) of the sample fractions containing the appropriate compound; $0.496 = 100/(80.7 \times 5 \times 0.5)$; $0.131 = 100/(305 \times 5 \times 0.5)$; $0.234 = 100/(171 \times 5 \times 0.5)$; 100 = factor for conversion to percent; 0.5 = effective sample concentration (in g/liter); 5 = cell pathlength (in cm); 80.7, 305, and 171 = approximate absorptivities of sulfanilic acid, Schaeffer's salt, and DONS, respectively.

2-Naphthol-6-Sulfonic Acid in FD&C Yellow No. 6 (29)

Using Schaeffer's salt-free colorant and 0.2% NaOH, prepare standard solutions containing 10 mg (100% pure dye basis) of colorant per liter of solution plus 0 μg, 5μg, 10μg, and 20 μg of added Schaeffer's salt. Determine the fluorescence of these solutions with a G. K. Turner Model 110 fluorometer equipped with a 7-60 primary filter and a 2A secondary filter. Prepare a calibration curve of fluorescence plotted against micrograms of Schaeffer's salt and use this to determine the Schaeffer's salt content of sample solutions similarly prepared.

4,4'-(Diazoamino)-Dibenzenesulfonic Acid (DAADBSA) in FD&C Yellow No. 6

High-pressure Liquid-chromatographic Procedure (1, 26, 28)

Apparatus and Reagents:
Liquid chromatograph—A DuPont Model 830 equipped with a 1 m $\times$ 2.1 mm-ID DuPont SAX column (strong anion exchange, No. 830950405), a

gradient elution accessory, and a Model 835 multiwavelength detector fitted with a 365 nm filter (No. 835052-907).

Primary eluant—0.01 M $Na_2B_4O_7$.

Secondary eluant—0.50 M aqueous $NaClO_4$ in 0.01 M $Na_2B_4O_7$.

Instrument Parameters:

Eluant flow rate: 1 ml/min; temperature: ambient; gradient: 0-100% secondary at 1%/min, slow-start exponential function 4.

Sample Preparation and Resolution: Weigh 1 g of each sample into separate 10-ml volumetric flasks. Dissolve each in primary eluant and dilute to volume with same. Before injecting the first sample run a blank (0-100%) gradient and then pump primary eluant through the column for 14 min. Immediately inject 5 μl of sample solution into the high-pressure liquid chromatograph (HPLC) using a microliter syringe. Start the 0-100% gradient at once and "hold at limit" until the chromatogram is complete (ca. 50 min). Pump primary eluant through the column for 14 min and then inject the next sample (DAADBSA elutes in ca. 17 min).

Gravity Column Procedure (8): Weigh 0.5 g of FD&C Yellow No. 6 into a 10-ml volumetric flask. Add two drops of 50% NaOH and 7 ml of 18% aqueous Na_2SO_4 and swirl to dissolve. Dilute to volume with Na_2SO_4 solution and mix well.

Slurry 20 g of BW Solka Floc (Brown Co., Berlin, NH) in 150 ml of water and pour the slurry into a 22-mm (ID) × 20-cm glass-chromatographic column fitted with a 100-200-ml reservoir and a 22-mm fritted disk. Apply 2 psi of air pressure to the top of the column until all the liquid has entered the packing. Add 100 ml of eluant (150 g of Na_2SO_4, 150 g of NaCl, and 5 ml of 50% NaOH to 1 liter with water) and again apply pressure until the solution has entered the column. Add 1 ml of sample solution and wash into the column with pressure. Rinse down the walls of the column with a small amount of eluant, force the washings into the column with pressure, then fill the reservoir with eluant and elute the column under 2 psi of pressure (DAADBSA elutes just before FD&C Yellow No. 6). A narrow band of subsidiary color may elute before the DAADBSA. Collect the DAADBSA band, measure its volume, and determine its absorbance at 410 nm using a 5 cm cell.

2-Aminoanthraquinone in D&C Blue No. 9 (4)

Transfer 0.5 g of sample to a 100-ml beaker. Add 10 ml of dimethylsulfoxide, cover with a watch glass, and boil gently for 10 min. Cool the sample to room temperature and filter through a Buchner funnel with suction. With the vacuum still on, place the filtering flask on a steam bath and evaporate the solution to 0.5 ml.

Using a syringe, transfer the sample solution as a streak across a 20 cm × 20 cm TLC plate coated with 0.38 mm of silica gel G. Dry the plate, rinse the flask with a small amount of acetone, and streak the washing onto the same plate.

Heat the plate 15 min at 110° C, cool, and place in a 10.5 in. × 10.5 in. ×

5.5 in.-high museum jar (with a glass top) to which 150 ml of diethyl ether had been added 10-20 min earlier. Develop until the solvent reaches the top of the plate.

Air dry the plate, scrape off the yellow band corresponding to 2-aminoanthraquinone, extract the compound from the silica gel with 15 ml of acetic acid, filter, dilute to 25 ml, and determine the sample's absorbance in a 1-cm cell at the maximum near 428 nm. Compare against a standard.

1,4-Dihydroxyanthraquinone in D&C Green No. 5 and Ext. D&C Violet No. 2 (15)

Transfer 0.1 g of sample into a 250-ml beaker. Add 100 ml of 0.1% NaOH, cover with a watch glass, and heat to boiling. Cool and transfer to a 250-ml separatory funnel, rinse the beaker with H_2O, and add the rinsings to the funnel. Add 5 ml of dilute HCl (8 + 92) and mix. Add 25 ml of isooctane and shake for 10-15 sec. Allow the layers to separate. If any quinizarin is present, the isooctane layer is yellow. Transfer the lower (aqueous) layer to a second funnel and extract with a second 25-ml portion of isooctane. Transfer the aqueous layer to a third funnel and extract with a fresh 25-ml portion of isooctane; discard the aqueous layer. Wash the isooctane in the first funnel with 25 ml of dilute HCl (1 + 199). Move this wash through the three funnels in the same manner as the initial aqueous solution. Continue to wash the isooctane layers with 25-ml portions of dilute HCl until all the colorant is removed. Discard the washes.

Draw the isooctane from the first separatory funnel into a 150-ml beaker and then pass the isooctane from the second funnel through the first funnel into the collection beaker. Similarly, pass the isooctane from the third funnel through the second and first funnels into the collection beaker.

Filter the isooctane through adsorbent cotton into a 100-ml volumetric flask. Rinse the filter into the flask with isooctane, make to volume with same, and determine the sample's absorbance versus a blank in a 1-cm cell at 249 nm. Compare against standards.

1,4-Dihydroxyanthraquinone and 1-Hydroxyanthraquinone in D&C Green No. 6 and D&C Violet No. 2 (14, 37)

Transfer 0.1 g of sample into a 250-ml beaker, add 100 ml of isopropyl ether, cover with a watch glass, and boil gently for 10 min. Cool and then transfer the solution to a 500-ml separatory funnel. If any undissolved dye is left in the beaker, add 50 ml of isopropyl ether and repeat the boiling; add the solution to the separatory funnel.

Add 50 ml of 5% NaOH solution to the funnel and shake for 10-15 sec. Transfer the aqueous phase to a clean funnel. Extract the isopropyl ether with additional 25-ml portions of 5% NaOH until an aqueous wash is colorless. (Quinizarin is intensely purple in aqueous alkaline solution.) Wash the combined NaOH extracts by shaking with successive 25-ml portions of isopropyl ether until an ether wash is colorless. Discard the ether washes.

Acidify the NaOH solution with HCl. Add 25 ml of isooctane and shake.

Transfer the aqueous phase to a clean separatory funnel and extract with 10 ml of isooctane. Repeat this process until the isooctane extractions are colorless.

Combine the isooctane extracts and wash twice with 20-25 ml portions of water. Allow the separatory funnel to stand a while after the second washing to permit as much water as possible to settle. Pass the isooctane layer through a plug of absorbent cotton (prewashed with isooctane). Rinse the separatory funnel with 20 ml of isooctane and pass this through the cotton. Transfer the isooctane to a 100-ml volumetric flask, make to volume, and determine spectrophometrically at 249 nm.

Pyrene in D&C Green No. 8 (32)

Dissolve 2 g of sample in 200 ml of hot H_2O and let the solution cool to room temperature. Filter through a tared Gooch crucible, wash with cold water until the washings are colorless, and then dry at $135°C$. Extract the isoluble residue with 50 ml of ethyl ether, filter the extract into a tared dish, evaporate to dryness at $40-50°C$, dry in a desiccator over sulfuric acid for 3 hr, and then weigh.

2,4-Dinitroaniline in D&C Orange No. 17 (12)

Paste a 1-g sample with 10 ml of concentrated sulfuric acid in a 500-ml beaker. Very slowly stir in 100 ml of methanol. Add 150 ml of water, mix, and evaporate to about 100 ml on a steam bath. Transfer to a 250-ml volumetric flask, cool, and dilute to volume with water. Filter 100 ml into a 500-ml separatory funnel. Add 30% sodium hydroxide until the sample solution is alkaline to litmus, cool, and extract with three 50-ml portions of ethyl ether. Wash the combined extracts with 20-ml portions of water. Transfer the ether solution into a beaker, using 10 ml of ethyl ether, and evaporate to dryness on a steam bath. Dissolve the residue in warm $0.1N$ ethanolic hydrochloric acid and transfer it to a 100-ml volumetric flask. Rinse the beaker into the flask with two 20-ml portions of ethanolic hydrochloric acid. Dilute to volume with the ethanolic hydrochloric acid and measure the absorbance at 290 nm, 335 nm, and 380 nm against a standard solution of 0.01 g/liter concentration.

$$\text{Percent 2,4-dinitroaniline} = \left[\frac{A_{335} - \dfrac{(A_{380} + A_{290})}{2}}{B_{335} - \dfrac{(B_{380} + B_{290})}{2}} \right] 0.25$$

where A is the absorbance of the sample at the wavelength specified, and B is the absorbance of the standard at the wavelength specified.

Lake Red C Amine (2-Chloro-5-aminotoluene-5-sulfonic Acid) in D&C Red Nos. 8 and 9 (6)

Weigh a 1-g sample and transfer to a 500-ml beaker with 5 ml of acetone. Add

100 ml of 2% barium chloride solution. Boil for 10 min and filter the hot mixture through Whatman No. 12 paper into a separatory funnel. Return the filter paper and the dye to the original beaker and repeat the extraction and filtering twice more.

Cool the combined filtrates, acidify with 5 ml of 1:1 hydrochloric acid, and extract with three 20-ml portions of benzene. Wash the combined benzene extracts with 20 ml of water and add the wash to the aqueous layer. Filter the aqueous layer through cotton into a beaker. Boil for 15-20 min to remove benzene. Cool. Adjust the pH to about 8 with ammonium hydroxide and dilute with water to 500 ml in a volumetric flask.

Measure the absorbance of sample against a standard solution at 247 nm.

Aminoazobenzene in D&C Red No. 17 (12)

Weigh a 1-g sample and transfer to an Erlenmeyer flask containing 100 ml of acetone. Heat to boiling on a steam bath. Slowly stir in 100-200 g of crushed ice and allow the mixture to stand for 15 min. Filter through a large fluted filter paper into a separatory funnel; wash the residue with 50 ml of 1:49 hydrochloric acid. Return the paper to the Erlenmeyer flask, and add 100 ml of acetone, and heat to boiling on a steam bath. Repeat the precipitation with ice and the filtration. Make the combined filtrates alkaline to litmus with 30% sodium hydroxide and extract with 50-ml portions of petroleum ether until the extracts are colorless. Wash the combined petroleum ether extracts with 20 ml of 2% sodium hydroxide and then extract with 10-ml portions of $4N$ hydrochloric acid until the acid extracts are colorless. Heat the combined acid extracts for 15 min on a steam bath, cool, and dilute to 250 ml with $4N$ hydrochloric acid. Determine aminoazobenzene spectrophotometrically at 500 nm.

m-Diethylaminophenol in D&C Red No. 19 (13)

Prepare a m-diethylaminophenol standard solution by dissolving 0.2 g of purified m-diethylaminophenol in 100 ml of 50% ethanol and diluting to 200 ml with water. Then transfer a 20-ml aliquot to a 1-liter flask, add 100 ml of N sodium hydroxide, and dilute to volume with water.

Place 20 g of potassium dihydrogen orthophosphate and 35 g of sodium chloride in a 500-ml round-bottomed flask. Weigh a 2.5-g sample into a 200-ml beaker and paste with 50 ml of water. Add 50 ml of water and transfer the mixture to the round-bottomed flask with several 10-ml portions of water. Dilute to 250 ml and adjust the pH to 6.6 ± 0.2 with dilute sodium hydroxide. Fit the flask with a steam distillation trap and condenser and collect 175 ml of distillate. Transfer the distillate to a 250-ml volumetric flask, add 10 ml of 10% sodium hydroxide, and dilute to volume with water. Determine the absorbance of the sample solution and the standard solution at 295 nm.

o-(2-Hydroxy-4-diethylaminobenzoyl) Benzoic Acid in D&C Red No. 19

Follow the procedure given for determination of subsidiary dyes in D&C Red

No. 19 (p. 245). The aroylbenzoic acid elutes in the first few fractions and can be measured at 365 nm.

m-Diethylaminophenol in D&C Red No. 37 (13)

Proceed as described for D&C Red No. 19, up to and including "...collect 175 ml of distillate." Transfer the distillate to a 500-ml round-bottom flask, rinse the receiver with several 10-ml portions of water, and add the washings to the flask. Add 30 g of sodium chloride and 15 g of potassium dihydrogen orthophosphate and adjust the pH to 6.5-6.7 with 30% aqueous sodium hydroxide. Distill, collect about 175 ml of distillate, and filter it into a 250-ml volumetric flask containing 10 ml of 10% sodium hydroxide. Wash the receiver with several 10-ml portions of water, filtering each wash into the flask. Dilute to volume with water and measure the absorbance as described above.

Phthalic Acid and *m*-Diethylaminophenol in D&C Red No. 37 (18)

Prepare a phthalic acid-diethylaminophenol standard by dissolving 50 mg of phthalic acid and 20 mg of *m*-diethylaminophenol in anhydrous methanol. Dilute to 100 ml with alcohol and store in a cool dark place.

Prepare silylating reagent by mixing equal volumes of *N*,O-bis (trimethylsilyl) acetamide and Tri-Sil/BSA (Pierce Chemical Co., Rockford, Ill.) Store under refrigeration in a rubber-capped serum vial.

Pipette 1 ml of the phthalic acid-diethylaminophenol standard into a silylation vial and evaporate to dryness in a stream of nitrogen. Weigh 0.1 g of sample into the vial. Into a second vial weigh another 1-g portion of sample. Cap both vials with septums, inject 1 ml of silylating reagent into each, and then heat for 10 min at 60°C.

Compare the sample against the standard by injecting 4 µl of each into a Hewlett-Packard Model 5750 (or similar) gas chromatograph equipped with a flame ionization detector and a 10 ft × 0.25-in. copper column containing 5% OV 11 (Supelco, Bellefonte, Pa.) on 60-80-mesh Chromosorb WAW (Johns-Manville, Celite Division, New York, N. Y.). Observe the following conditions: helium flow rate, 50 ml/min; injection port and flame ionization detector temperature, 230°C; column temperature, 150-250°C at 6°/min, then hold at 250°C until the chromatogram is complete (ca. 15 min).

o-(2-Hydroxy-4-diethylaminobenzoyl) Benzoic Acid (HDBA) in D&C Red No. 37 (18)

Dissolve 2 g of sample in 100 ml of methanol. Apply 0.5 ml of sample solution as a streak 0.5 cm (or less) wide, 2 cm from the bottom of a 20 cm × 20-cm LQIF thin-layer plate (Quantagram LQIF, Quantum Industries, Fairfield, N. J.). Develop using 10% methanol in chloroform until the solvent front travels 12 cm. When viewed under UV light the HDBA is detected as a stripe about 1 cm above the D&C Red No. 37. Extract from the plate with methanol and determine spectrophotometrically at 350 nm.

4-Methylimidazole in Caramel (39)

Dissolve 4 g of sample in 6-7 ml of water, adjust the pH to 8.5 with N NaOH, and add it to a column of Amberlite IRC-50 (H^+ form). Wash the column with water, then elute the 4-methylimidazole with aqueous $4N$ NH_3. Concentrate the eluate to 0.5 ml in a rotary evaporator, dissolve the residue in saturated aqueous NaCl, make alkaline with aqueous NH_3, and extract it four times with 20 ml of chloroform. Filter the combined chloroform extracts, remove the solvent at $30°$ in a rotary evaporator, add 5 ml of CH_2Cl_2, and remove any residual water by azeotropic evaporation. Dissolve the residue in 0.8 ml of tetrahydrofuran containing 1 mg of imidazole as an internal standard, add 0.2 ml of acetic anhydride, heat at $60°C$ for 5 min, and then evaporate to 0.2 ml in a stream of nitrogen. Chromatograph 2-3 μl of sample on a 2 m × 2 mm gas chromatographic column packed with 3% OV-17 on Gas Chrom Q (80-100 mesh). Temperature program from 80-205°C at 5°/min; use nitrogen as carrier gas at 20 ml/min. Calculate the methylimidazole from the ratio of the peak area of the acetyl derivative to that of the internal standard.

Free Gossypol in Cottonseed Flour (40)

Pretreat all aniline used in this procedure by distilling the aniline over zinc dust (using an air condenser). Discard the first and last 10% of the distillate. Store in a glass-stoppered brown bottle in a refrigerator. Redistill when the reagent blank exceeds an absorbance of 0.022.

Grind 50 g of cottonseed through a Bauer Bros. No. 148 laboratory mill equipped with a No. 6912 plate. Grind with the plates separated so that the hulls are broken. Separate the meats from the hulls and lint by screening, then grind the meats through a Wiley mill equipped with a 1-mm screen. Do not preheat the cottonseed, and avoid heating the sample during grinding. If the sample is a cake, pellet, or meal, directly grind 50 g through a Wiley mill.

Weigh 1 g of sample into a small beaker and add 25 ml of acetone. Stir for 2 min and filter into a test tube. To one half the filtrate add a pellet of solid sodium hydroxide and heat on a steam bath for 2-3 min, swirling frequently. The appearance of a deep orange-red to red color indicates the prescence of dianilinogossypol. In this case, use Method B. If the light-yellow filtrate does not turn red on treatment with sodium hydroxide, use Method A.

Method A: Weigh 1 g of sample and transfer to a 250-ml Erlenmeyer flask. Cover the bottom of the flask with glass beads 6 mm in diameter. Pipette 50 ml of 7:3 acetone-water into the flask, stopper, and shake on a mechanical shaker for 1 hr.

Filter the solution through dry, medium-porosity filter paper into a small glass-stoppered flask, discarding the first few milliliters of filtrate. Pipette two 10-ml portions of filtrate into separate 25-ml volumetric flasks. Dilute the first portion to volume with 8:2 isopropyl alcohol-water. To the second aliquot, add 2 ml of redistilled aniline and heat in a boiling bath for 30 min, along with a reagent blank containing 2 ml of aniline and 10 ml of 7:3 acetone-water. Remove both solutions from the bath, cool to room temperature, and dilute to volume with 8:2 isopropyl alcohol-water. Measure the absorbance of the two sample solutions and the blank against isopropyl alcohol at the maximum near 440 nm. Calculate the corrected absorbance (difference between the absorbance of the samples, corrected for the blank) and compare against standards similarly prepared.

Method B: Prepare an aqueous acetone aniline solution by mixing 700 ml of acetone, 300 ml of water, and 0.5 ml of redistilled aniline. Do not use after one day.

Weigh 1 g of sample into a 250-ml Erlenmeyer flask and cover the bottom of the flask with 6-mm-diameter glass beads. Pipette 50 ml of the aqueous acetone-aniline solution into the flask. Stopper the flask and shake on a mechanical shaker for 1 hr.

Filter the solution through dry, medium-porosity filter paper into a glass-stoppered flask. Within 3 hr, pipette 10-ml aliquots into two separate 25-ml volumetric flasks. Dilute the first aliquot to volume with 8:2 isopropyl alcohol-water, mix well, and allow to stand for 25-30 min. Treat the second aliquot and a reagent blank exactly as described in Method A. Determine the net absorbance and compare against standards similarly prepared.

Oxalic Acid in Ferrous Gluconate (41)

Dissolve 1 g of sample in 10 ml of water, add 2 ml of hydrochloric acid, and transfer it to a separatory funnel. Extract successively with 50 ml and 20 ml of ethyl ether. Combine the extracts, add 10 ml of water, and evaporate the ether on a steam bath. Add one drop of 36% acetic acid and 1 ml of 5% calcium acetate solution. Acceptable ferrous gluconate should show no turbidity within 5 min.

Reducing Sugars in Ferrous Gluconate (41)

Prepare a cupric tartrate solution by dissolving 34.66 g of cupric sulfate, $CuSO_4 \cdot 5H_2O$, in water and diluting to 500 ml. Then dissolve 173 g of potassium sodium tartrate, $KNaC_4H_4O_6 \cdot 4H_2O$, and 50 g of sodium hydroxide in water and dilute to 500 ml. Mix equal amounts of these two solutions when required.

Dissolve 0.5 g of sample in 10 ml of water, warm, and make the solution alkaline with 1 ml of 1:9 ammonium hydroxide. Pass hydrogen sulfide into the solution to precipitate iron and allow the mixture to stand for 30 min to coagulate the precipitate. Filter and wash the precipitate with two 5-ml portions of water. Acidify the combined filtrate and washings with 1:9 hydrochloric acid. Add 2 ml of 1:9 hydrochloric acid in excess. Boil the solution until the vapors no longer darken lead acetate paper and then concentrate to about 10 ml. Cool. Add 5 ml of 10.5% sodium carbonate solution and 20 ml of water. Filter the solution and adjust the volume of the filtrate to 100 ml. To 5 ml of this solution add 2 ml of cupric tartrate solution and boil for 1 min. Acceptable ferrous gluconate shows no red precipitate within 1 min.

Acid-soluble Substances in Talc (42)

Digest 1 g of sample with 20 ml of 1:3 hydrochloric acid at 50°C for 15 min.

Add water to restore original volume, mix, and filter. Add 1 ml of 10% (w/v) sulfuric acid to 10 ml of filtrate, evaporate to dryness, and ignite to constant weight.

Reaction and Soluble Substances in Talc (42)

Boil 10 g of sample with 50 ml of water for 30 min; periodically add water to maintain approximately the original volume. Filter. Evaporate one-half the filtrate to dryness, and dry at 105°C for 1 hr.

Water-soluble Iron in Talc (42)

Using hydrochloric acid, slightly acidify the second half of the filtrate obtained in the test described above. Add 1 ml of fresh 10% potassium ferrocyanide solution. If the sample is of acceptable quality, the liquid should not turn blue.

BIBLIOGRAPHY

1. BAILEY, J. E., COX, E. A. JAOAC 58, 609-613 (1975). Chromatographic Analysis of 4,4'-(Diazoamino)-Dibenzenesulfonic Acid in FD&C Yellow No. 6.

2. BAILEY, J. E., CALVEY, R. J. JAOAC 58, 1087-1128 (1975). Spectral Compilation of Dyes, Intermediates, and Other Reaction Products Structurally Related to FD&C Yellow No. 6.

3. BAILEY, J. E., COX, E. A. JAOAC 59, 5-11 (1976). 4,4'-(Diazoamino)-Bis(5-Methoxy-2-Methylbenzenesulfonic Acid): Preparation and Determination in FD&C Red No. 40.

4. BELL, S. J. JAOAC 52, 831-832 (1969). TLC Separation and Spectrophotometric Determination of 2-Aminoanthraquinone in D&C Blue No. 9.

5. DANTZMAN, J., STEIN, C. JAOAC 57, 963-965 (1974). Leuco Base Determination in Triphenylmethane Dyes, FD&C Blue No. 1, FD&C Green No. 3, and FD&C Violet No. 1.

6. ETTELSTEIN, N. JAOAC 35, 419-421 (1952). Report on Sulfonated Amine Intermediates in Coal-Tar Colors.

7. Food and Drug Administration, Washington, D. C., private communication.

8. FRATZ, D. D. BAILEY, J. E. JAOAC 59, 12-13 (1976). Quantitative Determination of 4,4'-(Diazoamino)-Dibenzenesulfonic Acid in FD&C Yellow No. 6 by Elution Chromatography.

9. FRATZ, D. D. JAOAC 59, 578-579 (1976). Quantitative Determination of 4,4'-(Diazoamino)-Bis(5-Methoxy-2-Methylbenzenesulfonic Acid) in FD&C Red No. 40 by Ion Exchange Chromatography.

10. GRAICHEN, C. JAOAC 33, 398-401 (1950). Report on Intermediates Derived from Phthalic Acid.

11. GRAICHEN, C. JAOAC *34*, 407-411 (1951). Report on Intermediates Derived from Phthalic Acid.

12. HARROW, L. S. JAOAC *33*, 390-396 (1950). Report on Non-Volatile Unsulfonated Amine Intermediates.

13. HARROW, L. S. JAOAC *34*, 133-135 (1951). Determination of *m*-Diethylaminophenol in D&C Red No. 19 and D&C Red No. 37.

14. HARROW, L. S., HEINE, K. S., Jr. JAOAC *35*, 751-754 (1952). The Determination of 1,4-Dihydroxyanthraquinone in D&C Violet No. 2 and D&C Green No. 6.

15. HOSKINS, ELIZABETH C. JAOAC *54*, 1270-1271 (1971). Determination of 1,4-Dihydroxyanthraquinone in D&C Green No. 5 and the Former Ext. D&C Violet No. 2.

16. JOHNSON, R. K. JAOAC *50*, 526-530 (1967). Uncombined Intermediates in FD&C Blue No. 1.

17. JONES, J. H., DOLINSKY, M., HARROW, L. S., HEINE, K. S., Jr., STAVES, M. C. JAOAC *38*, 977-1010 (1955). Studies on the Triphenylmethane Colors Derived from Ethylbenzylaniline Sulfonic Acid.

18. KABACOFF, B. L., MOHR, G., FAIRCHILD, C. M. J. Soc. Cosmet. Chem. *24*, 551-560 (1973). Chromatographic Determination of Trace Components in D&C Red No. 37.

19. KOCH, L. JAOAC *42*, 444-445 (1959). The Isolation and Estimation of Beta-Naphthol in D&C Red No. 36.

20. LINK, W. B. JAOAC *44*, 43-53 (1961). Intermediates in Food, Drug and Cosmetic Colors.

21. MARMION, D. M. JAOAC *54*, 131-136 (1971). Analysis of Allura Red AC Dye (A Potential New Color Additive).

22. MARMION, D. M., WHITE, R. G., CASHION, F. W., WHITCOMB, B. B. JAOAC *54*, 137-140 (1971). 6,6'-Oxybis (2-Naphthalenesulfonic Acid) in Schaeffer's Salt.

23. MARMION, D. M. JAOAC *54*, 141 (1971). 6,6'-Oxybis(2-Naphthalenesulfonic Acid) in FD&C Yellow No. 6.

24. MARMION, D. M. JAOAC *55*, 723-726 (1972). Uncombined Intermediates in FD&C Yellow No. 6.

25. MARMION, D. M. JAOAC *56*, 700-702 (1973). Uncombined Intermediates in FD&C Red No. 40.

26. MARMION, D. M. JAOAC *58*, 719-724 (1975). Determination of 4,4'-(Diazoamino) Dibenzenesulfonic Acid in FD&C Yellow No. 6.

27. MARMION, D. M. JAOAC *59*, 838-845 (1976). The Determination of 4,4'-Diazoaminobis(5-Methoxy-2-Methylbenzenesulfonic Acid) in FD&C Red No. 40.

28. MARMION, D. M. JAOAC *60*, 168-172 (1977). Determination of 4,4'-(Diazoamino) Dibenzenesulfonic Acid in FD&C Yellow No. 6 (Part II).

29. MOTEN, L., KOTTEMANN, C. JAOAC *52*, 31-33 (1969). TLC Determination of Uncombined 2-Naphthol-6-Sulfonic Acid (Sodium Salt) (Schaeffer Salt) in Color Additives.

30. *Official Methods of Analysis*, 11th ed. Association of Official Analytical Chemists Washington, D. C., 1970, p. 594.

31. Ibid., p. 595.

32. *Official Methods of Analysis*, 12th ed. Association of Official Analytical Chemists, Washington, D. C., 1975, p. 641.

33. SCHUMACHER, R. J. JAOAC *48*, 819-826 (1965). Organic Compounds in FD&C Blue No. 1.

34. SINGH, M. JAOAC *57*, 219-220 (1974). High-Pressure Liquid Chromatographic Determination of Uncombined Intermediates in FD&C Red No. 40.

35. SINGH, M. JAOAC *57*, 358-359 (1974). Determination of Uncombined Intermediates in FD&C Yellow No. 6 by High-Pressure Liquid Chromatgraphy.

36. SINGH, M. JAOAC *58*, 48-49 (1975). High-Pressure Liquid Chromatographic Determination of Uncombined Intermediates and Subsidiary Colors in FD&C Blue No. 2.

37. STEIN, C. JAOAC *50*, 1297-1298 (1967). Determination of Anthraquinone Intermediates in D&C Violet No. 2.

38. WITTMER, D. P., NUESSLE, N. O. HANEY, W. G. Jr., Anal. Chem. *47*, 1422-1423 (1975). Simultaneous Analysis of Tartrazine and its Intermediates by Reversed Phase Liquid Chromatography.

39. CARNEVALE, J. Food Technol. Aust. *27*, 165-166, 172 (1975). Improved Method for the Determination of 4-Methylimidazole in Caramel.

40. Food Additives Analytical Manual, Vol. 1. U.S. Dept. of Health, Education, and Welfare, Food and Drug Administration, Washington, D. C., November 1967.

41. Food Chemicals Codex, Publication 1406, 1st ed. National Academy of Sciences, National Research Council, Washington, D. C., 1966.

42. *The United States Pharmacopeia*, 17 ed. (XVII). Mack Publishing Co., Eston, Pa., 1965, p. 695.

Chapter 13 Homologous, Isomeric, and Other Related Colorants

Lower-sulfonated Colors in FD&C Blue No. 1

Solvent Extraction Procedures (8, 10): Prepare a salt-acetate solution as follows. Dissolve 125 g of sodium chloride and 13.6 g of sodium acetate in water, add 12 ml of glacial acetic acid, and dilute to 500 ml with water.

Prepare a 0.1% aqueous sample solution. To a 10-ml aliquot add 40 ml of the salt-acetate solution and extract successively in three separatory funnels, each containing 100 ml of isoamyl alcohol. Wash the alcohol extracts with 100-ml portions of the salt-acetate solution until the washings are colorless. Pass each wash successively through the funnels in the same order as described above. Dilute the alcohol layer with one or more volumes of hexane. Remove the dye from the alcohol-hexane mixture by washing with several 10-ml portions of water, passing each washing through the funnels as described above. To the combined aqueous extracts add ammonium acetate to a concentration of about 0.04N. Determine the subsidiary colors present spectrophotometrically by comparison with the spectra of a standard solution of Guinea Green B, CI No. 42085 (formerly FD&C Green No. 1).

Two other schemes have been used to separate lower-sulfonated colors in FD&C Blue No. 1. In the first, the dye mixture is dissolved in water and the solution is acidified. This solution is extracted with isoamyl alcohol, which in turn is washed with several portions of 1:99 hydrochloric acid. The trisulfonated colors plus some of the disulfonated compounds pass into the aqueous solution and the monosulfonated and disulfonated compounds remain in the amyl alcohol. After dilution with petroleum ether the disulfonated compounds are extracted with water.

If no trisulfonated compounds are present, another scheme can be applied. The dye mixture is first extracted with hot benzene; the unsulfonated compounds are dissolved and the mono- and disulfonated substances remain in the residue. The disulfonated compounds (with traces of monosulfonated substances) are extracted from the residue with hot water. The aqueous solution is acidified and extracted with isoamyl alcohol, the alcohol extract is diluted with petroleum ether, and the disulfonated compounds are extracted with water. The monosulfonated colors remain in the alcohol-ether solution.

Thin-layer-chromatographic procedure (1, 37): Dissolve 1 g of dye in water and dilute to 50 ml in a volumetric flask. Streak 0.1 ml of this solution about 2 cm from the edge of a 20 cm $\times$ 20 cm silica gel G (200 μ thick) chromatographic plate. Dry the plate for 5 min at 100°C and then develop using isoamyl alcohol-acetonitrile-methyl ethyl ketone-water-NH$_4$OH (50:50:15:5:5) as the eluant. Leach the colors from the plate with EtOH and determine spectrophotometrically.

5,7'-Disulfo-3,3'-Dioxo-$\Delta^{2,2'}$-Biindoline and 5-Sulfo-3,3'-Dioxo-$\Delta^{2,2'}$-Biindoline in FD&C Blue No. 2 (33)

Prepare chromatographic solvents as follows. Mix 1 volume of concentrated HCl with 8 volumes of 2.5% (w/v) aqueous hydroxylamine hydrochloride. In a separatory funnel mix 2 volumes of this reagent with 1 volume of butanol and 1 volume of chloroform. Shake for 1 min and allow the layers to separate. The lower (organic) layer is the mobile phase and the upper (aqueous) layer is the stationary phase.

Add 7 ml of stationary phase to 12 g of Johns-Manville Celite 545 and mix thoroughly. Transfer to a 40 cm $\times$ 2.5 cm-ID chromatographic tube and pack firmly with a plunger.

Dissolve 6 mg of FD&C Blue No. 2 in 5 ml of stationary phase, add 10 g of Celite, and mix thoroughly. Transfer the mixture to the chromatographic tube and pack firmly. Dry wash the beaker with about 1 g of Celite and pack the wash into the column.

Develop the chromatogram with mobile phase. The monosulfonated color elutes first, followed by the isomeric color. Collect each as separate fractions and dilute each with equal volumes of chloroform.

Extract the solution of monosulfonated color with three 5-ml portions of water and dilute the combined extracts to 25 ml with water. Extract the solution of isomeric color with three 15-ml portions of water and dilute the combined extracts to 50 ml with water. Using 1-cm cells, obtain the visible spectra of the solutions from 700 nm to 400 nm and compare against standards.

Lower Sulfonated and Isomeric Colors in FD&C Green No. 3 (39)

These colorants can be determined using the solvent extraction and the TLC procedures described for FD&C Blue No. 1 (see pp. 229-230).

Lower Iodinated Colorants in FD&C Red No. 3

Paper-chromatographic Procedure (43): Prepare an eluant by mixing 400 ml of methyl ethyl ketone, 100 ml of acetone, 100 ml of water, and 1 ml of concentrated ammonium hydroxide. Place two 3-in.-wide blotting-paper drapes in a 18 in. × 6-in. tank, one on each side, add the eluant to the tank, and allow time to equilibrate.

Prepare a 0.1% sample solution in 2:98 ammonium hydroxide. Apply a 0.1-ml aliquot as a 0.5 in. × 2.5-in. band to 3-in.-wide Schleicher & Schuell No. 2043 chromatographic paper, 1.5 in. from the bottom of the strip. Suspend the strip in the tank so that the lower edge dips 0.5 in. into the eluant. Develop for 6 hr protected from light. Remove the strip and dry in the dark. Extract each spot from the chromatogram with small amounts of 1:199 ammonium hydroxide. Dilute as needed, filter, and determine the individual colors spectrophotometrically. The order of elution, absorption maxima, and approximate absorptivities of the disodium salts are listed in the table that follows.

Order of Elution	Color	Absorption Maximum (in nm)	Absorptivity of Disodium Salts (in liters/g-cm)
1	2,4,5,7-Tetraiodofluorescein	527-530	108 [a]
2	2,4,7-Triiodofluorescein	517-520	140
3	2,4,5-Triiodofluorescein	516-519	116
4	{ 2,7-Diiodofluorescein	511-513	179
	{ 2,5-Diiodofluorescein	509-511	145
5	4,5-Diiodofluorescein	507-509	122
6	2-Iodofluorescein	501-503	193
7	4-Iodofluorescein	497-500	154
8	Fluorescein	491-493	228

[a]For monohydrate.

Column-chromatographic Procedure (11): Prepare ethanol wash solution by diluting 75 ml of 95% ethanol to 100 ml with water. Prepare each of the following eluants:

Eluant No. 1—350 ml of 25% aqueous sodium chloride solution containing 1.75 ml of concentrated ammonium hydroxide.

Eluant No. 2—500 ml of 2% aqueous sodium sulfate solution containing 2.5 ml of concentrated ammonium hydroxide.

Eluant No. 3—1000 ml of 1% aqueous sodium sulfate solution containing 5 ml of concentrated ammonium hydroxide.

Eluant No. 4—600 ml of 0.5% (v/v) ammonium hydroxide.

Eluant No. 5—400 ml of 60% ethanol containing 2 ml of ammonium hydroxide.

Use a glass chromatographic column equipped at the top with the eluant distribution system shown in Fig. 13. Fill the tube with 3 in. of water; add 0.5 g of dry Whatman Column Chromedia CF11 and allow to settle. Add a slurry of 16 g of Solka Floc BW-100 in 150 ml of water to the column. Open the outlet and drain the liquid nearly to the top of the packing. Add 100 ml

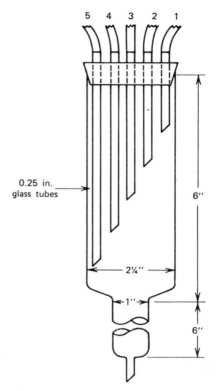

Figure 13 Eluant Distribution System for Column-chromatographic Determination of Subsidiary Colors in FD&C Red No. 3 (Numbers Above Inlet Tubes Refer to Serial Number of Respective Eluant)

of the ethanol wash solution and drain nearly to the top of the packing. Add 100 ml of eluant No. 1, washing the cellulose from the sides of the tube and keeping a level surface on the column. Drain the liquid to about 1 in. above the column packing. Wrap the 1-in.-diameter section of the column with aluminum foil to protect the sample from light.

Dissolve a 0.1-g sample in 100 ml of 1:199 ammonium hydroxide. Pipette 5 ml of the sample solution onto the column, mixing it with the eluant present and taking care not to disturb the surface of the packing. Drain the solution just to the packing surface. Carefully wash down the sides of the tube with eluant No. 1 and drain into the column. Add 20 ml of eluant No. 1 and about 3 g of the Chromedia to the column. When the cellulose has settled to form a protective cap over the surface, fill the column with eluant No. 1. Place any remaining eluant No. 1 in a flask and connect to tube No. 1. Place eluants Nos. 2-5 in suitable flasks and connect to the appropriate tubes. Elute the column successively with eluants Nos. 1-5. Collect the individual bands eluted and determine the colors spectrophotometrically. The order of elution and the identity of the subsidiary colors are as follows:

Eluant No. 1—Band 1, unidentified color; Band 2, fluorescein.

Eluant No. 2—Band 3, mixture of 2- and 4-monoiodofluoresceins; Band 4, mixture of 2,4-, 2,5-, and 2,7-diiodofluoresceins.

Eluant No. 3—Band 5, mixture of 4,5-diiodofluorescein and 2,4,7-triiodofluorescein; Band 6, 2,4,5-triiodofluorescein.

Eluant No. 4—Band 7, 2,4,5,7-tetraiodofluorescein.

Eluant No. 5—Band 8, unidentified color.

The amount of the individual colors can be estimated with reasonable accuracy if it is assumed that their absorptivities equal that of FD&C Red No. 3.

2-(2,4-Xylylazo)-1-naphthol-4-sulfonic Acid and 2-(5-Sulfo-2,4-xylylazo)-1-naphthol in FD&C Red No. 4 (11)

Extract 150 ml of n-butanol with three 50-ml portions of 10% sodium hydroxide and then wash the butanol with 50-ml portions of water until neutral (prepare fresh daily). Mix 100 ml of the washed butanol with 100 ml of carbon tetrachloride in a separatory funnel and add 100 ml of 2% hydrochloric acid containing 20 mg/ml of hydroxylamine hydrochloride. Shake the mixture for 3 min and then allow the layers to separate. Use the upper layer as the stationary phase and the lower layer as the mobile phase.

Mix 10 g of Celite 545 with 5 ml of the stationary phase, and firmly pack the mixture into a 45 cm × 2.5 cm chromatographic column.

Dissolve 0.05 g of sample in 25 ml of the stationary phase, mix a 5-ml aliquot of this solution with 10 g of Celite, and pack the mixture on the top of the column. Elute with mobile phase and collect the subsidiary color in the leading edge of the eluate.

If the subsidiary color is red, dilute it to 25 ml with the mobile phase and determine the color spectrophotometrically against a standard solution of 2-(2,4-xylylazo)-1-naphthol-4-sulfonic acid. If the eluate is yellow, transfer it to a separatory funnel and add an equal volume of chloroform. Wash with two 10-ml portions of 10% ammonium sulfate solution. Add a second volume of chloroform and extract with two 10-ml portions of 10% ammonium sulfate solution containing 10 mg/ml of sodium hydroxide. Adjust the aqueous extract to 25 ml with the extracting solution and determine the color spectrophotometrically against a standard solution of 2-(5-sulfo-2,4-xylylazo)-1-naphthol.

2-(4-Sulfo-1-Naphthylazo)-1-Naphthol-4-Sulfonic Acid in FD&C Red No. 4(11)

Prepare eluant by dissolving 50 g of sodium chloride in water containing 0.5 ml of concentrated ammonium hydroxide and then diluting this solution to 1 liter with water. Slurry Solka Floc BW-40 in water, pack two 1-in. chromatographic columns 12 in. high with the slurry, and wash each with 50 ml of eluant.

Dissolve 0.025 g of sample in 10 ml of water and add it to the first column. When the color settles on the column, wash it with eluant until most of the FD&C Red No. 4 is eluted. Strip the remaining color from the column with water and collect. Add 20 g of solid sodium chloride to each 100 ml of this solution and transfer it to the second column. Elute the remaining FD&C Red No. 4 with eluant and then elute the subsidiary color with water and determine it spectrophotometrically.

2-(3-Sulfo-2,6-Xylylazo)-1-Naphthol-4-Sulfonic Acid in FD&C Red No. 4 (42)

Prepare eluant by dissolving 800 g of anhydrous sodium sulfate in 5 liters of 0.2 M hydrochloric acid. Slurry powdered cellulose (Whatman Column Chromedia CF 11 or equivalent) in the eluant, pour sufficient slurry into a 24 in. × 2-in. chromatographic column to make a 15-in. bed, and wash the column with 100 ml of eluant.

Prepare a 0.5% aqueous sample solution and pipette a 20-ml aliquot onto the column. Elute the column with eluant; collect the fraction containing the subsidiary color, which elutes before FD&C Red No. 4. Saturate this fraction with sodium sulfate and pass the resulting solution through a 1-in.-diameter column packed with a 3-in. bed of cellulose. Elute the color with 100 ml of water and determine it spectrophotometrically against a standard solution.

2-(6-Sulfo-2,4-xylylazo)-1-naphthol-4-sulfonic Acid in FD&C Red No. 4 (13)

Slurry Solka Floc BW-40 in water, pack a 1-in.-diameter chromatographic column 12-in. high with the slurry, and wash it with 50 ml of a 20% sodium sulfate solution. Prepare a 0.5% aqueous sample solution, pipette a 10-ml aliquot onto the column, and elute with 20% sodium sulfate solution. FD&C Red No. 4 is quantitatively adsorbed. Collect the subsidiary color eluted and determine spectrophotometrically against a standard.

This compound can also be separated from FD&C Red No. 4 by paper chromatography using 200:88:2:40 butanol-water-ammonium hydroxide-ethanol or 300:150:5:80 butanol-water-acetic acid-ethanol as the eluant.

Subsidiary Colors in FD&C Red No. 4 (11)

Thin-layer Chromatography: Using a microliter syringe, spot about 4 mg of color as a band onto a 20 cm × 20 cm TLC plate coated with 250 μ of silica gel G. Elute with ethyl acetate-ethyl alcohol-diethylamine-water (55:20:10:10) until the solvent front nears the top of the plate. Leach the colorants from the plate with water or water-alcohol (1:1) and determine spectrophotometrically.

The order (top to bottom) in which the colorants appear on the plate is:

2-(5-Sulfo-2,4-xylylazo)-1-naphthol.

2-(2,4-Xylylazo)-1-naphthol-4-sulfonic acid.

2-(5-Sulfo-2,4-xylylazo)-1-naphthol.

2-(6-Sulfo-2,4-xylylazo)-1-naphthol-4-sulfonic acid.

FD&C Red No. 4.

2-(4-Sulfo-2,5-xylylazo)-1-naphthol-4-sulfonic acid.

2-(3-Sulfo-2,6-xylylazo)-1-naphthol-4-sulfonic acid.

Paper Chromatography: Prepare an eluant by mixing 70 volumes of methyl ethyl ketone, 30 volumes of acetone, 30 volumes of water, and 0.2 volume of concentrated ammonium hydroxide. Use a tank suitable for ascending chromatography and saturate its atmosphere with eluant. Prepare a 1% aqueous sample solution and apply 0.1 ml within a 18 cm × 0.7-cm rectangle, 2.5 cm from the bottom of a 20 cm × 20-cm sheet of Whatman No. 1 chromatographic paper. Allow the paper to dry at or below 50°C. Mount the sheet in the tank so that the eluant is 1 cm below the baseline of the sheet. Elute to a height of 17.5 cm or until the separation is satisfactory. Visually compare the chromatogram with knowns similarly prepared or extract the colors from the paper and determine spectrophotometrically.

3-Hydroxy-4-[(2-methoxy-5-methyl-4-sulfophenyl)azo]-2,7-naphthalenedisulfonic Acid (R-Salt + CSA) and 7-Hydroxy-8-[(2-methoxy-5-methyl-4-sulfophenyl)azo]-1,3-naphthalenedisulfonic Acid (G-Salt + CSA) in FD&C Red No. 40 (19)

Weigh about 200 g of Celite 545 (Fisher C-212) into a Petri dish and place in a desiccator containing 25 ml of General Electric SC-77 Dri-Film. Let stand until the Celite no longer wets when mixed with water ($\geqslant$3 hr).

Place 100 ml of BuOH, 100 ml of CCl$_4$, and 100 ml of dilute HCl (1 + 4) into a separatory funnel, shake for 3 min, and allow the layers to separate. The lower layer is the stationary phase and the upper layer is the eluant.

Mix 15 g of silane-treated Celite with 7.5 ml of stationary phase. Using a rammer, pack firmly into an 8 in. × 1 in. OD glass chromatographic tube. Wash the column with 15 ml of eluant. (This wash frequently flushes a yellow color from the column just ahead of the red subsidiary dye. The yellow color is leached from the packing itself and should be discarded.)

Prepare a 0.2% sample solution in eluant. Pipette 3 ml onto the column and let drain to the surface of the packing. Elute the column with eluant. Collect the desired fraction in a 25-ml volumetric flask and dilute to volume with eluant. Similarly prepare and elute a blank column to which no color has been added.

Record the visible spectra of the sample and blank from 680 nm to 480 nm in a 1-cm absorption cell against eluant. The G-salt + CSA dye has an absorption maximum near 503 nm. The R-salt + CSA dye has an absorption maximum near 512 nm.

$$\text{Percent } R\text{-salt} + \text{CSA dye (mw 599.48)} = A \times 8.41$$
$$\text{Percent } G\text{-salt} + \text{CSA dye (mw 599.48)} = A \times 9.04$$

where 8.41 is equal to 100/(49.4 × 1 × 0.24); 9.04 is equal to 100/(46.1 × 1 × 0.24); 100 is factor for conversion to percent; 1 is cell path length (in cm); 0.24 is effective sample concentration (in g/liter); 49.4 and 46.1 are absorptivities in liters/g-cm of R-salt + CSA dye and G-salt + CSA dye, respectively; and A is blank-corrected absorbance (also corrected for column blank where necessary) of the sample fraction at the appropriate absorption maximum.

6-Hydroxy-5-[(2-methoxy-5-methylphenyl)azo]-2-naphthalenesulfonic Acid (Cresidine + Schaeffer's Salt) and 4-[(2-Hydroxy-1-naphthyl)azo]-5-methoxy-2-methylbenzenesulfonic Acid (CSA + β-Naphthol) in FD&C Red No. 40 (19)

Add 100 ml of BuOH, 100 ml of CCl_4, and 100 ml of dilute HCl (2 + 98) to a 500-ml separatory funnel. Shake for 3 min and let the layers separate. The upper layer is the stationary phase and the lower layer is the eluant

Pipette 5 ml of stationary phase onto 10 g of Celite 545; mix well. Pack this mixture firmly into an 8 in. × 1 in.-OD glass chromatographic column. Pipette 5 ml of a 0.2% sample solution (in stationary phase) onto a second 10-g portion of Celite, mix well, and then pack firmly into the column with a rammer. Elute the column with eluant. Collect the desired fraction in a 25-ml volumetric flask and dilute to volume with eluant.

Immediately record the visible spectrum of the sample and blank (similarly prepared) from 680 nm to 480 nm in a 1-cm absorption cell against eluant.

Percent cresidine + Schaeffer's dye (mw 394.4) = $A \times 4.19$

Percent CSA + β-naphthol dye (mw 394.4) = $A \times 3.88$

where 4.19 is equal to 100/(59.7 × 1 × 0.40); 3.88 is equal to 100/(64.4 × 1 × 0.40); 100 is factor for conversion to percent; 1 is cell path length (in cm); 0.40 is effective sample concentration in g/liter; 59.7 and 64.4 are absorptivities in liters/g-cm of cresidine + Schaeffer's dye and CSA + β-naphthol dye, respectively; and A is blank-corrected absorbance (also corrected for column blank where necessary) of the sample fraction at absorption maximum near 508 nm.

Screening Procedures for Subsidiary Colors in FD&C Red No. 40

Paper-chromatographic Method (19): Apply 0.01 ml of 5% aqueous sample solutions as 1.5-in. bands on Whatman No. 1 chromatographic paper (10 in. × 10 in.). Dry for 15 min in a 50°C air oven and then develop for about 2 hr in a 10 in. × 10 in. glass chromatographic tank (Arthur H. Thomas Co., 3108-B05), using 130 ml of methyl ethyl ketone-acetone-water (70:30:30) as the eluant. Compare with standards run simultaneously on the same sheet. Figure 14 is a schematic representation of the resolution of a synthetic mixture containing subsidiary colors that could be present.

Thin-layer-chromatographic Procedure (4, 11): Spot 3 μl of a 2% sample solution and the appropriate standards on a 20 cm × 20-cm thin-layer plate coated with a 250-μ layer of silica gel G. Develop in an eluant composed of isoamyl alcohol-1,4-dioxane-acetonitrile-ethyl acetate-water-ammonium hydroxide (10:10:10:10:10:2). Compare sample and standards visually or extract from the plate with water and examine spectrophotometrically.

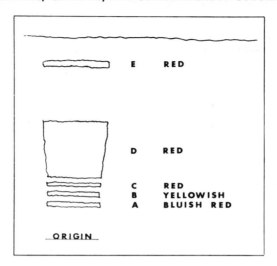

Figure 14 Paper-chromatographic Resolution of Dyes: A, *R*-salt + CSA Dye; B, *G*-Salt + CSA Dye; C, (6-Hydroxy-5-[(2-methoxy-5-methyl-4-sulfonphenyl)azo] - 8-(2-Methoxy-5-methyl-4-sulfophenoxy)-2-Naphthalenesulfonic acid); D, FD&C Red No. 40 and E, Cresidine + Schaeffer's Dye and CSA + β-Naphthol Dye

Lower-sulfonated colors appear above the main band, whereas higher-sulfonated colors appear below the main band.

Lower-sulfonated Colors in FD&C Yellow No. 5

Liquid-Liquid Extraction Method (7): Dissolve 0.2 g of sample in 100 ml of warm water. To 50 ml of this solution add 1 ml of concentrated hydrochloric acid and extract the solution successively in three separatory funnels, each containing 50 ml of amyl alcohol. Wash the alcohol extracts by shaking successively with 50-ml portions of 0.25N hydrochloric acid until the washings are practically colorless. Pass each acid portion through the funnels in the order used for the original alcohol extraction. Dilute the alcohol extracts in each funnel with 1-2 volumes of petroleum ether and extract the lower-sulfonated colors by washing with several 10-20-ml portions of water. Pass each portion through the funnels in the order reverse to that previously followed. Transfer the water solution to a 100-ml volumetric flask, add about 1 g of solid ammonium acetate, and dilute to volume with water. Determine the compounds present spectrophotometrically at 434 nm against a standard.

Column-chromatographic Method (11): Transfer 100 ml of BuOH, 100 ml of CCl4, and 100 ml of dilute (1 + 19) HCl to a separatory funnel. Shake the mixture for 3 min and then allow the layers to separate. Use the lower organic layer as the mobile phase and the upper aqueous layer as the stationary phase.

Mix 5 g of Celite 545 and 2.5 ml of stationary phase. Place a pledget of glass wool in the constriction of a 250 mm × 22 mm-ID glass chromatographic column then firmly pack all but a little of the mixture into the column.

Dissolve 0.04 g of color in 5 ml of stationary phase. Mix the solution with 5 g of Celite and then pack the mixture firmly into the column. Rinse the beaker with the reserved Celite and pack the rinse into the column. Elute with mobile phase and collect the lower sulfonated colors which elute first.

Transfer the eluant containing the subsidiary colors to a separatory funnel, add an equal volume of hexane, and extract the color with several small portions of water. Make to a known volume and examine spectrophotometrically.

Screening Procedures for Subsidiary Colors in FD&C Yellow No. 5

Paper Chromatography (5): This procedure separates the lower sulfonated colors as well as the ethyl ester of the parent compound from FD&C Yellow No. 5. It does not separate the lower sulfonated colors from each other.

Prepare the eluant by mixing 70 volumes of methyl ethyl ketone, 30 volumes of acetone, and 30 volumes of water. Use a tank suitable for ascending chromatography and saturate its atmosphere with the eluant. Prepare a 1% aqueous sample solution and apply 0.1 ml within a 18 cm × 0.7-cm rectangle, 2.5 cm from the bottom of a 20 cm × 20-cm sheet of Whatman No. 1 chromatographic paper. Allow the paper to dry at or below 50°C. Mount the sheet in the tank so that the eluant is 1 cm below the baseline of the sheet. Elute to a height of 17.5 cm or until the separation is satisfactory. Visually compare the chromatogram with knowns similarly prepared, or extract the colors from the paper and make a spectrophotometric determination.

Thin-layer Chromatography (11): Dissolve 2 g of sample in 100 ml of water. Spot 3 μl of this solution and the appropriate standards onto a silica gel thin-layer plate. Dry the plate and elute using 1,4-dioxane-isoamyl alcohol-water-ammonium hydroxide (10:10:4:1). Leach the colorants from the plate and determine spectrophotometrically.

Screening Procedures for Subsidiary Colors in FD&C Yellow No. 6

Paper Chromatography: The lower- and higher-sulfonated compounds can be determined in one step by paper chromatography according to the method given above for FD&C Yellow No. 5 except that a mixture of 350 volumes of methyl ethyl ketone, 150 volumes of acetone, 150 volumes of water, and 1 volume of concentrated ammonium hydroxide is used as the eluant. The higher-sulfonated compounds are separated from each other, but the lower-sulfonated compounds are not.

Thin-layer Chromatography (3): Spot 3 μl of a 2% aqueous sample solution onto an 8 in. × 8-in. chromatographic plate coated with 0.25 mm of silica gel G. Air dry the plate and then elute using isoamyl alcohol-acetone-water-ammonium hydroxide (65:50:20:5). If present, lower-sulfonated colors appear above the main band and higher-sulfonated colors appear below the main band.

Higher-sulfonated Dyes in FD&C Yellow No. 6

Extraction Procedure (9): Dissolve 0.1 g of sample in 100 ml of 1:25 hydro-chloric acid. Dilute 10 ml of this solution with 40 ml of 1:25 hydrochloric acid. Extract by shaking the solution successively in five separatory funnels, each containing 50 ml of isoamyl alcohol. Transfer the acid layer to a 100-ml volumetric flask. Wash the alcohol extracts with two 25-ml portions of 1:25 hydrochloric acid, passing each portion through the funnels in the same order used for the original extraction. Add the washings to the acid layer in the flask and dilute to 100 ml with water. Determine the higher-sulfonated colors spectrophotometrically against a standard.

Chromatographic Method (11): In a separatory funnel mix 100 ml of n-bu-tanol and 100 ml of carbon tetrachloride. Add 100 ml of 1:4 hydrochloric acid, shake for 3 min, and allow to settle. Use the lower layer as the station-ary and the upper layer as the mobile phase. Prepare silane-treated Celite by adding 25 ml of General Electric GS-77 Dri Film to the bottom of an empty desiccator, placing a dish containing 200 g of Celite 545 into the desiccator, and leaving it covered for at least 3 hr. Test completion of the silanization by mixing a small amount of the treated Celite with water; the Celite should not be wetted.

Mix 15 g of the treated Celite with 7.5 ml of the stationary phase, pack the mixture into a 45 cm $\times$ 2.5-cm chromatographic column and then wash the column with 15 ml of mobile phase. Dissolve 0.05 g of sample in 25 ml of the mobile phase. Transfer a 5-ml aliquot of this solution to the top of the column and elute with mobile phase. The higher-sulfonated colors will elute as one band before the parent compound. Determine their content in the fractions spectrophotometrically against a standard.

Lower Sulfonated Dyes in FD&C Yellow No. 6

Extraction Procedure (26): Dissolve 0.2 g of sample in 20 ml of water, add 1 ml of concentrated hydrochloric acid, and dilute to 50 ml. Extract by shaking the solution successively in three separatory funnels, each containing 50 ml of amyl alcohol. Wash the alcohol extracts with 50-ml portions of 5% aqueous sodium chloride solution until the washings are colorless. Dilute each alcohol layer with 100 ml of petroleum ether and extract the lower-sulfonated dye with several 10-ml portions of water. Pass each portion through the funnels in the order reverse to that previously followed. Combine the aqueous extracts in a 100-ml volumetric flask, add 1 ml of 2N ammonium acetate solution, and dilute to volume with water. Determine the lower-sulfonated colors present spectrophotometrically at 485 nm against a stand-ard solution of D&C Orange No. 4.

Chromatographic Method (11): Wash 150 ml of n-butanol with three 50-ml portions of 10% sodium hydroxide and then wash the butanol with 50-ml portions of water until neutral; prepare fresh daily. Mix 100 ml of the washed butanol with 100 ml of carbon tetrachloride in a separatory funnel and add 100 ml of 2% (v/v) hydrochloric acid containing 20 mg/ml of hydroxylamine hydrochloride. Shake the mixture for 3 min and then allow the layers to separate. Use the upper layer as the stationary phase and the lower layer as the mobile phase.

Mix 10 g of Celite 545 with 5 ml of the stationary phase and firmly pack the mixture into a 45 cm × 2.5-cm chromatographic column.

Dissolve 0.05 g of sample in 25 ml of the stationary phase, mix a 5-ml aliquot of this solution with 10 g of Celite, and pack the mixture on the top of the column. Elute with mobile phase, and collect the desired fractions. Dilute them to 25 ml with mobile phase and determine the amount of color spectrophotometrically.

Subsidiary Colors in Orange B

Column-chromatographic Procedure (11): Insert a glass-wool pledget in the constriction above the tip of a 59 cm × 3.3 cm-ID glass chromatographic column. Slurry Solka-floc BW-40 in water and add the mixture to the column to a height of about 45 cm. Wash the column with about 100 ml of 20% aqueous NaCl.

Transfer 0.2 g of sample into a small beaker and dissolve in 10 ml of water. Add 20 ml of 20% NaCl then, using small portions of 10% NaCl, quantitatively transfer the sample to the column. Drain the column just to the surface of the support and then cover the top of the support with a pledget of glass wool. Elute with 10% NaCl until the main band is halfway down the column and then elute with 2% NaCl.

Orange B elutes first followed by Orange K [1-(4-sulfophenyl)-3-carboxy-4-(4-sulfonaphthylazo)-5-hydroxypyrazole]. When most of the Orange K has eluted, wash 2-(4-sulfonaphthylazo) naphthionic acid or any other subsidiary color present from the column with water. Determine the colors spectrophotometrically.

Thin-layer-chromatographic Method (11): Using a microliter syringe, spot as a band about 3 mg of color onto a 20 cm × 20-cm, 250-μ thick cellulose plate. Allow the plate to dry thoroughly in the dark (ca. 20 min) and then develop using 1,4-dioxane-isoamyl alcohol-acetic acid-water (45:25:1:20) as the eluant. The order of elution from top to bottom of the plate is:

Orange B, 2-(4-sulfonaphthylazo)naphthionic acid,

1-(4-sulfophenyl)-3-carboxy-4-(4-sulfonaphthylazo)-5-hydroxypyrazole.

1[4-(2,5-Dimethoxyphenylazo)-2,5-Dimethoxyphenylazo]-2-Naphthol and 1,1'-(2,2',5,5'-Tetramethoxy-4,4'-Biphenylenebisazo)-di-2-Naphthol in Citrus Red No. 2(38)

Transfer 2.5 mg of sample (in chloroform) as a band 1 in. from the bottom of a 20 cm × 20-cm TLC plate coated with 0.38 mm of silica gel G. Allow the plate to dry and then develop it with chloroform until the solvent reaches the upper edge of the plate. Remove the plate from the tank, air dry it for about 10 min, and then leach the bands from the plate and determine them spectrophotometrically. The colors appear in the ascending order: (1) 1[4-(2,5-dimethoxyphenylazo)-2,5-dimethoxyphenylazo]-2-naphthol (dimethoxy color), (2) 1,1'-(2,2',5,5'-tetramethoxy-4,4'-biphenylenebisazo)-di-2-naphthol (benzidine color), and (3) Citrus Red No. 2.

Lower-sulfonated Colors in D&C Blue No. 4

This colorant is the ammonium salt corresponding to FD&C Blue No. 1 and can be analyzed by the methods given for that color.

Indirubin in D&C Blue No. 6 (41)

Weigh 0.3-0.4 g of sample into a 6-dram-capacity vial containing about 1 cm of Ottawa sand (MCB No. SX75). Add another 1 cm of sand, replace the cap, and shake. Transfer the sample and sand to a double-thickness cellulose extraction thimble containing 1-2 cm of sand. Further additions (ca. 3-4) of sand to the vial followed by shaking and transfer to the extraction thimble are made until sample transfer is essentially quantitative and the thimble is 3/4 full. Insert the thimble into the extraction section of a Soxhlet extractor with a 24/45 upper joint. Add 150 ml of glacial acetic acid to the flask and extract at a rate of four or more drops per second for 4 hr. Use boiling chips. After cooling, transfer the extract to a 200-ml volumetric flask. Rinse the flask with three 10-ml portions of acetic acid, adding the rinses to the extract. Make to volume with acetic acid. Dilute as needed and if a precipitate of indigo occurs filter through a 0.45-μm millipore filter.

Determine the blank-corrected sample absorbance at 615 nm and 533 nm against acetic acid.

$$\text{Percent indirubin} = \frac{(4.42 \times A_{533} - A_{615})}{177.32 \times w \times (1000/200) \times b} \times DF \times 100$$

$$= \frac{0.1128\,(4.42 \times A_{533} - A_{516}) \times DF}{w \times b}$$

where w is sample weight (in g), 1000/200 is factor for conversion of effective sample concentrations to g/liter, b = cell path length (in cm), DF = dilution factor, 100 = factor for conversion to percent, and $[(4.42 \times A_{533} - A_{615})]/177.32$ is the solution of the following simultaneous equations for c_1.

$$A_{533} = 40.866\,bc_1 + 11.601\,bc_2$$

$$A_{615} = 3.310\,bc_1 + 51.276\,bc_2$$

where c_1 and c_2 are the effective sample concentrations in (g/liter) of indirubin and indigo, respectively; 40.866 and 3.31 are the absorptivities in liters/g-cm of indirubin at 533 nm and 615 nm, respectively; and 11.601 and 51.276 are the absorptivities of indigo at the same wavelengths.

Monosulfonated Color in D&C Green No. 5 (16)

Transfer 10 g of sample to a 250-ml Erlenmeyer flask. Add 50 ml of a 10:2:1 glacial acetic acid-concentrated hydrochloric acid-water mixture and 10 ml of concentrated hydrochloric acid containing 1 g/ml of stannous chloride.

Boil gently until the volume is reduced to about 25 ml and then dilute the hot mixture with 100 ml of water. Transfer to a 500-ml volumetric flask and dilute to volume with water. Filter a 100-ml aliquot into a 500-ml extraction funnel; make it alkaline with 25 ml of 50% sodium hydroxide solution, cooling the funnel during the addition. Extract the liberated amine with two 100-ml portions of ethyl ether; combine the extracts and wash them with four 25-ml portions of water. Extract the amine with five 25-ml portions of 0.3N hydrochloric acid and transfer the washings to an iodination flask.

Boil to expel the dissolved ether, concentrate to 100 ml, then cool. Add 25 ml of 0.3N hydrochloric acid and about 100 g of crushed ice. Add 0.05N potassium bromide-bromate to the agitating solution until it remains yellow for at least 30 sec, and then add 5 ml in excess. Stopper the flask and let it stand in an ice bath for 10 min. Add 2-3 g of potassium iodide and titrate while cold with 0.05N sodium thiosulfate to a starch end point (add indicator internally near the end point). Perform a blank determination. Calculate the result as p-toluidine; 1 ml of 0.05N potassium bromide-bromate is equivalent to 1.34 mg of p-toluidine. Calculate the amount of monosulfonated dye present in the sample as follows:

$$\text{Monosulfonated dye, weight \%} = \frac{(W)(4.858)(100)}{w}$$

where W is amount of p-toluidine (in mg) and w is sample weight (in mg).

Sodium Salts of 1-(p-Toluidino)-4-(o-Sulfo-p-Toluidino) Anthraquinone and 1-Hydroxy-4-(o-Sulfo-p-Toluidino) Anthraquinone in D&C Green No. 5 (40)

Streak 1-1.5 mg of sample as an aqueous solution onto a 20 cm × 20-cm thin-layer plate coated with a 250-μm layer of MN300 cellulose. Allow the plate to air dry at room temperature for about 45 min and then develop with butyl acetate-dimethylformamide-water (10:5:1) in a tank that has been equilibrated with eluant for about 30 min. The colorants appear on the plate in descending order:

Unsulfonated D&C Green No. 5 (D&C Green No. 6),

monosulfonated D&C Green No. 5 (1-(p-toluidino)-4-(o-sulfo-p-toluidino)-anthraquinone),

1-hydroxy-4-(o-sulfo-p-toluidino)anthraquinone (Ext. D&C Violet No. 2).

Alizurol Purple (D&C Violet No. 2) in D&C Green No. 6 (11)

Spot about 1 mg of color as a chloroform solution onto a 20 cm × 20-cm thin-layer plate coated with silica gel G. Allow the plate to dry and then develop it using hexane-trichloroethylene-diethylamine (6:2:1) as the eluant. Leach the alizurol purple from the plate with chloroform and determine it spectrophotometrically.

Lower-sulfonated Colors in D&C Orange No. 4 (11)

Use the chromatographic method given for lower sulfonated dyes in FD&C Yellow No. 6 (p. 239) except prepare the mobile phase from 100 ml of washed butanol, 100 ml of carbon tetrachloride, and 100 ml of distilled water.

Related Bromofluoresceins in D&C Orange No. 5 (14)

Prepare a 1% ammoniacal sample solution and spot 5 μl on a 18 in. X 22-in. Whatman No. 1 chromatographic paper, about 1 in. from the edges. Immerse the 18 in. edge of the paper in an eluant composed of 1 g of sodium chloride and 1 ml of concentrated ammonium hydroxide dissolved in 100 ml of water (solution 1). Develop for 24 hr. Remove the sheet from the tank, air dry it in the dark, and then rotate it 90° and develop for an additional 48 hr in an eluant composed of 100 ml of solution 1, 300 ml of n-butanol, and 70 ml of ethanol. Leach the spots from the paper with 1:99 ammonium hydroxide and determine the individual compounds spectrophotometrically. See Fig. 15.

4-Toluene-azo-2-naphthol-3-carboxylic Acid in D&C Red Nos. 6 & 7 (18)

Place 0.25 g of sample in a 125-ml acetylation flask. Add 80 ml of methyl Cellosolve and 5 ml of concentrated hydrochloric acid, attach an air condenser, and reflux for 15-20 min. Transfer the solution to a separatory

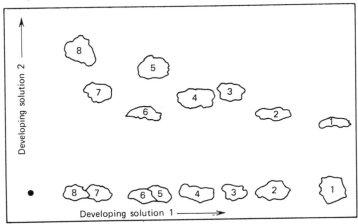

Figure 15 Paper-chromatographic Separation of Fluorescein and Nine Bromo-fluoresceins: (1) Fluorescein, (2) 2-Bromofluorescein and 4-Bromofluorescein, (3) 2,7-Dibromofluorescein, (4) 2,4-Dibromofluorescein and 2,5-Dibromofluorescein, (5) 2,4,7-Tribromofluorescein, (6) 4,5-Dibromofluorescein, (7) 2,4,5-Tribromo-fluorescein, (8) 2,4,5,7-Tetrabromofluorescein (black dot in lower left corner shows initial spot of sample containing 0.003 mg of each compound) (Reprinted with the permission of the Association of Official Analytical Chemists)

funnel. Rinse the flask into the funnel with four 10-ml portions of isopropyl ether and two 75-ml portions of water, shake, and then let the two phases separate. Drain the lower aqueous phase into a second separatory funnel, extract it with portions of isopropyl ether totaling 40 ml, combine the ether extracts, and discard the residual aqueous solution.

Extract the ether layer with 20-ml portions of water until the water extract is colorless. Filter the ether layer through a cotton pledget, rinse the cotton and the funnel with isopropyl ether, and dilute the filtrate to 100 ml with isopropyl ether. Compare the absorbance of the sample against a standard at the maximum near 507 nm.

4(4-Chloro-2-sulfo-5-tolylazo)-1-naphthol in D&C Red Nos. 8 and 9 (36)

Transfer a 0.1-g sample to a 600-ml beaker. Add 100 ml of ethanol, 20 ml of 10% sodium hydroxide solution, and 80 ml of water. Cover and then boil gently for 5 min. Remove from the heat and add 4 ml of concentrated HCl; and then add hydrochloric acid dropwise until the solution is acidic. Render basic by the dropwise addition of a 2% ethanolamine solution and add 2 ml in excess. Evaporate to 100 ml on a hot plate and then add 100 ml of a 2% ethanolamine solution, bring to a boil, cool, and filter.

Return the filter paper with the precipitate to the beaker and leach with a stream of 5:1 ethanol-10% sodium hydroxide solution. Add 80 ml of water and then dilute to 200 ml with the 5:1 ethanol-10% sodium hydroxide solution. Boil, adjust the pH, and filter as described above. Combine the filtrates.

Transfer the filtrates to a 500-ml separatory funnel and acidify with about 15 ml of hydrochloric acid. Extract the dye with 75 ml of isoamyl alcohol and discard the aqueous layer. Wash the alcohol layer twice by shaking it with 50-ml portions of a 5% sodium chloride solution. Add 50 ml of isooctane to the alcohol layer and extract it with 50 ml of an aqueous solution containing 1 g of ethanolamine and 2.5 g of sodium chloride. Repeat with 25-ml portions of the same solution until no more color is extracted. Combine the extracts in a 250-ml volumetric flask and dilute to volume with a solution of 2% ethanolamine in 5% sodium chloride. Determine the amount of dye present spectrophotometrically by comparison with a standard.

Determination of 1-Phenylazo-2-naphthol (PAN) in D&C Red No. 17 (24)

Prepare a 2.5% solution of the sample in chloroform and apply a 0.2-ml aliquot across a 20 cm × 20-cm glass plate coated with a silica gel G layer, 0.375 mm thick. Allow the plate to air dry. Develop with toluene in a lined chromatographic tank until the solvent front reaches the top of the plate. Remove the plate and air dry. If the orange-yellow line of PAN is not adequately resolved from the D&C Red No. 17, repeat the elution. If PAN is separated from the D&C Red No. 17 but mixed with another colorant, scrape the commingled colors from the plate, leach with chloroform, filter through sintered glass, evaporate the filtrate and any needed washings to 0.1-0.2 ml, and apply this solution to a 10 cm × 20-cm silica gel G plate, 0.375 mm thick. Develop the plate with methylene chloride. Leach the compounds from the plate with chloroform and determine spectrophotometrically.

Triethylrhodamine and Other Subsidiary Colors in D&C Red No. 19

Column-chromatographic Method (29): Prepare silane-treated Celite as described under Chromatographic Method for FD&C Yellow No. 6 (p. 239). Slurry 20 g of the treated Celite with 18 ml of N-decanol that has been equilibrated with 1:99 ammonium hydroxide and then pack the slurry tightly into a 1-in.-diameter chromatographic column.

Dissolve 0.05 g of sample in 5 ml of water, add one drop of concentrated ammonium hydroxide, and add the solution to the column. Elute with 1:99 ammonium hydroxide and collect 20-ml fractions. Determine the triethylrhodamine spectrophotometrically at 540 nm. Frequently, as many as nine distinct bands are separated; triethylrhodamine elutes just before the parent compound.

Thin-layer-chromatographic Method (2): Spot 0.2 ml of a 0.5% (w/v) ethanol sample solution as a band 3 cm from the bottom of a 20 cm × 20-cm TLC plate coated with 0.375 mm of silica gel G. (The plate should have been dried at room temperature and heated for 15 min at 125°C immediately prior to use.) Allow the plate to air dry and then place in an unlined tank that has been equilibrated for 10 min with acetone-chloroform-triethylamine-water (30:45:5:1). Develop the plate until the solvent front reaches the top of the plate. Remove the plate and allow it to air dry.

The top band is that of D&C Red No. 19 followed by the major subsidiary, triethylrhodamine. Minor subsidiaries follow in order of decreasing ethylation. Scrape the subsidiaries of interest from the plate, slurry with ethanol, filter, and determine spectrophotometrically.

Lower-brominated Subsidiary Colors in D&C Red No. 21 (15)

Weigh 2 g of sample and transfer it to a 100-ml volumetric flask. Add 50 ml of a solvent prepared by mixing 50 ml of S. D. No. 1 ethanol, 35 ml of water, and 15 ml of concentrated ammonium hydroxide. Shake to dissolve; warm if necessary. Dilute to volume with solvent. Apply 0.1 ml of sample solution to a 3 in. × 16 in. Schleicher & Schuell No. 2043 chromatographic paper and let it dry in the dark. Suspend the strip in a 18 in. × 6 in.-ID glass tank and develop in the dark for 24-48 hr by the ascending technique using n-butanol-water-ammonium hydroxide-ethanol (100:44:1:22.5). Remove the strip and dry in the dark. Wash each colored zone into a separate volumetric flask using 1:199 ammonium hydroxide, filter or centrifuge if necessary, and determine the individual substances spectrophotometrically. Their absorption maxima and approximate absorptivities are in the list that follows.

	Absorption Maximum (in nm)	Absorptivity (in liters/g-cm)
Fluorescein	492	258
4-Monobromofluorescein	498	206
2,4- and 2,5-Dibromofluorescein	506	188
4,5-Dibromofluorescein	504	163
2,4,5-Tribromofluorescein	512	157
2,4,5,7-Tetrabromofluorescein	518	149

Uranine in D&C Red No. 22 (17)

Slurry 20 g of Solka Floc SW-40-A in 800 ml of water, pack the slurry into a 120 cm × 2.2-cm chromatographic column, and wash the column with 50 ml of 5% sodium sulfate solution. Prepare a 0.3% aqueous sample solution and pipette a 5-ml aliquot into a beaker. Add 0.1 ml of 10% sodium hydroxide solution and 1 g of anhydrous sodium sulfate, dilute to 20 ml with water, and then add, with stirring, 0.5 g of Solka Floc. Transfer the mixture to the top of the column with small portions of 5% sodium sulfate solution and elute the uranine with this solution. Render the eluate alkaline with sodium hydroxide and determine the uranine spectrophotometrically at 490 nm.

1-(Phenylazo)-2-naphthol in D&C Red No. 31 (11)

Extract 0.1-gram of sample in a Soxhlet extractor with approximately 100 ml of benzene until the leachings are colorless or have a slight persistent bleed. Transfer the extract to a separatory funnel and wash with 1% NaOH solution until the washings are colorless. Wash the benzene layer with several 50-ml portions of water to remove the excess NaOH. Evaporate the benzene solution to dryness. Dissolve the residue in alcohol and dilute with alcohol to 100 ml or 200 ml, depending on the intensity of the color. Determine the amount of subsidiary color present spectrophotometrically.

Chromotrope 2R in D&C Red No. 33 (31)

Slurry Solka Floc BW40 in water and pack it into a 100 cm × 2-cm chromatographic column to a settled height of 50 cm, and then wash the column with 20% sodium chloride solution. Prepare a 0.1% aqueous sample solution. To a 5-ml aliquot add 20 ml of a 20% sodium chloride solution; transfer it to the column with the sodium chloride solution. Elute the column with 10% sodium chloride solution containing 1% ammonium hydroxide. Chromotrope 2R elutes first followed by the parent compound. (An unknown blue dye frequently remains at the top of the column.) When the Chromotrope band is separated by about 10 cm from the main band, change the eluant to a 5% sodium chloride solution containing 1% ammonium hydroxide. Neutralize the appropriate fractions with acetic acid and determine Chromotrope 2R spectrophotometrically at the absorption maximum near 508 nm.

Triethylrhodamine and Other Subsidiary Colors in D&C Red No. 37

Use the TLC method described for D&C Red No. 19 (p. 245).

Subsidiary Dyes in D&C Yellow No. 10 (30)

Weigh about 200 g of Celite 545 (Fisher C-212) into a Petri dish and place in a desiccator containing 25 ml of General Electric SC-77 Dri Film. Let stand until the Celite no longer wets when mixed with water (3-24 hr).

Prepare a 2.5% solution of triisooctylamine in n-butanol and equilibrate with an equal volume of (1 + 9) HCl. Allow the layers to separate.

Thoroughly mix 18 ml of the *n*-butanol layer from the solution described above with 20 g of the silane-treated Celite and pack the mixture into a 1 in.-diameter glass column. Wash the column with mobile phase (the aqueous-acidic layer from that described above) and allow the column to drain just to the surface of the packing.

Transfer 1 ml of a 1% sample solution in mobile phase to the column and elute with mobile phase. Collect 20-ml fractions and determine the individual colorants spectrophotometrically.

When present, the subsidiary colors emerge in the following order:

2-(2-Quinolyl-6-sulfonic acid)-1,3-indandione-5-sulfonic acid.

2-(2-Quinolyl-8-sulfonic acid)-1,3-indandione-5-sulfonic acid.

2-(2-Quinolyl-6,8-disulfonic acid)-1,3-indandione.

2-(2-Quinolyl-6-sulfonic acid)-1,3-indandione.

2-(2-Quinolyl-8-sulfonic acid)-1,3-indandione.

Quinizarin Green (D&C Green 6) in D&C Violet No. 2 (11)

Spot as a band about 1 mg of sample as a chloroform solution onto a 20 cm × 20-cm TLC plate coated with silica gel G. Allow the plate to air dry and then develop using hexane-trichloroethylene-diethylamine (30:10:5) until the solvent front nears the top of the plate.

The colors appear in the descending order:

Quinizarin Green.

D&C Violet No. 2

Related Colorants in Annatto (12)

Paper-chromatographic Method (20-23) (see McKeown, G. G. under Resolution of Mixtures, p. 266).

Thin-layer-chromatographic Methods:

Method A—(see Francis, B. J., and Ramamurthy, M. K., and Bhalerao, V. R. under Resolution of Mixtures, p. 263 and 272).

Method B—Chloroform extracts of annatto (6): Coat 2 in. × 15-in. glass plates with silica gel containing 12% gypsum as a binder. Dry the plates for 1 hr at 100°C.

Streak 0.2 ml of a 0.2-1% chloroform solution of pigment onto the plate and elute using an eluant of acetic acid-chloroform-acetone (1:50:50).

Method C—Vegetable oil or propylene glycol extracts of annatto (28): Apply sample to a 20 cm × 20-cm cellulose thin-layer plate (Eastman Chromagram Sheet, 6065) and elute briefly with cyclohexane to separate the oil from the pigments. Air dry the plate then elute in a paper-lined, eluant-equilibrated tank containing cyclohexane-chloroform-acetic acid (65:5:1).

Method D—Aqueous-alkaline extract of annatto (28): Apply sample to an untreated 10 cm X 20-cm Merk silica gel thin-layer plate. Air dry and then elute in a paper-lined, eluant-equilibrated tank using chloroform-absolute ethanol-acetic acid (68:2:1).

Fractionation of Caramel by Gel Filtration (25)

Caramel is separated in an aqueous medium into high- and low-molecular-weight components on Sephadex G-25 and G-50.

Separation of α- and β-Carotene (see Usher, C. D. et al. under Isolation of Colorants, Dairy Products, p. 301)

Lumiflavin in Riboflavin (34)

Shake 20 ml of chloroform with 20 ml of water for 3 min. Allow the layers to separate. Drain the chloroform and repeat the extraction twice with 20-ml portions of water. Filter the washed chloroform through dry filter paper. Shake the filtrate for 5 min with 5 g of powdered anhydrous sodium sulfate. Allow the mixture to stand for 2 hr and then decant or filter the clear chloroform.

Shake 0.025 g of riboflavin with 10 ml of washed chloroform for 5 min and filter. A color in the filtrate similar to that of potassium dichromate solution indicates the presence of lumiflavin. Acceptable riboflavin should have no more color than 0.0003 N potassium dichromate solution when viewed under identical circumstances.

Curcuma Zedoaria and Curcuma Aromatica in Turmeric (32)

Prepare a 1% benzene solution of the essential oils steam-distilled from turmeric. Spot 10 μl of this solution onto a 10 X 20 cm glass plate coated with 500 μm of silica gel G (the plate should have been activated for 1 hr. at 110°C), then elute in a solvent-saturated tank using ethyl acetate-n-hexane (3:17).

Spray the air-dried plate with concentrated sulfuric acid-nitric acid (50:0.5), allow it to stand 1 minute, then examine visually under daylight and ultraviolet light (365 nm). Then spray with anisaldehyde-sulfuric acid-ethanol-glacial acetic acid (0.5:0.5:9.0:0.1) and similarly examine.

Compare versus standards prepared simultaneously.

Related Colorants in Saffron

Extraction/TLC procedure (27)

Boil 1.00 g of sample in 20 ml of water, evaporate the filtered extract to a few drops, and chromatograph the solution on Whatman No. 1 paper versus safflower and other knowns using the following eluants:

1 ml 0.88 ammonia + 99 ml water

2.5% aqueous sodium chloride

80g phenol + 20g water

5 ml 0.88 ammonia +95 ml water + 2g trisodium citrate

Thermomicro Separation (TAS)/TLC method (35).

Weigh the powdered sample into a glass cartridge fitted with a capillary tube, heat the tube at an appropriate temperature (ca. $200°C$), and collect the distillate on a Silica Gel HF_{254} plate. Elute with benzene-chloroform (80:20) and examine under short-wavelength UV light.

BIBLIOGRAPHY

1. BELL, S. J. JAOAC 56, 947-949 (1973). Lower Sulfonated Subsidiary Colors in FD&C Blue No. 1.

2. BELL, S. J. JAOAC 57, 961-962 (1974). Thin Layer Chromatographic Determination of Subsidiary Dyes in D&C Red No. 19 and D&C Red No. 37.

3. BELL, S. J. JAOAC 58, 717-718 (1975). Thin Layer Chromatographic Separation and Spectrodensitometric Determination of Higher and Lower Sulfonated Subsidiary Dyes in FD&C Yellow No. 6.

4. BELL, S. J. JAOAC 59, 1294-1311 (1976). Preparation and Spectral Compilation of FD&C Red No. 40 Intermediates and Subsidiary Dyes.

5. British Standards No. 3210, 1960. Methods for the Analysis of Water-Soluble Coal Tar Dyes Permitted in Foods.

6. DENDY, D. A. V. East Afric. Agric. Forest J. 32, 126-132 (1966). Annatto, The Pigment of Bixa Orellana.

7. DOLINSKI, M. JAOAC 35, 421-423 (1952). Lower Sulfonated Dyes in FD&C Yellow No. 5.

8. DOLINSKY, M. JAOAC 36, 798-802 (1953). Report on Lower Sulfonated Dyes in FD&C Blue No. 1.

9. DOLINSKY, M. JAOAC 37, 805-808 (1954). Report on Subsidiary Dyes in FD&C Colors. I. Higher Sulfonated Dyes in FD&C Yellow No. 6.

10. DOLINSKY, M. JAOAC 38, 359-365 (1955). Lower Sulfonated Dye in FD&C Blue No. 1.

11. Food and Drug Administration, Washington, D.C., private communication.

12. FREISE, F. W., Pharm. Zentralhalle Deutschland 76, 4 (1935). Approximate Analysis of Bixa Orellana Seeds.

13. GRAICHEN, C., HEINE, K. S., Jr. JAOAC 37, 905-912 (1954). Studies on Coal-Tar Colors. XVI. FD&C Red No. 4.

14. GRAICHEN, C., MOLITOR, J. JAOAC 42, 149-160 (1959). Studies on Coal-Tar Colors. XXII. 4,5-Dibromofluorescein and Related Bromofluoresceins.

15. HANIG, I., KOCH, L. JAOAC 46, 1010-1013 (1963). Quantitative Paper Chromatography of D&C Red No. 21 (Tetrabromofluorescein).

16. KOCH L. JAOAC 29, 237-240 (1946). Report on Subsidiary Dyes in D&C Colors.

17. KOCH, L. JAOAC 41, 249-250 (1958). Report on Subsidiary Dyes in D&C Colors: Uranine in D&C Red No. 22 (Eosine).

18. KOCH, L. JAOAC 46, 344-346 (1963). Subsidiary Dyes in D&C Colors (4-Toluene-Azo-2-Naphthol-3-Carboxylic Acid in D&C Red Nos. 6 and 7).

19. MARMION, D.M. JAOAC 54, 131-136 (1971). Analysis of Allura Red AC Dye (A Potential New Color Additive).

20. MC KEOWN, G. G. JAOAC 44, 347-351 (1961). Paper Chromatography of Bixin and Related Compounds.

21. MC KEOWN, G. G., MARK, E. JAOAC 45, 761-766 (1962). The Composition of Oil-Soluble Annatto Food Colors.

22. MC KEOWN, G. G. JAOAC 46, 790-796 (1963). Composition of Oil-Soluble Annatto Food Colors. II. Thermal Degradation of Bixin.

23. MC KEOWN, G. G. JAOAC 48, 835-837 (1965). Composition of Oil-Soluble Annatto Food Colors. III. Structure of the Yellow Pigment Formed by the Thermal Degradation of Bixin.

24. MOLITOR, J. C. JAOAC 50, 1198-1199 (1967). Determination of 1-Phenylazo-2-Naphthol in D&C Red No. 17.

25. OERSI, F. Nahrung 13, 53-57 (1969). Fractionation of Caramel by Gel Filtration.

26. Official Methods of Analysis, 11 ed., Association of Official Analytical Chemists, Washington, D. C., 1970, p. 597.

27. PARVENEH, V. J. Assoc. Publ. Analysts 10, 31-32 (1972). Assessment of the Purity of Saffron Colour.

28. REITH, J. F., Gielen, J. W. J. Food Sci. 36, 861-864 (1971). Properties of Bixin and Norbixin and the Composition of Annatto Extracts.

29. RICHIE, C. D., WENNINGER, J. A., JONES, J. H. JAOAC 42, 720-724 (1959). Studies on Coal-Tar Colors. XXII. D&C Red No. 19: Identification and Determination of Triethylrhodamine and o-(2-Hydroxy-4-diethylaminobenzoyl) Benzoic Acid in Commercial Samples of D&C Red No. 19.

30. RICHIE, C. D., WENNINGER, J. A., JONES, J. H. JAOAC 44, 733-739 (1961). Studies on Coal-Tar Colors. XXV. D&C Yellow No. 10.

31. SCLAR, R. N. JAOAC 36, 930-936 (1953). Studies on Coal-Tar Colors. XIII. D&C Red No. 33.

32. SEN, A. R., GUPTA, P. S., DASTIDAR, N. G. Analyst 99, 153-155 (1974). Detection of *Curcuma zedoaria* and *Curcuma aromatica* in *Curcuma longa* (Turmeric) by Thin-Layer Chromatography.

33. SINGH, M. JAOAC 53, 250-251 (1970). Determination of 5,7'-Disulfo-3,3'-Dioxo-$\Delta^{2,2'}$-Biindoline, Disodium Salt, and 5-Sulfo-3,3'-Dioxo-$\Delta^{2,2'}$-Biindoline, Sodium Salt in FD&C Blue No. 2.

34. *Specifications for Identity and Purity of Food Aditivies*, Vol. 2 Food Colors, Food and Agriculture Organization of the United Nations, Rome, 1963, p. 58.

35. STAHL, E., WAGNER, C. J. Chromatog. *40*, 308 (1969). TAS-Method for the Microanalysis of Important Constituents of Saffron.

36. STEIN, C. JAOAC *50*, 1199-1201 (1967). Determination of 4(4-Chloro-2-sulfo-5-tolylazo)-1-naphthol in D&C Red Nos. 8 and 9.

37. STEIN, C. JAOAC *52*, 34-40 (1969). Subsidiary Colors in FD&C Blue No. 1.

38. STEIN, C. JAOAC *53*, 26-28 (1970). TLC and Spectrophotometric Determination of 1[4-(2,5-Dimethoxyphenylazo)-2,5-dimethoxyphenyl-azo]-2-naphthol and 1,1'-(2,2',5,5'-tetramethoxy-4,4'-biphenylenebi-sazo)-di-2-naphthol in Citrus Red No. 2.

39. STEIN, C. JAOAC *53*, 677-681 (1970). Subsidiary Colors in FD&C Green No. 3.

40. STEIN, C., COX, E. A. JAOAC *56*, 1188-1190 (1973). Determination of Sodium Salt of 1-(*p*-Toluidino)-4-(*o*-sulfo-*p*-toluidino) Anthraquinone and the Sodium Salt of 1-Hydroxy-4-(*o*-sulfo-*p*-toluidino) Anthra-quinon in D&C Green No. 5.

41. Unpublished data.

42. WENNINGER, J. A., JONES, J. H., DOLINSKY, M. JAOAC *43*, 805-809 (1960). Studies on Coal-Tar Colors. XXIV. FD&C Red No. 4.

43. WOZNICKI, E. J., Private communication.

PART C RESOLUTION OF MIXTURES AND ANALYSIS OF COMMERCIAL PRODUCTS

Chapter **14** Resolution of Mixtures

Frequently, no single dye is capable of producing a desired shade, so mixtures or "secondary colors" are used. The determination of the nature and the amount of individual colorants in such mixtures presents a special problem. If the mixtures are not too complicated and if the component colorants have sufficiently different spectra not masked or distorted by the presence of excipients, NMR, IR and visible spectrometry can be used to analyze them directly. Rarely, though, is the analyst blessed with such ideal conditions, and most often separation of the mixtures into their component parts is necessary for a successful analysis.

The literature is teeming with examples of the analysis of such mixtures using most every separations technique available. The method to use, of course, is dictated by the needs of the analyst and the equipment available to him. Electrophoresis, TLC, and paper chromatography are relatively simple methods and require a minimum of equipment, applied time, and technique but yield only semiquantitative results even after extensive calibration. These methods are best used on a "go-no-go" basis versus an acceptable standard. Conventional gravity-column chromatography provides greater precision and accuracy but usually at the expense of longer analysis times and more attention on the part of the analyst. Solvent-solvent extraction is simple but in most cases inadequate, whereas counter-current distribution is a powerful enough tool, but its use generally requires too much sophistication and time. To date, little use has been made of gas chromatography since few colorant mixtures are amenable to separation by this method. The technique that presently offers the most promise as both a rapid and quantitative method is high-pressure liquid chromatography (HPLC). Although still in its infancy, HPLC has already proven to be a powerful weapon for the determination of impurities in color additives and it's undoubtedly only a question of time before it is routinely applied to the separation of mixtures of colors. Unfortunately, the instrumentation needed is quite costly and much remains to be learned about its use.

Procedures exist for separating groups of dyestuffs having similar properties or applications such as the carotenoids, the water-soluble food colors, and lipstick dyes, but no one method has yet been written that separates all the permitted color additives and most certainly none ever will. The majority of extant methods are deficient since they either deal with one or more colors no longer permitted or fail to consider newer related colorants or both. However, these methods can frequently be modified to meet one's needs or used as the starting point for developing a better one. A number of them are summarized in the following bibliography. Others can be found under the discussion of the isolation of colorants from commercial products where they are used as a means of identification.

BIBLIOGRAPHY

ALDRED, J. B. J. Assoc. Public Analysts 3, 79-82 (1965). The Identification of Violet BNP and Its Distinction from Other Violet Colors. Thin-layer chromatography is used to separate various violet colors. The method uses Kieselgel G plates and an eluant composed of iso-BuOH-EtOH-H$_2$O (2:2:1, v/v). The R_f values given are: Violet BNP (CI 42580), 0.38; Violet 5BN (CI 42650), 0.46; FD&C Violet No. 1 (CI 42640), 0.46; Methyl Violet (CI 42535), 0.58; and Acid Violet 4BN (CI 42561), 0.42.

ANWAR, M. H., NORMAN, S., ANWAR, B., LAPLACA, P. J. Chem. Ed. 40, 537-538 (1963). Electrophoretic Study of Synthetic Food Dyes. Thoroughly wet a cellulose acetate strip with a buffer solution consisting of equal parts of 0.1 M sodium acetate and isopropyl alcohol adjusted to pH = 4.6 with acetic acid. Blot between absorbent paper and spot a solution of the dye mixture on the strip. Separate by electrophoresis in the buffer solution using an applied voltage of 270 V. Fading and oxidation of triphenylmethane dyes can be minimized by conducting the separation in the dark under an inert atmosphere. See Fig. 16 for order of separation.

BAINBRIDGE, W. C. JAOAC 29, 240 (1946). Report on the Analysis of Mixtures of D&C Red No. 7 and D&C Red No. 10. Transfer a 0.25-g sample to a 250-ml Erlenmeyer flask. Add 100 ml of methyl Cellosolve and 5 ml of hydrochloric acid. Connect the flask to an air condenser and gently reflux the mix until it is in solution. Buffer the solution with 10 g of sodium bitartrate dissolved in 75 ml of boiling water and titrate with 0.1N titanous chloride solution. Record the results (in ml) as the combined titer for the two dyes. Dissolve a second 0.25-g sample in 100 ml of methyl Cellosolve and transfer the solution to a 500-ml separatory funnel. Transfer any residual dye with 10-ml portions of ethyl ether. Bring the ether volume to 150 ml and add 250 ml of 10% sodium chloride solution. Withdraw the lower layer and extract it with a second 75-ml portion of ethyl ether. Withdraw and discard the lower layer. Combine the ether extracts and wash them with 30-ml portions of water until the washings are colorless; discard the washings. Transfer the ether layer to a 250-ml Erlenmeyer flask with two 10-ml portions of ethyl ether. Evaporate the solution to dryness, add 100 ml of methyl Cellosolve, and heat into solution. Buffer with 10 g of sodium bitartrate in 75 ml of boiling water, heat to boiling, and titrate with 0.1N titanous chloride solution. Calculate as D&C Red No. 10. Subtract the milliliters of titer for the D&C

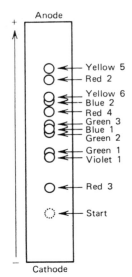

Anode

Yellow 5
Red 2
Yellow 6
Blue 2
Red 4
Green 3
Blue 1
Green 2
Green 1
Violet 1

Red 3

Start

Cathode

**Figure 16 Electrophoretic Separation of Food Colors
[From J. Chem. Ed. _40_ 537 (1963)]**

Red No. 10 from the milliliters of titer for the two dyes and calculate the difference as D&C Red No. 7.

BANDELIN, F. J., TUSCHHOFF, J. V. J. Am. Pharm. Assoc. 49, 302-304 (1960). Paper Chromatography of Some Certified Dyes. Common certified dyes are separated on paper using 2% aqueous NH_4OH containing 2% iso-BuOH.

BARRETT, J. F., RYAN, A. J. Nature _199_, 372-373 (1963). Thin-Layer Chromatography of Some Food Colors on Silica Gel. Describes the use of TLC on silica gel for the separation of sulfonated dyestuffs including FD&C Red Nos. 3 and 4, FD&C Yellow Nos. 5 and 6, and FD&C Blue No. 2. The solvents studied were 9:1 EtOH-NH_4OH; 5:2:1 acetoacetic ester-MeOH-NH_4OH; 5:2:1 acetoacetic ester-C_5H_5N-NH_4OH; 10:10:1 AmOH-EtOH-NH_4OH, and 7:3:1 EtOAc-C_5H_5N-H_2O.

BAYER, J. Acta Pharm. Hung. 31B, Suppl. 51-58 (1961). Evaluation of Paper Chromatograms of Pharmaceutical Preparations by the Densitometer. Use of a densitometer to quantitate the components of drugs separated by paper chromatography. Includes the determination of FD&C Yellow No. 5.

BOLLINGER, H. R., KOENIG, A., SCHWIETER, U. Chimia _18_, 136 (1964). Thin-Layer Chromatography of Carotenes. Six carotenes are separated on activated MgO layers with petroleum ether (boiling range 90-110°)-benzene (50:50). For clearer separations of ϵ-, α-, and β-carotenes use 9:10 ether-benzene and for mixtures of δ-, γ-carotene and lycopene use 10:90 ether-benzene.

Solvents: S_1 = Light petroleum (boiling range 90-110°)-benzene (9:1).

S_2 = Light petroleum (boiling range 90-110°)-benzene (5:5).

S_3 = Light petroleum (boiling range 90-110°)-benzene (1:9).

Thin layer: "Darlington" light magnesium oxide, activated for 1 hr at 120°.

Compound	S_1	S_2	S_3
		R_f	
ε-Carotene	0.47	0.70	0.84
α-Carotene	0.26	0.66	0.80
β-Carotene	0.11	0.49	0.74
δ-Carotene	0.00	0.20	0.55
γ-Carotene	0.00	0.11	0.41
Lycopene	0.00	0.00-0.02	0.13

BROWN, J. C. JSDC 85, 137-146 (1969). The Chromatography and Identification of Dyes. A general description of TLC, paper chromatography, and electrophoresis as tools for the separation of dyestuffs.

CAMACHO, I., DUARTE, M. I. Rev. Colomb. Cienc. Quim. Farm. 1, 5-32 (1971). Identification of Dyes Used in Lipstick in Columbia. Thin-layer and paper chromatography are used to separate and identify lipstick colorants.

CALZOLARI, L., COASSINI, L., LOKAR, L. Rass. Chim. 15, 49-60 (1963). Partition Paper Chromatography of Food Dyes.

CANUTI, A., MAGRASSI, B. L. Chim. Ind. (Milan) 46, 284-286 (1964). Food Colors. I. Application of Thin-Layer Chromatography for Determining Added Artificial Food Colors. Artificial acidic food colors permitted in Italy are separated using BuOH-H$_2$O-EtOH-NH$_4$OH (50:25:25:10) as the eluant.

CELAP, M. B., JANJIC, T. J., JEVTIC, V. D. Mikrochim. Ichnoanal. Acta 4, 647-751 (1965). Application of the Ring-Oven Method to the Determination of Dyes. The Weisz ring-oven method was applied to the separation and determination of various dyes including FD&C Red No. 3, FD&C Blue No. 2 and FD&C Yellow No. 5.

CERESA, G. Ann. Sper. Agrar. (Rome) 13, 545-571 (1959). Identification of Synthetic Dyes Used in the Food Industry. The 13 food dyes permitted by Italian legislation are separated by one-dimensional paper chromatography using EtOH-BuOH-H$_2$O (50:25:25) as eluant. Those dyes not separated by this mixture are resolved using normal HCl or by adding 10 ml of concentrated NH$_4$OH to 100 ml of the mixture described above.

CHAPMAN, W. B., OAKLAND, D. J. Assoc. Publ. Analysts 6, 124-128 (1968). Differentiation of Blue Colouring Matters in Food and Drugs With Particular Reference to Blue VRS (CI Acid Blue 1) and Patent Blue V (CI Acid Blue 3). Thin-layer chromatography and paper electrophoresis were used to differentiate 14 blue colors. Using TLC, colors applied as 0.1% aqueous solution (5 μl) to layers of Kieselgel G (250 μm) activated at 105° for 2 hr, which were then developed with fresh isopropyl alcohol-concentrated aqueous NH$_3$ (4:1) for about 2 hr, or to layers of Cellulose CC41 dried overnight, which were developed with isobutyl alcohol-H$_2$O-ethanol-concentrated aqueous NH$_3$(25:25:50:2) for 1-1.5

hr. Using paper electrophoresis, colors were applied in aqueous solution to Whatman 3MM paper (25 cm $\times$ 10 cm) and dried. The paper was saturated with electrolyte, $0.1N$ aqueous NH_3 or 0.25 M $Na_2B_4O_7$ buffer (pH = 9.2), and a 6-mA current was passed for about 2 hr.

CHIANG, H. C., CHEN, C. H. J. Pharm. Sci. *59*, 266-267 (1970). Polyamide – Silica Gel Layer Chromatography of Yellow Food Dyes. Various yellow colorants including FD&C Yellow Nos. 5 and 6 were separated on mixed polyamide-silica gel plates using either MeOH-23% NH_4Cl-CHCl$_3$ (30:20:1.3) or iso-BuOH-EtOH-0.45% NaCl (3:5:1) as the eluant. Plates are prepared by dissolving 8 g of polyamide chip (Nylon 6, type 1022B, UBE Industrial Ltd., Osaka, Japan) in 80 ml of 90% formic acid and then adding 20 ml of distilled water and warming and stirring the mixture to form a homogeneous solution. The mixture is then cooled to room temperature and 52 g of Silica Gel G (E. Merck) is added. Coated glass plates are air dried for 3 hr and then heated at $100°C$ for 30 min. Separations are better than those obtained on plates made from either polyamide or silica gel G alone.

CHIANG, H. C. J. Chromat. *40*, 189-190 (1969). Polyamide-Silica Gel Thin-Layer Chromatography of Red Food Dyes. FD&C Red Nos. 3 and 4, D&C Red No. 19 and other red colorants are separated on plates coated with a mixture of polyamide and silica gel G. Plates are made by dissolving 7 g of polyamide (ϵ-polycaprolactam CM 1007S, Toyo Rayon Co., Tokyo, Japan) in 100 ml of warm 75% formic acid, adding 52 g of Silica Gel G and coating the mixture onto plates that are air dried for 3 hr and then heated at $100°C$ for 30 min. Eluants studied include iso-PrOH-5% NH_4Cl (8:3), ether-iso-PrOH-5% NH_4Cl (1:2:2), CHCl$_3$-iso-PrOH-5% NH_4Cl-glacial AcOH (1:5:2:1), n-BuOH-EtOH-5% Na citrate (6:4:3) and CHCl$_3$-iso-PrOH-5% NaCl-glacial AcOH (5:25:5:1).

CHIANG, H. C., LIN, S. L. J. Chromatog. *44*, 203-204 (1969). Polyamide-Kieselguhr Thin-Layer Chromatography of Yellow Food Dyes. Eight yellow colorants, including FD&C Yellow Nos. 5 and 6, are separated on TLC plates made from Nylon 6 and Kieselguhr G. The plates are made by dissolving 10 g of polyamide chip (Nylon 6, type 1022 B, UBE Industries Ltd., Osaka, Japan) in 80 ml of 90% HCO_2H, adding 20 ml of water and warming ($<40°C$) and stirring the solution until homogenous. The mixture is then cooled and 40 g of Kieselguhr G (E. Merck) is mixed in. Glass plates are then coated, air dried for 3 hr, and heated at $100°C$ for 30 min. Using MeOH-$(CH_3)_2$CO-H_2O-30% AcONa-ethylenediamine (10:10:20:5:2) as eluant, R_f values were 0.66, 0.11, 0.05, 0.91, 0.38, 0.31, 0.73, and 0.53, respectively, for Naphthol Yellow S, Yellow AB, Yellow OB, FD&C Yellow No. 5, FD&C Yellow No. 6, Metanil Yellow, auramine, and picric acid. With EtOH-H_2O-Et$_2$O-5% NH_4Cl-ethylenediamine (15:15:10:5:2) the respective R_f values were 0.69, 0.35, 0.23, 0.88, 0.81, 0.53, 0.76, and 0.60.

CIELESZKY, V., SOHAR, J. Koloriszt. Ertesito 6, 358-373 (1964). The Use of Chromatographic Methods for the Separation, Identification, and Purification of Synthetic Food Dyes.

COTTER, R. L. Paper No. 41, 1975 Pittsburgh Conference. The Use of High Pressure Liquid Chromatography for the Analysis of Food, Drug and Cosmetic Colorings. Describes the use of a reverse-phase column (Micro Bondapak C-18, Waters Associates Inc., Milford, Mass.) for resolving mixtures of colorants and detecting impurities in colorants.

CRIDDLE, W. J., MOODY, G. J., THOMAS, J. D. R. J. Chromatog. *16*, 350-359 (1964). Thin Film Electrophoresis. Part I. The Electrophoretic Behavior of Coal-Tar Food Colours on Paper and Thin Films. Twenty-six colorants permitted in foods in the United Kingdom were subjected to electrophoresis for 1 hr at 200 V on thin layers of Kieselguhr, alumina G, silica gel G, and Whatman No. 1 paper. Thin films were prepared from slurries of 30 g of adsorbent in 60 ml of H_2O spread on 20 cm $\times$ 17.5 cm plates and dried at $105°C$. The electrolytes used were normal HOAc, $0.1N$ NH_4OH and buffer solutions of pH-4, 6, 8, and 9.2.

CRIDDLE, W. J., MOODY, G. J., THOMAS, J. D. R. Nature *202*, 1327 (1964). Use of Thin Films for Electrophoresis of Coal-Tar Food Colours. Ten colorants, including FD&C Yellow No. 5, FD&C Red No. 3 and FD&C Blue No. 2, were separated by electrophoresis using thin layers of alumina, Kieselguhr, and silica gel. Results using Kieselguhr and an eluant of 0.05 M borax (pH = 9.18) at a potential of 200 V were compared with those similarly obtained on Whatman No. 1 paper.

CROSSLEY, J., THOMAS, J. D. R. Analyst *83*, 462-465 (1958). The Separation of Some Coal-Tar Food Colours by Paper Electrophoresis.

Apparatus: E. E. L. electrophoresis apparatus.

Substrate: Whatman No. 1 filter paper

Electrolytes:

1. Normal acetic acid.
2. pH-4 Buffer: 6 ml of $0.1N$ NaOH + 750 ml of 0.1 M monopotassium phthalate diluted to 1.5 liters.

Electrolyte	Distances Moved (in mm)					
	1	2	3	4	5	6
Current density (in mA per 5 cm)	0.6	1.7	1.7	2.0	2.0	1.7
Time (in hr)	2	1.75	2	2	1.5	2
FD&C Red No. 3	0	0	0	0	9	0
FD&C Yellow No. 5	130	74	23	15	103	83
FD&C Blue No. 2	52	35	9	8	38	10
Ponceau MX			0	0	0 and 16	
Ponceau 4R			26	30	100	
Ponceau 3R			0	0	11	

3. pH-6 Buffer: 85.5 ml of $0.1N$ NaOH + 750 ml of 0.1 M monopotassium phthalate diluted to 1.5 liters.
4. pH-8 Buffer: 702 ml of $0.1N$ NaOH + 750 ml of 0.1 M monopotassium phthalate diluted to 1.5 liters.
5. 1% Sodium tetraborate.
6. $0.1N$ NH_4OH.

CUZZONI, M. T. Farmaco (Pavia) Ed. pract. *15*, 752-758 (1960). Food Additives Permitted in Italy. Electrophoretic Determination of Synthetic Dyes.

DAMIANI, C. Ind. Aliment. *4*, 41-48 (1965). Identification of Water-Soluble Food Colors by Paper Chromatography Using a Pyridine Based Eluant. Mixtures of colorants are separated using EtOH-BuOH-Pyridine-Water (5:35:30:30) as the eluant.

DAVIDEK, J., JANICEK, G. Qualitas Plantarum et Materiae Vegetabiles *16*, 253-257 (1968). Thin Layer Chromatographic Separation of Fat Soluble and Water Soluble Food Dyes. Recommends a number of chromatographic systems. For fat-soluble colorants: aluminum oxide plates, petroleum ether-CCl_4; or paraffin-impregnated starch plates, MeOH-H_2O-AcOH (16:3:1). For water-soluble dyes: polyamide powder plates, NH_4OH-MeOH-H_2O (5:15:80).

DAVIDEK, J., DAVIDKOVA, E. J. Chromatog. *26*, 529-531 (1967). The Use of Polyamide in Analyses of Water Soluble Food Dyes. IV. Thin-Layer Chromatographic Separation of Water Soluble Food Dyes. Various combinations of NH_4OH-MeOH-H_2O were used to resolve colorants on polyamide powder. The best separation was achieved using NH_4OH-MeOH-H_2O (5:15:80). Plates were prepared by homogenizing 12 g of polyamide powder (Chemical Fabrics Lovosice Workshop Rudnik, Czechoslovakia) with 40 ml of MeOH, coating plates with a 0.2-mm layer of the mixture and then drying them at $40°C$ for 30 min.

Color	R_f
Amaranth	0.77
Azorubin	0.78
Echrot	0.42
Cochenillerot	0.34
FD&C Red No. 3	0.21
FD&C Yellow No. 6	0.72
FD&C Yellow No. 5	0.88
Naftolgelb	0.62
FD&C Blue No. 2	0.70
Brillantschwarz	0.60

DAVIDEK, J., JANICEK, G. J. Chromatog. *15*, 542-545 (1964). Chromatography of Fat-Soluble Food Dyes on Thin Starch Layers With Stationary Non-Polar Phases.

DAVIDEK, J., POKORNY, J., JANICEK, G. Z. Lebensm. Forsch. *116*, 13-19 (1961). Detection and Determination of Fat Soluble Food Colors with the Aid of Thin-Layer Chromatography on Aluminum Oxide. Of the solvents studied, petroleum ether, CCl_4, and mixtures of these gave the best results.

DE GORI, R., CANTAGALLI, P. Boll. lab. chim. provinciali *8*, 23-26 (1957). Extraction and Identification of Synthetic Coloring Added to Food. Naphthol Yellow S (Ext. D&C Yellow No. 7), FD&C Yellow No. 5, and croisidine are separated on Whatman No. 1 paper using EtOH-H_2O-BuOH-NH_4OH (1:1:2:1) as eluant. The R_f values are 0.35, 0.05, and 0.89, respectively.

DE GORI, R., GRANDI, F. Boll. lab. chim. provinciali 9, 168-177 (1958). Separation and Identification of the Artificial Dyes Authorized for Alimentary Use by Decree of the High Commissioner of Hygiene and Sanitation of 22 December 1957. A discussion of the separation of a variety of colorants, including FD&C Yellow No. 5, FD&C Yellow No. 6, D&C Yellow No. 10, and FD&C Blue No. 2 on SS2043A paper using EtOH-BuOH-H_2O (20:25:25).

PLA-DELFINA, J. M. J. Soc. Cosmet. Chemists *13*, 214-244 (1962). Systematic Identification of Food, Drug and Cosmetic Azo Dyes.

DICKES, G. J. J. Assoc. Public Analysts *3*, 49-52 (1965). Separation of Synthetic Water-Soluble Coloring Matters by Thin-Layer Chromatography. Separations were performed on Kieselgel G layers (250 μ, 20 cm × 10-cm glass plates heated at 160 °C for 1.5 hr), using iso-PrOH-NH_4OH-H_2O (10:1:1) or saturated KNO_3 as the eluant.

DOBRECKY, J., CARNEVALE BONINO, R. C. D'A DE. Revta Asoc. bioquim. argent. *32*, 12-15 (1967). Chromatographic Separation of Food Dyes Permitted in Argentina. FD&C Red Nos. 2 and 3, FD&C Yellow Nos. 5 and 6, FD&C Blue No. 2, and several other colorants are separated by radial paper chromatography using 0.1 M HCl as the eluant.

DOBRECKY, J., CARNEVALE BONINO, R. C. D'A. DE. Revta Asoc. bioquim. argent. *32*, 16-19 (1967). Paper Chromatographic Separation of Dyes Permitted for Foods, Drugs and Cosmetics in the U.S.A. Food colorants are separated on paper by two-dimensional chromatography using the organic phase of a mixture of BuOH-AcOH-H_2O (4:1:5) and then a solution containing 2 g of EDTA and 5 ml of 25% aqueous NH_3 in 100 ml of H_2O.

DOBRECKY, J., CARNEVALE BONINO, R. C. D'A. DE. Revta Asoc. bioquim. argent. *32*, 139-143 (1967). Paper Chromatographic Separation of Dyes Permitted by the European Economic Community. Fourteen dyes permitted in foods and drugs were separated by two-dimensional paper chromatography using BuOH-AcOH-H_2O (4:1:5) and 0.1N HCl as the solvents. Red and blue dyes in the series were separated by circular-paper chromatography using a 2% solution of EDTA in 5% aqueous NH_3 as solvent.

DOBRECKY, J., CARNEVALE BONINO, R. C. D'A. DE. Revta Asoc. bioquim. argent. *36*, 143-145 (1971). Separation, Identification and Determination of Six Dyes Not Permitted (in Argentina) in Medicines or Foods. Ponceau 2R, Ponceau SX (FD&C Red No. 4), Rhodamine B (D&C Red No. 19), Naphthol Yellow S (Ext. D&C Yellow No. 7), Malachite

Green, and Auramine were separated by two-dimensional paper chromatography on Whatman No. 1 paper. The first developing solvent was 2% EDTA (disodium salt) in 5% aqueous NH$_3$. The second was H$_2$O.

DOBRECKY, J., CARNEVALE BONINO, R. C. D'A. DE. Rev. Farm. *114*, 21-22 (1972). Paper Chromatography of Colouring Agents Used in Drugs and Cosmetics.

EGGER, K. Chromatographic Symposium II 1962. Société Belge des Sciences Pharmaceutiques, Bruxelles, 1963, p. 75. Eleven carotenoids, including canthaxanthin, β-apo-8'-carotenal, and β-carotene, were separated on thin layers of paraffin-impregnated Kieselguhr using various combinations of acetone and 95% ethanol.

EGGER, K., VOIGT, H. Z. Pflanzenphysiol, *53*, 64-71 (1965). Carotenoid Separation on Thin Layers of Polyamide. Thirty-one carotenoids were chromatographed on thin layers of polyamide using nine solvent systems. The best eluants were isooctane-MeOH-MeCOEt (80:10:10) and MeCOEt-MeOH-H$_2$O (30:30:10).

ESPADA, A. M. Inform. Quim. Anal. *24*, 63-67 (1970). Chromatographic Identification of Dyes Used in Carbonated Beverages. Eight colorants, including FD&C Red No. 4, FD&C Yellow Nos. 5 and 6, FD&C Blue No. 1, and caramel, were separated and identified by paper chromatography using Whatman No. 1 paper and an eluant containing 4.8 ml of 28% NH$_4$OH, 6 ml of BuOH, 4 g of NaCl, and 100 ml of H$_2$O.

FOPPEN, F. Chromatog. Rev. *14*, 133-298 (1971). Tables for the Identification of Carotenoid Pigments. Includes paper, thin-layer, and column chromatography data.

FOUASIN, R. Rev. fermentations inds. aliment. *7*, 195-219 (1953). A Systematic Method for Separation and Identification of Synthetic Colors Used in Foods. Based on a study of the chromatographic properties of more than 80 colorants, a scheme of qualitative analysis was devised that first separates the colorants into groups using immiscible solvents, and then subdivides them using various acid and alkaline eluants.

FRANCIS, B. J. Analyst *90*, 374 (1965). The Separation of Annatto Pigments by Thin-Layer Chromatography with Special Reference to the Use of Analytical-Grade Reagents. The Separation of Annatto Pigments on silica gel as reported by Ramamurthy and Bhalerao [(Analyst *89*, 740-744 (1964)] was found to be dependent on the acetic acid content of the amyl acetate used as the eluant.

GRAHAM, R. J. T., NYA, A. E. International Symposium on Chromatography and Electrophoresis, 5th Bruxelles, 1969, p. 486-490. Twenty-eight British food colors were chromatographed on silica gel thin layers using BuOH-EtCOMe-NH$_4$OH (d 0.88)-H$_2$O (5:3:1:1) as the eluant.

GRAHAM, R. J. T., NYA, A. E. J. Chromatog. *43*, 547-550 (1969). The Partition Chromatography of Food Dyes on Polycarbonate-Coated Foils. Twenty-eight food dyes permitted in Britain, including FD&C Blue No. 2, FD&C Yellow Nos. 5 and 6, and FD&C Red Nos. 2, 3, and 4, were separated on precoated polycarbonate foils (10 cm × 10 cm) using butanol aqueous NH$_3$ (d 0.88) (99:1).

GREENSHIELDS, R. N., HUNT, P. C., FEASEY, R., MAC GILLIVRAY, A. W. J. Inst. Brew. London 75, 542-550 (1969). Preliminary Investigation of the Electrophoretic Properties of Caramels.

GRIFFITHS, M. H. E. J. Food Technol. 1, 63-72 (1966). Systematic Identification of Food Dyes Using Paper Chromatography. Procedures are presented for the separation and identification of a variety of food colors permitted in the United Kingdom, the United States, and the European Economic Community. The technique of "double spotting" is recommended as a means of overcoming irregularities in R_f values caused by the impurities derived from the foodstuffs. "Double-spotting" consists in placing a spot of the unknown dye on top of the spots of knowns so that both will be equally affected by impurities present.

GROB, E. C., PFANDER, H., LEUENBERGER, U., SIGNER, R. Chimia 25, 332-333 (1971). Separation of Carotenoid Mixtures by Counter-Current Extraction. β-Carotene, cryptoxanthin, canthaxanthin, and zeaxanthin were separated by counter-current distribution using the solvent system methanol-H_2O (19:1) and light petroleum (boiling range 50-70°C). Separations were performed under an atmosphere of nitrogen using the apparatus developed by Signer and Arm [Analyt. Abstr. 15, 3034 (1968)].

HANSENS, M., DE RUDDER-TACK, Y. Pharm Tijdschr. Belg. 44, 125-131 (1967). Paper Chromatographic Determination of Synthetic Water-Soluble Food Colors. Samples (0.25%) were prepared in 50% EtOH, spotted on paper, and developed with 2% tri-Na citrate in 5% NH_4OH.

HAYES, W. P., NYAKU, N. Y., BURNS, D. T. J. Chromatog. 71, 585-587 (1972). Separation and Identification of Food Colours. III. Improved Resolution of Selected Dye Pairs. [For Parts I and II, see Hoodless et al., J. Chromatog. 54, 393-404 (1971); 56, 332-337 (1971).] Systems were devised for dye pairs previously unresolved. For Orange GGN and FD&C Yellow No. 6, use BuOH-H_2O-AcOH (10:5:1); for FD&C Green No. 1 and Green S, use iso-BuOH-EtOH-H_2O-concentreated aqueous NH_3 (60:20:2:1); and for FD&C Blue No. 1 and Light Green Yellowish, use ethyl acetate-MeOH-concentrated aqueous NH_3 (10:3:3). All separations were on thin layers of cellulose powder (Applied Science Laboratories, microcrystalline).

HEILINGOETTER, R. Kosmet, Aerosole 44, 970 (1971). Chromatography of Hair Dyes.

HOODLESS, R. A., PITMAN, K. G., STEWART, T. E., THOMSON, J., ARNOLD, J. E. J. Chromatog. 54, 393-404 (1971). Separation and Identification of Food Colours. I. Identification of Synthetic Water-Soluble Food Colours Using Thin-Layer Chromatography. A TLC method is described for the separation and identification of 49 synthetic food colors that are or have been used in food products. The R_f and R_x (with respect to Orange G) values are tabulated and a scheme for the rapid identification of the components of a mixture of dyes is proposed. Cellulose and silica gel plates were used with a variety of solvents.

HOODLESS, R. A., THOMSON, J., ARNOLD, J. E. J. Chromatog. 56, 332-337 (1971). Separation and Identification of Food Colours. II. Identification of Synthetic Oil-Soluble Food Colours Using Thin-Layer Chromatography. Cellulose layers (0.25 mm) are immersed in a 10% solution of liquid paraffin in light petroleum (boiling range 80-100°C) for 1 min and then either air dried or dried in an oven at 80°C. Then 1-2 μl of dye solution is applied and the chromatogram is developed with 2-methoxy-ethanol-MeOH-H$_2$O (11:3:6). Ten oil-soluble dyes, including four that are permitted in certain countries, are separated by this procedure.

IRIMESCU, I., COCIUMIAN, L., IDU, S. M. Z. Med. Labortech. 8, 85-93 (1967). Improved Circular Chromatographic Method. Orange GGN, FD&C Blue No. 2, Amaranth, and FD&C Yellow No. 5 are separated by paper chromatography using BuOH-C$_5$H$_5$N-H$_2$O (3:2:5).

JENSEN, A. Wiss. Veroeffentl, Deut. Ges. Ernaehrung 9, 119-127 (1963). Paper Chromatography of Carotenes and Carotenoids. A review.

JENSEN, A., JENSEN, S. L. Acta. Chem. Scand. 13, 1863 (1959). Separation of Twenty Five Different Carotenoids on 20% Kieselguhr Paper Using Mixtures of Petroleum Ether and Acetone.

JONES, J. H., CLARK, G. R., HARROW, L. S. JAOAC 34, 135-147 (1951). A Variable Reference Technique for Analysis by Absorption Spectro-photometry. A solution of an unknown is placed in the sample compart-ment of a double-beam spectrophotometer. The sample's composition is determined by continually varying the composition of a reference solu-tion until spectral balance is obtained. The composition of the reference solution is conveniently changed by equipping the spectrometer's reference compartment with a flow-through cell connected through a circulating pump to a titration vessel into which suspected knowns are added from burettes.

KAMIKURA, M. Shokuhin Eiseigaku Zasshi 7, 338-342 (1966). Thin Layer Chromatography of Synthetic Dyes. IV. Separation and Identification of Water-Soluble Dyes. 1. On the Developing Solvent and Condition of Activation of Silica Gel. Silica gel chromatography plates used for separa-tion of water-soluble dyes were prepared under three conditions of acti-vation: no activation, 60° activation for 60 min, and 100° activation for 60 min. Using eight developing solvents, the influence of the conditions of activation on the separation of water-soluble dyes was studied. Of the eight developing solvents, MeCOEt-H$_2$O (20:1) and MeCOEt-Me$_2$CO-H$_2$O (10:0.1:0.4) gave clear separation for xanthene dyes, including D&C Yel-low No. 7, FD&C Red No. 3, D&C Red No. 22, D&C Red No. 28, Rose Bengal, and Acid Red.

KOCH, L. JAOAC 26, 245-249 (1943). Systematic Group Separation of Mix-tures of FD&C, D&C and Ext. D&C Colors by Use of Immiscible Solvents.

KRAUZE, S., PIEKARSKI, L. Acta Polon. Pharm. 16, 395-402 (1959). Electrophoretic Separation and Determination of Dyes. Various dyes, including FD&C Yellow No. 5, D&C Orange No. 4, and D&C Red No. 19, were studied by paper electrophoresis using a potential of 400 V and pH = 12 phosphate buffer.

LEGRAND, P. Ann. fals. fraudes 52, 5-14 (1959). Identification by Micro-electrophoresis of Small Quantities of Synthetic Coloring Matter for Foods.

LEHMANN, G., HAHN, H. G., MARTINOD, P. Fresenius Z. Anal. Chem. *227*, 81-89 (1967). Quantitative Determination of Substances Separated on Thin Plates. After conventional thin-layer chromatography, the component of interest is scraped from the plate onto smooth parchment and then transferred quantitatively to a special microchromatographic column where it is eluted and measured spectrophotometrically.

LIN, S. C., LIN, Y., CHIANG, H. C. T'ai-Wan Yao Hsueh Tsa Chih *19*, 45-47 (1967). Polyamide Thin-Layer Chromatography of Food Colors. Eleven colorants were separated by polyamide TLC using three solvent mixtures: $CHCl_3$-Me_2CO-5% NaCl (0.4:9:3), $CHCl_3$-Me_2CO-5% Na salicylate (0.4: 9:3), and $CHCl_3$-Me_2CO-5% Na benzoate (0.4:9:3).

MAC DONELL, H. L. Anal. Chem. *33*, 1554-1555 (1961). Porous Glass Electrophoresis. Food colors, inks, and amino acids were separated by electrophoresis on porous glass slides (Corning, No. 7930).

MC KEOWN, G. G. JAOAC *37*, 527-529 (1954). The Separation of Amaranth and Tartrazine. Slurry alumina (Fisher Scientific adsorption alumina 80-200 mesh) in water and pack it into a 15 cm $\times$ 1.5 cm-ID column to a height of 7.5 cm. Wash the column with water until the eluate is clear. Activate the column with 1:100 hydrochloric acid. Next, dissolve a 0.01-g sample in 20 ml of 1:100 hydrochloric acid and pass it through the column. Wash the column with 100 ml of water. Colors are fixed on the top of the column as lakes. Elute FD&C Yellow No. 5 (Tartrazine) with 200-300 ml of 3% sodium acetate solution. Elute Amaranth with 100-200 ml of 0.4% sodium hydroxide. Next, adjust the eluate fractions to pH = 6 with hydrochloric acid and examine them spectrophotometrically.

MC KEOWN, G. G., THOMSON, J. L. JAOAC *37*, 917-920 (1954). A Separation of Triphenylmethane Food Colors by Column Chromatography. Slurry 80-200-mesh adsorption alumina (Fisher Scientific) in water and pack it into a 15 cm $\times$ 1.5 cm-ID column to a height of 10 cm. Wash it with water until the eluate is clear. Next, prepare an aqueous sample solution containing about 0.5 mg of each color. Adjust the solution to about $0.1N$ with dilute acetic acid and pass it through the column. Wash the column with 50 ml of water and then develop it with 350 ml of 1.5% aqueous pyridine. The basic solvent alters the color of some triphenylmethane dyes. To observe the position of the colors after development, pass 100 ml of water through the column followed by 100 ml of 1:100 acetic acid. The dyes then regain their original color.

MC KEOWN, G. G. JAOAC *44*, 347-351 (1961). Paper Chromatography of Bixin and Related Compounds. Prepare a strip of filter paper 6.5 in. $\times$ 22.5 in. from Whatman 3MM paper and mark a starting line 2.5 in. from one end. Impregnate by dipping into a 50% (v/v) solution of *N,N*-dimethylformide (DMF) in acetone and let dry in air for 10 min with the paper suspended in a vertical position. After drying, rapidly spot 2-μl volumes of the solutions to be analyzed. Develop the chromatogram by descending flow, using cyclohexane-chloroform-DMF-acetic acid (85:10: 3:2) for 3 hr, or until a satisfactory separation is obtained.

Compound	R_f
Stable norbixin	0.009
Labile norbixin	0.014
Stable bixin	0.09
Labile bixin	0.14
Oil Yellow AB	0.24
Stable methylbixin	0.38
Labile methylbixin	0.56

MC MILLION, C. R., DUNNING, H. A. B., Jr. J. Am. Pharm. Assoc. *48*, 249-251 (1959). A Chromatographic Technique for the Identification of Fluorescein and Phenolphthalein Derivatives. Methods are described for the chromatographic separation and identification of fluorescein and phenol derivatives on paper strips or cellulose columns by using 0.5 M Na_3PO_4 as the mobile phase.

MACCIO', I; Anales direc. nacl. quim 9, *52-54* (1956). Partition Chromatography of Inverted Phase. Separation of Some Fat-Soluble Dyes. The paper strip, cut from Whatman No. 1 filter paper was submerged for 1 min in a petroleum ether solution of Vaseline and then the petroleum ether was allowed to evaporate in the air for 3 hr. The resulting chromatogram was developed by placing the treated paper strip in a closed test tube for 18 hr. The following mixtures could be separated with a solvent containing 80% MeOH and 20% H_2O: (1) Sudan IV, Sudan III, and butter yellow; (2) Sudan IV, Sudan III, and Yellow OB, and (3) Sudan IV, Sudan II, and butter yellow. It is not possible to separate butter yellow from Yellow OB, or a mixture of Sudan II and Sudan III, by using this solvent mixture. With solvent mixtures containing EtOH or MeOH and H_2O and HCl, it is possible to separate mixtures of butter yellow and Yellow OB. None of these solvent mixtures separate Sudan II and Sudan III.

MACCIO', I. Anales direc. nacl. quim. *17*, 10-12 (1956). Chromatographic Study of New Dyes Derived from Coal Tar Allowed by the Food Authority for Food Use. The following dyes were separated by using a solvent consisting of 80% EtOH and 5% glacial AcOH in water (R_f values after each): FD&C Red No. 3 (0.88), Rose Bengal (0.88), Orange I (0.76), Ponceau 2R (0.24), Ext. D&C Yellow No. 7 (0.34), FD&C Blue No. 2 (0.08), FD&C Yellow No. 5 (0.07), Guinea Green B (0.90), and Patent Blue (0.93). Separation required 7 hr in a glass container 10 cm in diameter by 50 cm high. Separation was done on Whatman No. 1 paper 8 cm × 30 cm in size, using a 0.1% aqueous solution of the dye.

MARMION, D. M. JAOAC 57, 495-507 (1974). Applications of Nuclear Magnetic Resonance to the Analysis of Certified Food Colors. Individual colorants in secondary mixtures are identified and quantitated by NMR. Spectra are obtained in mixed deuterated solvent (water: dimethylsulfoxide; D_2O: DMSO-d_6, 2:1 v/v) at 100-105°C.

MASSART, D. L., DE CLERCQ, H. Anal. Chem. *46*, 1988-1992 (1974). Applications of Numerical Taxonomy Techniques to the Choice of Optimal Sets of Solvents in Thin Layer Chromatography. The problem of making a rational selection of a restricted set from a large number of available chromatographic systems for the separation of a particular group of substances is discussed. The systems are classified according to their mutual resemblance by numerical taxonomy techniques. From the resulting groups with dissimilar separation characteristics, one system per group can be chosen according to criteria such as availability and cost. In this way, a combination of systems with desirable characteristics and yielding relatively little correlated information should be obtained. This is illustrated by the selection of a combination of three solvent/stationary phases from a set of ten for the separation and identification of 26 yellow, orange, and red synthetic food dyes. The selection criterion in the groups, obtained by numerical taxonomy classification, is the information content. The resulting best combination is given and is found to permit unambiguous identification of all 26 dyes.

MIGLIETTA, E. Boll. lab. chim. provinciali *11*, 216-229 (1960). Chromatography and Spectrophotometry of Some Certified Dyes. Chromatographic values with four eluants and color curves are reported. A standardization of chromatographic and spectrophotometric characteristics of various certified dyes is proposed.

MITCHELL, L. C. JAOAC *36*, 943-946 (1953). The Separation of Certain Anthraquinone Dyes by Paper Chromatography. Spot a dimethylformamide solution of the sample on 8 in. × 8-in. Whatman No. 1 filter paper. Uniformly impregnate the paper with 1:99 refined soybean oil-ethyl ether by rapidly spraying it from top to bottom in horizontal strips. Elute the sheet in a 12 cm × 25 cm × 25-cm glass tank with 4:1 methyl Cellosolve-water until the solvent front approaches the top of the paper (ca. 2.5 hr). Compare against standards simultaneously prepared.

Color	R_f (approx.)
D&C Green No. 5	0.98
"Monosulfonated" D&C Green No. 5	0.86
D&C Green No. 6	0.14
Ext. D&C Violet No. 2	0.90
D&C Violet No. 2	0.24

MITRA, S. N., MATHEW, T. V., GUPTA, P. K. J. Inst. Chem. (India) *40*, 177-178 (1968). A Note on Paper Chromatography of Food Colours. Sample solutions (0.01%) were spotted on 40 cm × 10-cm Whatman No. 1 paper. (See table at the top of page 269.)

MITRA, S. N., CHATTERJI, R. K. J. Proc. Inst. Chemists *27*, 169-176 (1955). Separation of Permitted Coal-Tar Food Colors by Paper Chromatography. Paper chromatographic methods for the separation of FD&C Red No. 3, FD&C Yellow No. 5, FD&C Blue No. 2, Amaranth, and Orange I are described. The most suitable systems found were 5% NaCl or 0.25-1N HCl. Separations can be improved by two-dimensional chroma-

R_f Values of Food Colors Using Two Solvents

Color	R_f Solvent A	Solvent B
Ponceau 4R	0.183	0.437
FD&C Red No. 3	0.560	0.018
Carmoisine	0.10	0.085
Fast Red E	0.09	0.07
Amaranth	0.05	0.15
Red 6B	—	0.056
Red FB	—	0.0
Acid Magenta	—	0.89
FD&C Yellow No. 5	—	0.471
FD&C Yellow No. 6	—	0.21
FD&C Blue No. 2	0.113	—

Solvent A—Iso-amyl alcohol: 95% ethanol: $NH_4OH:H_2O$ (4:4:1:2); 18 hr; descending.

Solvent B—2% Sodium citrate in 5% NH_4OH; 3 hr; descending.

chromatography using iso-BuOH-5% NaCl as the second system.

MORI, I., KIMURA, M. J. Pharm. Soc. 74, 179 (1954). Electromigration of Food Colours. A number of systems were evaluated for the separation of various color additives using Toyo filter paper No. 50 as the support.

Conditions

	Electrolyte	V	mA/cm	hr
I	30% Acetic acid	700	0.5	4
II	10% Acetic acid	700	0.6	1
III	1% Borax	500	1.0	4
IV	0.1% NH_4OH	700	0.4	1
V	5% $NaHCO_3$	200	2.5	4

Relative Separation

	I	II	III	IV	V
Amaranth	38	30(71)[a]	23[b],60[c]	80	21
D&C Red No. 22	4	0	21	22(38)	
D&C Yellow No. 7	12	−8	46	74	
FD&C Blue No. 2	54	49	48,75[d]	60	14
FD&C Yellow No. 5	72	80	85		43
Naphthol Green B	23	21	28	28	
D&C Red No. 28	0		33		

[a] Parentheses indicate fluorescence.
[b] Blue Spot.
[c] Yellow Spot.
[d] Small Spot.

MORI, H., YOKOYAMA, T., HAMADA, K. Eisei Shikenjo Kenkyu Hokoku *81*, 57-60 (1963). Separation of Some Japanese Official Cosmetic Coal-Tar Dyes by Column Chromatography. Mixtures of red dyes were separated on a cellulose powder-alumina (1:1) column using a mixture of BuOH-EtOH-0.5N NH_4OH (6:2:3).

MÜLLER, K., TÄUFEL, K. Ernährungsforschung *1*, 354-361 (1956). Paper Chromatographic Separation and Identification of Food Dyes Allowed in the German Democratic Republic.

NAFF, M. B., NAFF, A. S. J. Chem. Ed. *40*, 534-535 (1963). TLC on Microscope Slides. D&C Yellow No. 7 and related fluoresceins were separated on silica gel G using toluene-acetic acid (65:35).

NETTO, I. Ann. fals fraudes *50*, No. 580 (1957). R_f Values of Food Colors. Substrate was Whatman No. 1 paper and solvent was 1N HCl.

Color	R_f
Ponceau 3R	0.031-0.040
Amaranth	0.063-0.077
FD&C Blue No. 2	0.099-0.131
FD&C Yellow No. 5	0.320-0.340
Ext. D&C Yellow No. 7	0.534-0.574
Guinea Green B	0.791-0.860

NEY, M., BERGNER, K. G., SPERLICH, H., MIETHKE, H. Deut. Lebensm. Rundschau *61*, 148-150 (1965). The Food Color Patent Blue. Patent Blue V (CI 42051), Patent Blue VF (CI 42045), Patent Blue AE (FD&C Blue No. 1, CI 42090), and Wool Green BS (CI 44090) were separated by paper chromatography using two solvents: BuOH-EtOH-25% NH_4OH-H_2O (4:4:1:3) and HOAc-pyridine-H_2O (55:25:20).

NÜTSU, Y. Bunseki Kagaku *13*, 1239-1242 (1964). High-Voltage Paper Electrophoretic Analysis of Water-Soluble Coal Tar Dyes for Food. Xanthene and triphenylmethane dyes were separated on filter paper using pH = 3-11.6 buffers and an applied voltage of 3000 V. The cooling agent was hexane. Migration distance was dependent on both the pH of the buffer and the structure of the dye.

PANOPOULOS, G., MÉGALDOIKONOMAS, J. Chim. anal. *36*, 68-69 (1954). Application of Chromatography to Identify the Dyes Used in Coloring Food Products.

PARIS, R. R., ROUSSELET, R. Ann. pharm. franc. *16*, 747-756 (1958). Characterization of Dyes of Vegetable Origin by Paper Chromatography. The R_f values and recommended solvents are given for a number of natural dyestuffs, including caramel, carotene, chlorophyll, indigo carmine, and saffron.

PARKÁNYI, C. Chemie *10*, 45-47 (1958). Square Capillary Analysis on Paper. Egacide Orange G and GG, FD&C Yellow No. 5, Amaranth, and Metanil Yellow O were separated on filter paper using water, aqueous 10% NH_3, or 5% HCl.

PARRISH, J. R. J. Chromatog. *33*, 542-543 (1968). Chromatography of

Food Dyes on Sephadex. Separations by TLC on Sephadex G-25 (super-fine grade) were used to predict separations on columns of the same material.

R_f Values of Dyes on Sephadex G-25

DYE (CI No.)	Eluants		
	I[a]	II[b]	III[c]
Blue VRS (42045)	0.48	0.41	0.31
FD&C Red No. 4 (14700)	0.47	0.27	0.20
Ponceau 4R (16255)	0.46	0.27	0.21
FD&C Yellow No. 5 (19140)	0.42	0.27	0.13
Ponceau 3R (16155)	0.40	0.12	0.06
FD&C Blue No. 2 (73015)	0.36	0.13	0.06
Amaranth (16185)	0.34	0.15	0.06
Ext. D&C Yellow No. 7 (10316)	0.33	0.29	0.16
Carmoisine (14720)	0.27	0.08	0.03
Orange G (16230)	0.25	0.07	0.04

[a] Water.
[b] 0.1% Sodium sulfate solution.
[c] 4% Sodium sulfate solution.

PEARSON, D. J. Assoc. Public Analysts 2, 30-34 (1964). R_f Values of Permitted Synthetic Water-Soluble Coloring Matters. Data are reported using a solvent composed of 80 g of PhOH + 20 g of H_2O.

PEARSON, D., CHAUDHRI, A. B. J. Assoc. Public Analysts 2, 22-30 (1964). R_f Values of Some Nonpermitted Synthetic Water-Soluble Coloring Matters. R_f Values were recorded for seven solvent systems.

PEARSON, D., WALKER, R. J. Assoc. Public Analysts 3, 45-48 (1965). R_f Values of Permitted Synthetic Water-Soluble Coloring Matters. Twelve solvent mixtures, including those proposed by the Association of Public Analysts, the British Standards Institute, and the British Food Manufact-turing Industries Research Association, were used to study 29 dyes by ascending chromatography.

PEARSON, D. J. Assoc. Public Analysts 11, 52-56 (1973). Identification of Oil-Soluble Food Colours. Reversed-phase paper chromatography using liquid paraffin as the stationary phase was used to study 16 colors and eight solvent systems. Spectrophotometric absorption maxima for solutions of the colors in light petroleum are also reported.

PEEREBOOM, J. W. Chem. Weekblad 57, 625 (1961). R_f Values (Thin Layer) of Fat-Soluble Dyestuffs.

PEEREBOOM, J. W. C., BEEKES, H. W. J. Chromatog. 20, 43-47 (1965). Thin-Layer Chromatography of Dyestuffs on Polyamide and "Silver Ni-trate" Layers. A study is reported of the separation of fat-soluble dyestuffs on layers of silica gel G, Kieselguhr G, aluminum oxide G, poly-amide, and silver nitrate-impregnated silica gel.

PENNER, M. H. J. Pharm. Sci 57, 2132-2135 (1968). Thin-Layer Chromato-graphy of Certified Coal-Tar Colour Additives. Nineteen dyes used in pharmaceutical preparations are separated on 0.25-mm layers of micro-crystalline cellulose (Avicel) using the following solvent systems; ethyl acetate-BuOH-pyridine-H_2O (5:5:6:5); ethyl acetate- BuOH-aqueous NH_3(d = 0.88) (4:11:5); ethyl acetate-PrOH-aqueous NH_3 (d = 0.88)-H_2O (7:7:4:4); and PrOH-ethyl acetate-aqueous NH_3 (d = 0.88) (13:15: 12).

PIETSCH, H. P., MEYER, R. Nahrung 9, 154 (1965). Thin-Layer Chromato-graphic Separation of Artificial Organic Food Dyes with Kieselgel D.

PIEKARSKI, L., KRAUZE, S. Roczniki Panstwowego Zakladu Hig. 10, 495-500 (1959). Determination of Dye Mixtures After Their Chromato-graphic Separation. Nine water-soluble dyes permitted in Poland for use in food are separated by two-dimensional paper chromatography using BuOH-EtOH-H_2O (2:1:1) and N NH_4OH.

PINTER, I., KRAMER, M., KLEEBERG, J. Elelmiszervizsgalati Kozlemen 14, 169-175 (1968). Thin-Layer Chromatographic Method for Detecting Various Cosmetic Dyes in Mixtures.

POPOV, A., MITSEV, I. Izvest. Inst. Org. Khim., Bulgar. Akad. Nauk 2, 5-11 (1965). Identification of Erythrosine in the Presence of Other Red Dyes. Mixtures of erythrosine and resorcinolphthalein food dyes are separated by paper chromatography using 10% NH_4OH-20% NaOAc-tert-BuOH (with 5% H_2O) (65:20:15).

POPOVICI, V., SCHWEIGER, A., SPITZER, A. Farmacia 13, 569-573 (1965). Paper Chromatography of Some Dyes Used in Pharmacy.

PUCHE, R. C. T. Rev. asoc. bioquim. argent. 22, 228-236 (1957). Paper Partition Chromatography of Synthetic Dyes Authorized for Coloring Foods. Food colors were spotted on S&S No. 0859 paper and developed with BuOH saturated with 10% HCl (I) or 0.5 ml of xylidine and 5 ml of concentrated HCl in 10 ml of H_2O (II). The following dyes were chro-matographed (name, R_f with solvent I, R_f with solvent II): FD&C Red No. 3, 0.00, 0.60; Rose Bengal, 0.00, 0.42; Bordeaux Red, 0.07, 0.21; Ponceau Red 2R, 0.18, 0.28; Orange I, 0.22, 0.41; FD&C Yellow No. 5, 0.47, 0.17; FD&C Blue No. 2, 0.94, 0.42; Guinea Green B, 0.89, 0.50.

RAI, J. Chromatographia 5, 211-213 (1971). Separation and Identification of Water-Soluble and Fat-Soluble Food Dyes by Thin-Layer Chromato-graphy. The R_f values are tabulated for 21 fat-soluble dyes in four solvents and 21 water-soluble dyes in five solvent systems using silica gel G as the adsorbent.

RAMAMURTHY, M. K., BHALERAO, V. R. Analyst 89, 740-744 (1964). A Thin-Layer Chromatographic Method for Identifying Annatto and other Food Colours. A simple technique is described for separating and identifying 11 yellow food colors: fat-soluble annatto, water-soluble annatto, curcumin, Oil Orange S, ethyl bixin, Oil Orange E, Yellow OB, Yellow AB, FD&C Yellow No. 5, FD&C Yellow No. 6, and β-carotene. Annatto and curcumin can be separted from other fat-soluble and water-soluble dyes on glass slides coated with H_2SiO_3 containing plaster of Paris or on silica gel G using amyl acetate as the developing solvent. The fat- and water-soluble dyes can be separated further on glass slides coated with CaCO$_3$ containing starch treated with liquid paraffin using MeOH-H$_2$O-NH$_3$ (20:5:1) as the eluant. (see Francis, B. J.).

RAO, T. S. S., SASTRY, L. V. L., SIDDAPPA, G. S. Indian J. Technol. 3, 332-334 (1965). Separation of Synthetic Food Colors by Chromatography and Electrophoresis. The R_f values (using ascending chromatography) are given for FD&C Red No. 3, Amaranth Carmoisine, Fast Red E, Red 6B, Ponceau 4RS, FD&C Yellow No. 5, FD&C Yellow No. 6, Blue VRS, Brilliant Black, and FD&C Blue No. 2 in 15 solvent systems. Five systems were studied using circular chromatography. Electrophoresis data are given for the dyes in pH = 8.6 borate, pH = 4.5 phthalate, pH = 8.6 1% Na$_2$B$_4$O$_7$, and pH = 7 phosphate buffers.

RAO, V. K., SARMA, P. S. N. J. Sci. Ind. Research 21D, 61-63 (1962). Paper Chromatography of Food Colors. An eluant consisting of 1.6% ethylenediamine hydrate and 2% iso-BuOH in H$_2$O is used in the paper-chromatographic separation and identification of coal-tar dyes permitted under the Indian Prevention of Food Adulteration Act of 1954.

ROY, B. R., SUNDARARAJAN, A. R., MITRA, S. N. J. Sci. Ind. Res. 18, 38-40 (1959). Analysis of Synthetic Food Colours Prescribed in India. Various chromatographic schemes are outlined for the paper chromatographic separation of food colors.

SADINI, V. Chimica 37, 381-394 (1961). Identification of Synthetic Dyes Permitted in Foods in the Countries Included in the European Economic Community. The chromatographic behavior of 23 dyes permitted in foods in the countries included in the European Economic Community is studied by descending chromatography using 21 solvent mixtures.

SADINI, V. Rass. Chim. 13, 13-18 (1961). Further Developments on Chrommatographic Research on the Coloring Matter of Food. A review with 100 references of research carried out during 1959 and 1960 on the chromatography of colors and pigments, natural and synthetic, both found in and added to foods.

SADINI, V. Rass. Chim. 12, 27-35 (1960). Partition Chromatography on Paper of Food-Additive Dyes. The 13 dyes that may be used as food additives in Italy were chromatographed on paper using 150 solvent systems.

SAENEZ, I., RUIZ, L., LAROCHE, C. Bull. Soc. Chim. France 1594-1597 (1963). Thin-Layer Chromatography of Synthetic Dyes.

SASAKI, H., IWATA, T. Shokuhin Eiseigaku Zasshi 13, 120-126 (1972). Analytical Studies on Food Dyes. IV. Direct Densitometry of Paper Chromatograms and Other Dyes by Transparent Methods. Paper chromatograms of 12 water-soluble and four oil-soluble dyes (including Amaranth and FD&C Yellow No. 6) were directly scanned using a photoelectric densitometer. The integrated readings were proportional to the root square of dye amounts or concentration.

SASAKI, H., FUKUSHIRO, S. Shokuhin Eiseigaku Zasshi 13, 127-132 (1972). Analytical Studies on Food Dyes VI. Determination of Monoazo Food Dye Mixtures by Direct Densitometry of Transparent Types. Paper chromatograms of two or four component mixtures of Amaranth, FD&C Yellow No. 5, New Coccine, and FD&C Yellow No. 6 were scanned with a densitometer. Recovery was 95.7-100.7% or 84.8-112.1% for the two- or four-component mixtures, respectively.

SASTRY, L. V. L., SEBASTIAN, K., KRISHNAPRASAD, C.A. J. Food Sci. Technol. *7*, 132-134 (1970). Estimation of Total Dye Content of Food-Colour Preparations. The component dyes were separated by chromatography on Whatman No. 3 paper with various solvent systems and extracted from the chromatogram with 0.1M HCl in 70% ethanol. The extinction of each extract was then measured at the appropriate wavelength of maximum absorption. The recoveries of pure dyes ranged from 97% to 101%, but that of indigo carmine was only 5-29%.

SCHNEIDER, H., HOFSTETTER, J. Deut. Apotheker-Zgt. *103*, 1423-1424 (1963). Thin Layer Chromatography of Enamel-Like Pigments Used in Drugs. Thin-layer chromatography of H_2O-insoluble enamel-type pigments can be performed after the Al salt is converted to the free acid by stirring for 2 hr with N HCl. An aliquot is spotted on a 0.5 mm cellulose plate and eluted with a 4:1 solution of 2.5% Na citrate-25% NH_3. Identification is by color and position of the spot compared with standards. The technique can be applied to most substances approved as food coloring, except for lightly colored sugar coatings, where no positive determination could be made because of sugar interference.

SCLAR, R., FREEMAN, K. JAOAC *38*, 796-809 (1955). Chromatographic Procedures for the Separation of Water-Soluble Acid Dye Mixtures. General methods are given for the chromatographic separation of FD&C Colors. Paper chromatography— Place one drop of fresh 1% aqueous sample solution on a 6 cm X 22-cm strip of Whatman No. 1 filter paper. The sample should be reasonably free of salt, sugar, starches, and other substances. Allow it to dry. Suspend the strip in an air-tight tank containing 0.5-1 cm of 2:3 water-ethylene glycol monomethyl ether acetate (methyl Cellosolve acetate). Alternatively, 1:4 water-methyl Cellosolve acetate can be used. Develop for 2-3 hr. If all FD&C colors are present, 10 colored zones may be observed. From the bottom up, the zones are as follows:

1. FD&C Yellow No. 5.
2. FD&C Red No. 2.
3. FD&C Blue No. 2.
4. FD&C Red No. 1 plus FD&C Red No. 4.
5. FD&C Yellow No. 6.
6. FD&C Blue No. 1 plus FD&C Green No. 2 plus FD&C Green No. 3.
7. FD&C Yellow No. 1.
8. FD&C Violet No. 1 plus FD&C Green No. 1.
9. FD&C Orange No. 1.
10. FD&C Red No. 3.

FD&C Reds No. 1 and 2, Greens No. 1 and 2, Yellow No. 1, Orange No. 1, and Violet No. 1 are no longer certifiable as FD&C colors.

Column chromatography—Prepare 30%, 25%, and 20% solutions of sodium chloride. To prepare solutions containing both ethyl alcohol and sodium chloride, mix the required reagents with water. For example, to prepare 20% ethyl alcohol plus 15% sodium chloride, mix 12 parts of 95% ethanol, 30 parts of 30% sodium chloride solution, and 18 parts of water. Slurry Solka-Floc BW-40 (manufactured by Grefco Inc., Dicalite Division, New York) in water and pour the amount indicated into a 2.5 cm-ID glass column. Allow it to stand for 12 hr. Treat sample as directed in the following table.

General Method for FD&C Colors

Pass the following solutions through the columns in the order given, allowing as little mixing as possible between successive solutions	Collect the following bands as they separate and emerge from the tubes; treat as indicated	ml (approx.)
Column No. 1 — 50 g of Cellulose		
1. 50 ml of 5% ethanol plus 15% sodium chloride (prewash)	Water plus prewash (discard)	350
2. Unknown colors dissolved in 50 ml of 20% sodium chloride		
3. 250 ml of 5% ethanol plus 1% sodium chloride	FD&C Green No. 3 plus FD&C Green No. 2[a] plus FD&C Blue No. 1	45
	Tailing plus ethanol solution (discard)	16
	FD&C Yellow No. 1[a] plus FD&C Yellow No. 5 (separate on column No. 4)	55
	Tailing plus ethanol solution (discard)	22
	FD&C Yellow No. 6	91
	Ethanol solution (discard)	25
4. 250 ml of 0.5% sodium chloride	FD&C Blue No. 2 plus FD&C Red No. 2[a] (separate on column No. 5)	132
	Dilute tailing (discard)	55
5. Sufficient water to remove remaining colors	Remaining colors	112

Column No. 2 — 40 g of Cellulose

1.	50 ml of 20% Ethanol plus 15% sodium chloride (prewash)	Water plus prewash (discard)	394
2.	Solution of remaining colors from column No. 1 (in 112 ml) adjusted to 20% sodium chloride		
3.	250 ml of 20% ethanol plus 15% sodium chloride	FD&C Green No. 1[a] plus FD&C Violet No. 1[a] (separate on column No. 6)	96
		Slight tailing of FD&C Violet No. 1[a] plus trace of FD&C Orange No. 1[a] (discard) FD&C Orange No. 1[a]	62
4.	50 ml of 20% Sodium chloride	Ethanol solution (discard)	140
5.	Sufficient water to Remove remaining colors	Remaining colors	42

Column No. 3 — 30 g of Cellulose

1.	50 ml of 20% NaCl (prewash)	H_2O + prewash (discard)	—
2.	Solution of remaining colors from column No. 2 (in 42 ml) made to 20% NaCl		
3.	200 ml of 1% NaCl + 1% NH_4OH	FD&C Red No. 4 (in basic form)	75
		Ammoniacal solution (discard)	—
4.	50 ml of 20% NaCl + 1% acetic acid		
5.	200 ml of 0.0625% NaCl	FD&C Red No. 1	—
6.	50 ml of H_2O + one or two drops of concentrated NH_4OH	Salt solution (discard)	—
7.	Sufficient H_2O to remove remaining color	FD&C Red No. 3	

Column No. 4 — 20 g of Cellulose
(Separation of FD&C Yellow No. 1 from FD&C Yellow No. 5)

1.	50 ml of 20% NaCl (prewash)	Water plus prewash (discard)	—
2.	Solution of FD&C Yellow No. 1 + FD&C Yellow No. 5, salted and freed of alcohol[b]		
3.	200 ml of 20% NaCl	FD&C Yellow No. 1	—
		Salt solution (discard)	—
4.	Sufficient H_2O to remove remaining color	FD&C Yellow No. 5	—

Column No. 5 — 30 g of Cellulose
(Separation of FD&C Blue No. 2 from FD&C Red No. 2)

1. 50 ml of 20% NaCl (prewash)	H_2O + prewash (discard)	—
2. Solution of FD&C Blue No. 2 + FD&C Red No. 2, salted and freed of alcohol[b]		
3. 200 ml of 1% NaCl + 1% NH_4OH	FD&C Red No. 2 (in basic form)	—
	Salt solution (discard)	—
4. Sufficient H_2O to remove remaining color	FD&C Blue No. 2	—

Column No. 6 — 30 g of Cellulose
(Separation of FD&C Green No. 1 from FD&C Violet No. 1)

1. 50 ml of 10% alcohol + 15% NaCl (prewash)	H_2O + prewash (discard)	—
2. Solution of FD&C Green No. 1 + FD&C Violet No. 1, salted and freed of alcohol[b]		
3. 250 ml of 10% alcohol + 1% NaCl	FD&C Green No. 1	140
4. 50 ml of 20% NaCl	Alcohol solution (discard)	—
5. Sufficient H_2O to remove re-remaining color	FD&C Violet No. 1	60

[a]These dyes are no longer certifiable as FD&C colors.
[b]Add sufficient sodium chloride to the fraction to bring the concentration to 20%. Extract with an equal volume of peroxide-free ethyl ether. If the dye tends to pass into the ether, add 0.5 ml of 10% sodium hydroxide for each 100 ml of dye solution. Neutralize the aqueous phase, if necessary, and remove the ether by passing air through it.

SHELTON, J. H., GILL, J. M. T. J. Assoc. Public Analysts *1*, 88-91 (1963). Paper Chromatographic Identification of Food Dyes. The method of Yanuka and colleagues is extended to the separation and identification of those food colors permitted in the United Kingdom and not included in the original paper.

SPALDING, R. C. J. Assoc. Public Analysts *2*, 111-112 (1964). R_f Values of Certain Synthetic Coloring Matters. The R_f values of 28 dyes obtained by overnight descending chromatography are compared to those obtained with 12-cm ascending runs. The values did not necessarily correspond.

STAHL, E., BOLLINGER, H. R., LEHNERT, L. Wiss. Veroeffentl. Deut. Ges. Ernaehrung *9*, 129-134 (1963). Thin-Layer Chromatography of Carotene and Carotenoid Mixtures. Some 30 carotene derivatives are separated, identified, and classified as to functional groups using $Ca(OH)_2$, Mg phosphate, and silica gel G. Solvent mixtures for separating carotenes were 5% CH_2Cl_2 in mixed hydrocarbons, aldehydes, esters, and CCl_4; for highly polar carotenoids, C_6H_6 or 20% EtOAc in CH_2Cl_2 was used. Separation took place under CO_2 to prevent decomposition. Ubiquinones as interfering substances were separated on paraffin-impregnated 50:50 Kieselguhr-silica gel with paraffin-saturated 9:1 MeOH-iso-PrOH and visualized with 5% phosphomolybdic acid.

STEWART, I., WHEATON, T. A. J. Chromatog. 55, 325-336 (1971). Continuous Flow Separation of Carotenoids by Liquid Chromatography. A liquid chromatographic system is described for the separation of complex mixtures of carotenoids. Carotenes are separated on magnesium oxide and xanthophylls are separated on zinc carbonate. The separation of complex mixtures required gradient elution. A variety of solvent combinations were tried.

SYNODINOS, E., KOTAKIS, G., KOKKOTI-KOTAKIS, E. Chim. Chronika 28, 77-79 (1963). Separation of Synthetic Dyes by Thin-Layer Chromatography. The following synthetic dyes used in Greece for coloring food, drugs, and cosmetics were separated by TLC on $CaCO_2$ using BuOH-EtOH-H_2O (2:1:1) with 10% NH_3 as the solvent system. The R_f values include: Ponceau BR (0.71), Amaranth (0.63), FD&C Red No. 3 (0.79), FD&C Red No. 4 (0.74), FD&C Yellow No. 5 (0.54), FD&C Blue No. 2 (0.65), and FD&C Yellow No. 6 (0.72).

SYNODINOS, E., KOTAKIS, G., KOKKOTI-KOTAKIS, E. Riv. Ital, Sostanze Grasse 40, 674-676 (1963). Separation of Synthetic Dyes by Thin-Layer Chromatography. Seven dyes approved by the Greek Food Regulations were studied. Plates were prepared using H_4SiO_4(I), Celite (II), MgO (III), $CaCO_3$(IV), rice starch (V), and gypsum, alone or in mixtures. The eluant was 2:1:1 BuOH-EtOH-H_2O containing 10% concentrated NH_3. The dyes studied were Ponceau 3R (1), Amaranth (2), FD&C Red No. 3 (3), FD&C Red No. 4 (4), a mixture of 1 and 4, a mixture of 1, 2, 3, and 4, FD&C Yellow No. 5, FD&C Yellow No. 6, FD&C Blue No. 2, a mixture of FD&C Yellow No. 5, FD&C Yellow No. 6 and FD&C Blue No. 2, chlorophyllin a and b, and Carmine; 1 and 4 can be separated on a plate made of IV and V. The best separation between chlorophyllin a and b is obtained using a plate made with V, I, II, and III.

SZOKOLAY, A. Z. Lebensm. Forsch. 120, 295-299 (1963). Paper Chromatographic and Spectrophotometric Detection of Fat-Soluble Synthetic Food and Cosmetic Dyes. Twelve oil-soluble synthetic food colors are separated by two-dimentional paper chromatography using dioxane-H_2O-NH_4OH (70:20:5) as the eluant.

SZOKOLAY, A. Cslka Hyg. 14, 289-292 (1969). Identification by Thin-Layer Chromatography of Food Dyes Permitted in Czechoslovakia. Mixtures of FD&C Red No. 4, FD&C Red No. 3, Amaranth, FD&C Blue No. 2, cochineal red and Ponceau 6R were readily separated on starch-bound silica gel using ethyl acetate-MeOH-4.6N aqueous NH_3 (25:8:5) as the eluant. Mixtures of FD&C Yellow No. 5 and azorubine were separated using BuOH-EtOH-H_2O-aqueous NH_3 (10:5:4:2) as the eluant.

TAKESHITA, R., ITOH, N., SAKAGAMI, Y. J. Chromatog. 57, 437-440 (1971). Separation and Detection of Basic Dyes by Polyamide Thin-Layer Chromatography. The chromatographic behavior of fifteen colorants used illegally in foods and drugs is studied in a variety of solvents.

TAYLOR, K. B. Nature *185*, 243-244 (1960). Chromatography of Xanthene Dyes. Commercial halogenated fluoresceins were successfully chromatographed on 20 cm $\times$ 40-cm Whatman No. 1 paper using 0.88 ammonia-EtOH-H$_2$O (5:10:85).

TERASHIMA,T. Shokuhin Eiseigaku Zasshi *2*, 44-51 (1961). High-Voltage Paper-Electrophoretic Analysis of Water-Soluble Coal-Tar Dyes. I. The Migration Distance of Dyes. Sixty-nine dyes, including many that are used in foods and drugs, were studied by paper electrophoresis using 5N AcOH or 0.1N NaOH plus 10% propylene glycol as electrolyte at 50 V/cm for 30 min.

TERASHIMA, T. Shokuhin Eiseigaku Zasshi *8*, 46-52 (1967). High-Voltage Paper-Electrophoretic Analysis of Water-Soluble Coal-Tars Dyes. VII. Systematic Separation Method of Dyes. 36 water-soluble food dyes are systematically classified.

TILDEN, D. H. JAOAC *35*, 423-435 (1952); JAOAC *36*, 802-810 (1953); JAOAC *37*, 812-818 (1954). Report on Paper Chromatography of Coal-Tar Colors. A study of the usefulness of paper chromatography for the separation and identification of color additivies is discussed.

TONET, N. Mitt. Geb. Lebensm. Hyg. *60*, 201-205, (1969). Use of High-Voltage Electrophoresis as a Supplementary Technique for the Identification of Water-Soluble Dyes. Seventy-five synthetic and several natural dyes were studied by electrophoresis at 4500 V using 20% acetic acid or 0.1M aqueous NH$_3$-3.3 mM acetic acid buffer adjusted to pH = 10.3 as electrolyte.

TURNER, T. D., JONES, B. E. J. Pharm. Pharmac. *23*, 806-807 (1971). Identification of Blue Triphenylmethane Food Dyes by Thin-Layer Chromatography. To separate FD&C Blue No. 1, Blue VRS, FD&C Green No. 3, Green S, and Patent Blue V, apply 1 μl of a 0.01% sample solution to a thin layer of DEAE-cellulose precoated on a plastic sheet (Macherey-Nagel) and elute with M NH$_4$I or 0.05M or 0.2M ammonium benzoate. Azo dyes have very low mobilities in these solvents.

VERMA, M. R., DASS, R. J. Sci. Ind. Res. *15C*, 186-192 (1956). Identification of Certifiable Food Colors. I. Determination of R_f Values of Single Food Colors. The R_f values of 45 dyes used in the food industry have been determined in a number of eluants.

VERMA, M. R., DASS, R. J. Sci. Ind. Res. *16*B, 131 (1957). R_f Values of Fat-Soluble Dyes.

VILLANUA, L., CARBALLIDO, A., OLMEDO, R. G., VALDEHITA, M. T. Anales Bromatol. *13*, 59-106 (1961). Synthetic Food Colors. VI. Characteristics, Properties, Spectrophotometry, and Circular Paper Chromatography of Prohibited Water-Soluble Dyes.

VILLANUA, L., CARBALLIDO, A., OLMEDO, R. G., VALDEHITA, M. T. Anales Bromatol. *13*, 263-285 (1961). Synthetic Food Colors. VII. Systematic Scheme for the Identification of Water-Soluble Dyes.

VILLANUA, L., CARBALLIDO, A., OLMEDO, R. G., VALDEHITA, M. T. Anales. Bromatol. *16*, 377-394 (1964). Synthetic Food Colors. IX.

Characteristics, Properties, Spectrophotometry, and Chromatography of Some Water-Soluble Artificial Dyes.

WALTHIER, J., JENEY, E. Olaj, Szappon, Kozmet, *17*, 85-88 (1968). Thin Layer Chromatography in the Analysis of Synthetic Food Colours.

WANG, K. T. Nature *213*, 212 (1967). Polyamide Layer Chromatography of Some Synthetic Food Colors. Ten colorants, including FD&C Red No. 4, Amaranth and FD&C Yellow Nos. 5 and 6, were separated on polyamide layers using five solvent systems.

WEISS, L. C. JAOAC *34*, 453-459 (1951). Chromatographic Properties of Oil-Soluble Coal-Tar Colors. Systems are presented for the column-chromatographic separation of 22 oil-soluble colorants.

Chapter 15 Analysis of Commerical Products

The determination of colorants in foods, drugs, and cosmetics is probably the most challenging and certainly the most needed analysis in the field of color additives today. The challenge arises from the inherent difficulties associated with isolating the colorants, knowing when recovery is complete, and whether low recovery reflects inadequate analytical procedures or product-related colorant decomposition. The simultaneous presence of more than one color additive as well as the presence of natural colorants in the product compound the problem. The need for the determination is manifold. Government requires that only limited amounts of specified colorants be used in products, and so it must police industry. Industry wants to know what its competition is doing, and both are interested in colorant stability after incorporation into various matrices.

A number of techniques have been used to determine color additives in manufactured goods. The simplest, of course, is to measure them spectro-photometrically *in situ*, an approach that is viable if the colorant or colorants present are not interfered with by the presence of natural dyestuffs or by each other. Still and carbonated soft drinks, powdered gelatin desserts, certain hard candies, and colored films can often be measured this way after only a minimum of sample preparation such as degassing, dilution, or dissolution. The chief concern in making such measurements is in being certain that the matrix does not affect the colorant's spectrum either qualitatively or quantitatively, a point that is best established using the technique of known additions.

Unfortunately, few products can be handled so simply. Items such as chocolate pudding, lavender lipstick, fruit-stripe gum, nail polish, and multi-colored cold capsules can be an analyst's nightmare since the colorants present are often a mixture of both soluble dyestuffs and insoluble pigments or lakes that are widely different in chemical nature, difficult to isolate from their matrices, and a real chore to resolve from each other.

Most colorants can be isolated from their matrices by one of three techniques: leaching, solvent-solvent extraction, or adsorption onto an active substrate. Leaching has been used successfully to remove colorants from the surface of oranges, sausages, and tablets as well as from packaging films and spices. In the simplest cases the product is merely soaked in an appropriate solvent and then filtered or centrifuged to isolate the colorant-bearing liquid. Further cleanup is done as needed. Solvent-solvent extraction using simple immiscible solvent pairs, or solvent pairs, one of which acts as a carrier for a complexing agent or an ion-exchange resin, is a widely and effectively used method of isolation. The procedure pioneered by Dolinsky and Stein and developed by Graichen and Molitor in which the liquid anion exchange resin Amerlite LA-2 (Rohm and Haas Co., Philadelphia, Pa.) dissolved in butanol or hexane is the extracting medium is among the most general and most useful of the methods employed today. Adsorption techniques have been developed using a variety of materials such as wool, powdered leather, cellulose, alumina, and polyamide powder. In using adsorption techniques the pH of the sample solution is adjusted as needed, then the solution is treated with adsorbent either by adding it to the sample or by passing the solution down a column packed with adsorbent. The adsorbent is freed of sample matrix and then stripped of colorant by washing with appropriate solvents.

As might be expected, no one method is capable of analyzing all kinds of samples, and only a thorough knowledge of specific procedures will enable one to develop techniques suitable for solving individual problems. Some of these methods are summarized in the following bibliography. Where appropriate, procedures are grouped according to the class of product to which they apply. The remainder of the procedures are listed as miscellaneous.

BAKED GOODS

BENDER, A. E., MACFARLANE, A. J. Analyst 90, 536-540 (1965). Determination of β-Carotene in a Roller-Dried Food. Three procedures involving enzyme treatment, saponification, and direct solvent extraction are compared for effectiveness in determining β-carotene in a roller-dried food. The enzyme method appears to be the best.

Enzyme Procedure: Weigh 10 g of powdered food and 0.3 g of Bacterase (Assocatied British Maltsters Ltd., Stockport, Cheshire) and mix in a 150-ml beaker. Add 40 ml of pH = 7 buffer (0.4 M disodium hydrogen orthophosphate plus 0.2 M citric acid) at 50°C, mix to a smooth paste, and incubate the mixture for 5 min at 50°C. Add ammonia at pH = 8.5 and incubate the mixture at 50 °C for an additional 5 min. Transfer the sample to a 250-ml separatory funnel using a minimum of water and extract it with successive 50-ml portions of extraction solvent (10% v/v of 99%, 74° over proof industrial methylated spirits in peroxide-free, analytical reagent-grade diethyl ether) until all color has been removed. Combine the extracts, centrifuge if necessary, and determine the absorption at the absorption maximum near 452 nm.

CASILLO, R., POLITO, A. Selezione Tec. Molitoria 14, 108-113 (1963). Addition of Carotene and Xanthophil to Farinaceous Products (method for cold extraction with benzene and for extraction preceded by hydrolysis with alcoholic potash).

Total carotenoids are determined by extracting with cold benzene and then measuring spectrophotometrically at 465 nm. β-Carotene and xanthophil are separated from the total carotenoids by passing the sample solution over activated alumina. Xanthophil is retained at the upper part of the column and elutes after β-carotene. Xanthophil is measured at 442 nm.

DI STEFANO, F., RENZI, D. Rend. ist. super. sanità 19, 294-297 (1956). Detection of Riboflavine in Artifically Colored Food Pastes. The sample is extracted in the cold by centrifugation with water and then exposed for 45 min to UV radiation to convert riboflavine to lumichrome. A blue fluorescence under Wood's light indicates the product originally contained riboflavine.

HAYES, W. P., NYAKU, N. Y., BURNS, D. T., HOODLESS, R. A., THOMSON, J. J. Chromatog. 84, 195-199 (1973). Separation and Identification of Food Dyes. V. Examination of Ponceau 6R Dyes: Extraction of Dyes from Confectionary Products (cakes, cake mixtures, and pastries). Weigh 5 g of sample into a glass evaporation dish and place it in a 100°C oven for 30 min. Add sufficient light petroleum (boiling range 40-60°C) to cover the sample (ca. 30 ml); stir the mixture. Allow the solids to settle and decant off the light petroleum. Repeat this procedure twice and then allow the residual light petroleum to evaporate. Grind the sample gently so as not to form too fine a powder, add 4 g of Celite 545 to the sample, and mix.

Place a plug of glass wool in the end of a chromatographic tube (250 mm × 15 mm) and transfer the powdered sample to it. Pour 30 ml of acetone on top of the column and when the solvent has percolated through the whole length of the column, apply a slight air pressure to aid uniform packing. Discard the eluate. Carefully pour 50 ml of a mixture of methanol, water, and 25% v/v aqueous tetramethylammonium hydroxide solution (40:9:1) through the column. Adjust the pH of the eluate to approximately 6 by the addition of dilute hydrochloric acid. Add 5 ml of 1% aqueous polyoxyethylene sorbitan monooleate solution and reduce the volume of the mixture to about one-half on a steam-bath with the aid of a current of air blown over the surface of the liquid. Add an equal volume of water to the solution and allow it to cool.

Place a plug of glass wool in a 15 mm × 500-mm chromatographic tube and add a suspension of equal amounts of cellulose powder (microgranular/CT, without additives, Whatman Inc., Clifton, N. J.) and silica gel in water to the tube to give a column about 20 mm high. Rinse the walls of the tube with a small volume of acetone to aid the settling of the column packing and then place sand on top of the packed column to form a layer about 6 mm deep. Pour the solution of extracted dye through the column and wash the column three times with 5-ml portions of acetone, five times with 5-ml portions of a mixture of chloroform, absolute ethanol, water and 90% formic acid (100:90:10:1), three times with 5-ml portions of acetone, and finally three times with 10-ml portions of water. Elute the dyes with a minimum volume of acetone-ammonia solution (40 ml of acetone, 9 ml of water, 1 ml of ammonia, specific gravity =

0.88), rejecting the eluate until the dyes are eluted. Remove the ammonia by blowing a current of air over the surface of the eluate and then reduce its volume to about half on a steambath. Add an equal volume of water and adjust the pH to approximately 6 with hydrochloric acid. Pour the solution through a column of cellulose powder-silica gel in a second 15 mm × 500 mm chromatographic tube prepared as described above and wash the column with the same volumes of solvents in the sequence as described above for the first column. Elute the dyes with the minimum volume of acetone-ammonia solution. Remove the ammonia by blowing a current of air over the surface of the liquid and then evaporate the solution almost to dryness on a steam bath. Dissolve the residue in a few drops of 0.1N hydrochloric acid and use this solution for TLC.

HILDENBRAND, K. Deut. Lebensm. Rundschau 63, 372-373 (1967). Identification of Riboflavine and Quinoline Yellow in Bakery and Confectionery Products. Riboflavine and Quinoline Yellow are isolated from food by fixation on wool fibers, then separated by chromatography on paper or polyamide powder.

LEHMANN, G., COLLET, P. Z. Lebensm. Forsch. 144, 104-106 (1970). Analysis of Dyes. IX. Detection of synthetic dyes in pastries, dough products, and grain. The chopped sample is dried at 110°C, ground in a mortar with sand, Celite, and acetone, then filtered. The cake is repeatedly extracted with acetone to remove fat and water. The residue is dried, finely ground, and transferred to a chromatographic tube, where it is extracted with concentrated aqueous NH_3-methanol (1:19). The natural and synthetic fat-soluble dyes in the acetone extract and the acid dyes eluted from the chromatographic tube are separated and identified chromatographically.

SINGH, M. JAOAC 53, 23-25 (1970). Stability of Color Additives: FD&C Red No. 2 in Baked Goods. Colorants are extracted from cookies, cakes, dog biscuits, and various other baked goods using the Amberlite LA-2 amine ion exchange resin method of Graichen and Molitor (see p. 323).

BEVERAGES

BAUERNFEIND, J. C., OSADCA, M., BUNNELL, R. H. Food Technol. 16, 101-107 (1962). β-Carotene, Color and Nutrient for Juices and Beverages. For manufacturing control, parallel assays are run on unfortified and β-carotene-fortified juice and the amount of added β-carotene is determined by difference (Procedure 1). When the unfortified sample is not available, added β-carotene is determined by column chromatography (Procedure 2).

Procedure 1: To a 40-ml glass-stoppered centrifuge tube add 0.5 g of $CaCO_3$ and 0.25 g of Hyflo Super Cel. Add 2 ml of a 1:1:1 mixture of water, methanol, and n-propanol, thoroughly wetting the adsorbent mixture. Using a 2-ml blowout volumetric pipette, transfer 2 ml of the orange-juice concentrate, preblended for 5 min in a Waring Blender, into the tube. Follow with 20 ml of Skellysolve B, stopper, and shake 5 min in a horizontal mechanical shaker. Centrifuge briefly. Add 2 g of anhydrous Na_2SO_4. Shake and centrifuge briefly. Add 4 g of anhydrous Na_2SO_4. Shake for 5 min and centrifuge for 5 min until the Skellysolve B supernatant is clear. Dilute as needed and determine spectrophotometrically.

Procedure 2: Extract 2 ml of orange-juice concentrate as described under Procedure 1. Transfer 10 ml of the extract to a 125-ml Erlenmeyer flask, blanket with a steady stream of nitrogen, and evaporate to dryness on a 40°C water bath. Dissolve the residue in 5 ml of Skellysolve B. Pack a 10 mm-ID chromatographic column with 8 cm of Merck No. 71707 reagent aluminum oxide using a vacuum of 15-20 in. of mercury. Just before applying the sample adjust the vacuum to obtain a solvent flow rate of about two drops per second.

Transfer the residue dissolved in Skellysolve B onto the alumina column. Rinse the flask two times with 5 ml of Skellysolve B and transfer the rinsings into the column. Elute the column with 35 ml of 2% acetone in Skellysolve B into a glass-stoppered graduated cylinder so that the final volume is 40 ml. (The colorless initial eluate is discarded to keep the volume in this range.) The carotenes (a-, β-, and ζ-) are eluted as a red-orange band; a deep yellow band that follows should be at least 2 cm from the bottom of the column after all the carotenes have been eluted. Determine spectrophotometrically.

BENK, E. Deut. Lebensm. Rundschau 57, 324-329 (1961). Detection of Added β-Carotene in Orange Juice and Orange Juice Products. To determine total carotenoids—Extract the sample with petroleum ether-MeOH, wash the extract repeatedly with aqueous MeOH and then with H_2O, and measure spectrophotometrically at 450-470 nm.

To determine β-carotene—Transfer the ether extract to a chromatographic column containing H_2O-deactivated Al_2O_3. Elute β-carotene with 40% C_6H_6 in petroleum ether and measure spectrophotometrically.

BENK, E. Essenze Deriv. Agrumari 35, 113-118 (1965). The Detection of β-Carotene and Carotene Compounds Added to Fruit Juices and Fruit Juice Stock. Methods are described for quantitative separation and detection of carotene compounds. Includes a chromatographic separation on partially activated Al_2O_3.

BENK, E. Rieschstoffe Aromen 12, 205-206 (1962). Detection of Coloring of Fruit Juice, Beverages, and Lemonades by Sugar Color.

Method A: Based on the change in color produced when the 5-hydroxymethyl-2-furaldehyde present in caramel is treated with resorcinol. Mix 5-10 g of syrup or juice with washed beach sand to form a soft mass. Extract the mass with Et_2O, evaporate the extract to dryness, and add a few drops of fresh 1% resorcinol in fuming HCl to the residue. A cherry-red color indicates the presence of large amounts of 5-hydroxymethyl-2-furaldehyde, a pale red or brown color indicates low or trace amounts, and an olive or dark green color indicates that none is present.

Method B: Reaction with *p*-toluidinebarbituric acid. To 30 ml of a 0.5% aqueous solution of syrup or 30 ml of beverage add 1 ml of Carrez Solution 1 (150 g of $K_4Fe(CN)_6$/liter) and 1 ml of Carrez Solution 2 (300 g of ZnOAc/liter). Dilute to 100 ml, filter, and add *p*-toluidine to the filtrate. To one aliquot of filtrate add 1 ml of water (blank) and to a second aliquot, add 1 ml of 0.5% barbituric acid. After 4-5 min, measure the absorbance of the red color formed versus the blank.

BENK, E., WOLFF, I., TREIBER, H. Deut. Lebensm. Rundschau 59, 39-42 (1963). Detection of Added Carotenoids in Orange Juice by Thin-Layer Chromatography. Carotenoids extracted from juice with Et_2O are separated on a column of Al_2O_3 then identified by TLC using SiO_2-coated glass plates, and petroleum ether-C_6H_6-Me_2CO-AcOH (80:20:2:1) as the eluant.

DAGHETTA, A., BRUSS, O. Ann. sper. agrar. *11*, 117-120 (1957). Determination of β-Carotene in Fruit Juices. The method described is that reported by Wall and Kelley (see this list) using Al_2O_3 as an adsorbent in place of magnesia.

DE GORI, R., GRANDI, F., SANTUCCI, F. Boll. lab. chim. provinciali *10*, 248-255 (1959). Determination of Some Dyes for Liquors. Mixtures of colors, including FD&C Blue No. 2, FD&C Yellow No. 6 and FD&C Yellow No. 5, are determined in liquors spectrophotometrically without prior isolation or separation.

DI GIACOMO, A., RISPOLI, G. Essenze Deriv. Agrum. *36*, 167-176 (1966). Countercurrent Distribution Determination of Synthetic Carotenoids Added to Orange Juice. Samples are prefractionated by column chromatography and then resolved in a 100-tube Craig apparatus using a two-phase system of petroleum ether-MeOH. The various carotenoids are identified by TLC or spectrophotometry at 340-550 nm.

DI GIACOMO, A., RISPOLI, G. Riv. Ital. Essenze-Profumi Pianti Offic. Aromi-Saponi Cosmet.-Aerosol. *48*, 631-636 (1966). Determination of Caramel Added to Orange Juice and Beverages. Samples are chromatographed on Sephadex.

HIGBY, W. K. Food Technol. *17*, 95-98 (1963). Analysis of Orange Juice for Total Carotenoids, Carotenes, and Added β-Carotene.

ILLI, J. Mitt. Gebiete Lebensm. Hyg. *54*, 434-437 (1963). The Isolation of Artificial Dyestuffs From Foods With the Aid of Acidic (Anionotropic) Activated Aluminum Oxide. Dilute the sample (soft drink or cordial) with water and pour the mixture onto an acidic (anionotropic) aluminum oxide column. Wash concomitant substances such as sugars from the column with 70% ethanol. Elute the color with 10 ml of 70% ethanol, to which 1 ml of 10% ammonium hydroxide has been added. Evaporate the eluate to 1-2 ml and resolve the colors by paper chromatography.

LEHMANN, G., COLLET, P., MORAN, M. Z. Lebensm. Forsch. *143*, 191-195 (1970). Detection of Artificial Dyes in Wines and Fruit Juices. The sample is acidified with 98% formic acid-methanol (2:3) (solvent A) and stirred with polyamide powder, and the mixture is transferred to a prepared microcolumn, which is washed with solvent A and then with H_2O until the washings are neutral. The percolate and washings, containing anthocyanins, are rejected. The adsorbed dyes are washed from the column with concentrated aqueous NH_3-methanol (1:19), then with methanol, and the solution is diluted with H_2O. The dyes are further purified by another treatment on a polyamide column. The final solution is evaporated to a small volume, and the dyes are identified by TLC on a 0.25-mm layer of cellulose powder with 2.5% aqueous Na citrate-concentrated aqueous NH_3-methanol (20:5:3) as solvent. For the identification of basic dyes, the sample is passed through a prepared column of polyamide powder and the column is washed with H_2O. The dyes and part of the anthocyanins are eluted from the column with methanol, and the eluate is passed through an anion-exchange column of DEAE-cellulose, which retains the natural colors. The basic dyes in the percolate are then adsorbed on a cation-exchange column of carboxymethylcellulose, from which they are eluted with solvent A. The eluate is evaporated to 0.2 ml and the dyes are identified by TLC.

MAGLITTO, C., GIANOTTI, L., MATTAREI, C. Boll. Lab. Chim. Provinciali *15*, 354-359 (1964). Rapid Extraction of Pigments and Their Detection by Thin−Layer Chromatography. I. Research of cuprous chlorophyllins in preserves, of malvin from hybrid wines, and of vegetable extracts added to brandies.

For juices and preserves— To 50 g of sample add 50 g NaCl and 0.5 ml HCO_2H. Extract with 25 ml of 3:2:5 Me_2CO-Et_2O-iso-PrOH. Evaporate the extract to dryness, add 10 ml of solvent, add NaCl, and reextract. Concentrate the residue for chromatography.

For wines, syrups and brandies— To 50 ml of sample add 13 g NaCl. Extract with three 25-ml portions of the mixed solvent as described above, concentrate the combined extracts under reduced pressure to 5 ml, and chromatograph.

MARTIN, G. E., FIGERT, D. M. JAOAC 57, 217-218 (1974). Qualitative Determination of Coal-Tar Dyes in Alcoholic Products by Thin-Layer Chromatography. Pipette 50 ml of flavor extract or 100 ml of alcoholic beverage into a 250-ml beaker. Add sufficient HCl to reduce pH to about 2. Add about 12 in. of wool yarn and boiling chips. Place solution on hot plate and boil until volume has been reduced to 25 ml. Remove wool from beaker and rinse thoroughly with cold water. If wool is white, no synthetic aromatic dye is present. If wool retains color, add approximately 25 ml 10% NH_4OH to wash dye from wool. Let stand for 15 min, remove wool, express dye solution with glass stirring rod and discard yarn.

Place beaker on hot plate and boil until solution is reduced to about 2 ml. Depending on dye content, transfer 0.5-2.0 μl of solution to two 20 cm × 20-cm glass thin-layer plates coated with 0.1 mm of cellulose (EM Laboratories, Elmsford, N. Y.). Similarly transfer standard solutions (0.1 g/100 ml) to the same plates. Chromatograph in separate 10 in. × 12 in. × 4-in. glass tanks using mobile phases A and B.

Mobile phase A: Ethyl acetate-n-butanol-pyridine-H_2O (25:25:30:25).

Mobile phase B: n-Propanol-H_2O-triethylamine (62:28:10).

MATTIONI, R. Boll. Lab. Chim. Provinciali 15, 539-545 (1964). Determination of Caramel in Beverages. To beverages containing not more than 20% EtOH add 1 g Na_2SO_4 and 2 ml BuOH. Agitate a few minutes and then add one drop of a fresh saturated solution of phloroglucinol in concentrated HCl. Shake well. A dark-pale red color in the BuOH layer indicates the presence of caramel.

RATHER, H. Riechstoffe Aromen 12, 33-41 (1962). The Determination of Carotene and Carotenoids as Coloring Adjuvants in Orange Juice and Concentrates. Colorants are isolated by solvent-solvent extraction using alcohol, Et_2O, petroleum ether mixtures and then chromatographed on activated Al_2O_3 using C_6H_6-petroleum ether as eluant.

REEDER, S. K., PARK, G. L. JAOAC 58, 595-598 (1975). A Specific Method for the Determination of Provitamin A Carotenoids in Orange Juice. Blend equivalent of 20 ml of single-strength orange juice with 20 ml of petroleum ether, 60 ml of isopropanol, and 50 mg of butylated hydroxytoluene (BHT) for 1 min under reduced light. Transfer mixture to a 500-ml separatory funnel. Add 100 ml of diethyl ether and 20 ml of NaCl-saturated water and shake for 2 min. Discard lower layer and wash upper layer with 4 ml X 100 ml of 10% aqueous NaCl. Transfer organic layer to a 250-ml Erlenmeyer flask. Add 10 ml of methanolic KOH and a stirring bar and flush with nitrogen. Stopper and stir gently magnetically for 45 min. Transfer to a 250-ml separatory funnel and wash free of alkali with 10% aqueous NaCl. Reduce extract to dryness in rotary vacuum evaporator at 35°C. Dissolve residue in 2 ml of benzene-n-hexane (3:5) plus 0.01% BHT (eluant No. 1). Filter extract through 0.2 μm of regenerated Sartorius cellulose membrane into small vial. Analyze immediately or store at -10°C.

Pack a 2.1 mm-ID X 3 ft stainless-steel column with basic alumina (Woelm B 18, 18-30 μm). Deactivate by eluting with 20 ml of 15% isopropanol in hexane followed by eluant No. 1 until baseline is stabilized. Using a loop injector, inject 120 μl of extract onto the column and elute under pressure with eluant No. 1 at 2 ml/min. Monitor eluant at 440 nm. Retention times for α- and β-carotene are 4.7 min and 6 min, respectively.

TEWARI, S. N., SHARMA, S. C., SHARMA, V. K. Chromatographia 7, 36-37 (1974). Paper-Chromatographic Technique for the Detection of Colouring Matter in Liquors and Wines. The sample (10 ml) is evaporated to dryness and the residue is dissolved in 50% aqueous ethanol (0.5 ml). An aliquot of this solution is spotted on Whatman No. 1 paper, together with appropriate standards, and a chromatogram is developed with butanol-acetic acid-H_2O (4:1:5).

VALCHER, S. Boll. Lab. Chim. Provinciali 13, 530-542 (1962). Identification of Artificial Colors in Wines and Other Liquids.

WALL, M. E., KELLEY, E. G. Ind. Eng. Chem., Anal. Ed. 15, 18-20 (1943). Determination of Pure Carotene in Plant Tissue, Rapid Chromatographic

Method. Extract 1 g of ground, dehydrated sample for 1 hr in a Soxhlet apparatus using 200 ml of acetone-Skellysolve B (30:70). Evaporate to 25-50 ml on a steam bath. Pack a 23 mm × 200-mm glass column 3/4 full with a mixture of 3 parts Hyflo Super-Cel and 1 part Micron Brand No. 2641 activated MgO. Using vacuum, wash the column with 50 ml of Skellysolve B, pass the sample through the column, and then elute α- and β-carotenes with 3-5% acetone in Skellysolve B. Most noncarotene pigments remain at the top of the column.

YUFERA, E. P., MALLENT, D. Rev. Agroquim. Technol. Alimentos 4, 499-500 (1964). Detection of Orange Juice Adultered by Addition of β-Carotene and Synthetic Carotenoids. Carotenoids are separated on silica gel G using petroleum ether-iso-PrOH (95:5). The R_f values are:

β-apo-8′-carotenal = 0.22

canthaxanthin = 0.12

β-carotene = 0.75

Me ester of β-apo-8′-carotenic acid = 0.35

YUFERA, E. P., MALLENT, D. Rev. Agroquim. Tecnol. Alimentos 6, 215-220 (1966). Detection of Adulterants in Citric Juices. VIII. Methods for the Characterization of Natural and Synthetic Carotenoids. Mixtures are separated on Kieselgel G using petroleum ether-iso-PrOH (95:5) for one-dimensional chromatography and petroleum ether-iso-PrOH-EtOAc (80:40:5) and petroleum ether-iso-PrOH-acetone (95:5:10) for two-dimensional chromatography.

WILD, R., DOBROVOLNY, H. Brauwissenschaft 29, 93-100 (1976). Detection of Tagetes Extracts in Orange Products by High-Pressure Liquid Chromatography. Total carotenoids are extracted from juices, concentrates, or oils and are then fractionated by column chromatography on alumina. The xanthophyll ester fraction is analyzed by HPLC using a 10-μm Bondapak C_{18} column and methanol as the eluant.

CANDY AND CONFECTIONS

LEHMANN, G., HAHN, H. G. Gordian 69, 310-322 (1969). Isolation of Food Dyes from Predominantly Sugar Containing Preparations by the Polyamide Chromatography Method. Colorants are isolated from foods by polyamide-column chromatography and identified by TLC.

LEHMANN, G., ARACHAL, T., MORAN, M. Z. Lebensm. Forsch. 153, 155-157 (1973). Analysis of Dyes. XIV. Detection of Fat Soluble Dyes in Fats and Chocolate. Oil or fat is dissolved in light petroleum (boiling range 40-60°C); chocolate is extracted with warm light petroleum and insoluble matter is removed by filtration. The light petroleum solution is shaken with dimethylformamide (I) and the I phase is separated, washed with light petroleum (to remove residual fat), and mixed with an equal volume of H_2O. Residual light petroleum is removed by distillation under reduced pressure. A portion of the I solution is applied to a column (25 cm × 15 mm) packed with polyamide powder MN SC6. The column is washed with H_2O to remove I, auramine, and riboflavine. Artificial and natural dyes are then eluted with suitable solvents, namely, methanol-$CHCl_3$ (for chocolate dyes), or methanol-aqueous NH_3 (19:1) and identified by TLC [e.g., on Kieselgel G with $CHCL_3$-methanol (1:4) as solvent].

LEHMANN, G., COLLECT P. Z. Lebensm. Forsch. *143*, 418-420 (1970). Analysis of Dyes. VI. Detection of Snythetic Dyes in Marzipan and Persipan. The sample (0.5-2 g) is treated with hot H_2O (25 ml), polyamide powder (1 g) is added, the mixture is transferred to a microcolumn, and the liquid allowed to run through. Fat and basic and fat-soluble dyes are eluted with acetone, and sugar is eluted with hot H_2O. Acid dyes are then washed from the column with 0.1% NaOH solution in 70% aqueous methanol.

MARION, D. M. JAOAC *54*, 131-136 (1971). Analysis of Allura* Red AC Dye (A Potential New Color Additive); Hard Candy (Orange Sour Balls) Containing Allura* Red AC Dye and FD&C Yellow No. 6. Prepare the following eluting mixtures: eluant No. 1 = 200 g NaCl and 50 ml SD No. 30 alcohol diluted to 1 liter with water; eluant No. 2 = 10 g of NaCl and 50 ml of SD No. 30 alcohol diluted to 1 liter with water. Slurry 30 g of Whatman Column Chromedia CF11 in 200 ml of eluant No. 1 and pour into 40 cm × 2.5-cm glass column (Corning No. 38450). Wash with 100 ml eluant No. 1.

Dissolve 150-g sample in 500 ml of water. Pipette 20 ml into a 50-ml beaker. Add 4 g of NaCl and 2 ml of SD No. 30 alcohol; stir to dissolve. Wash sample onto column with two 10 ml portions of eluant No. 1. Elute FD&C Yellow No. 6 from column with eluant No. 2. Elute Allura* Red AC dye from column with water. Determine both colors spectrophotometrically.

MATHEW, T. V., MITRA, S. N., ROY, A. K. J. Proc. Inst. Chemists *36*, 301-304 (1964). Isolation and Identification of Coal-Tar Colors in Sweetmeat (Halwa) by Thin-Layer Chromatography. Colorants were isolated by leaching 50 g of sample with 100 ml of 90% EtOH followed by 100 ml of 1% aqueous NH_4OH. The combined extracts are filtered, acidified with acetic acid, and then boiled for 15 min with three white defatted wool strands to adsorb colorant. The colorant is removed from the wool by heating for 15 min with 1% aqueous ammonia, concentrated, and then chromatographed on Al_2O_3 thin-layer plates containing 5% $CaSO_4$. The eluant was iso-AmOH-EtOH-NH_4OH-H_2O (4:4:1:2).

STINSON, E. E., WILLITS, C. O. JAOAC *46*, 329-330 (1963). Separation of Caramel Color from Salts and Sugar by Gel Filtration. Slurry 475 g of 50-270-mesh Sephadex G-25 (Pharmacia, Uppsala, Sweden) with water and transfer to a 120 cm × 5 cm-ID chromatographic column. Allow excess water to drain to the top of the Sephadex. Dilute 150 ml of syrup to 200 ml, apply it to the column, and wash the column with 3 liters of distilled water at approximately 12 ml/min. The first 680 ml of eluant is

*Registered trademark of Buffalo Color Corporation.

colorless. The colorant appears in the next 560 ml followed by organic salts, sucrose, and sodium chloride.

PIEKARSKI, L., KRAUZE, S. Acta Polon. Pharm. *18*, 103-109 (1961). Dyes Used for Coloring Dragées. Triturate 1-2 g of sample with 5-10 ml of water; filter. Heat the filtrate on a steam bath for 10 min with 0.5 ml of 10% $KHSO_4$ and a few threads of degreased (petroleum ether) wool. Wash the wool with cold water and then heat it for 10 min in 5 ml of 1% NH_4OH. Centrifuge and then evaporate the supernatant liquid to dryness. Dissolve in a few drops of water and chromatograph using BuOH-EtOH-H_2O (2:1:1).

Alternately, extract a solution of dragées with 10 ml of pH = 3 buffer and 10 ml of quinoline. Extract the organic layer with Et_2O and 1-2 ml of H_2O, then evaporate the aqueous layer to dryness, dissolve in a minimum amount of water, and chromatograph as described above.

COSMETICS

ALBORNOZ, A. L. Rev. Fac. Farm. Univ. Central Venezuela 5, 57-66 (1964). Paper Chromatography of Dyes in Lipsticks Made in Venezuela. Extract 20 mg of sample with 1 ml of 10% aqueous NH_3. Evaporate the extract to near dryness and chromatograph the residue on Whatman No. 1 paper using EtOH-H_2O-AcOEt-NH_4OH (25:60:12:3).

BARKER, A. M. L., CLARKE, P. D. B. J. Forens. Sci. Soc. *12*, 449-551 (1972). Examination of Small Quantities of Lipsticks. Extract a 3 mm X 3-mm area of cloth with a few drops of acetone-trichloroethylene (1:1). Chromatograph the extract against standards on a thin-layer plate of Alumina F$_{254}$ using isoamyl alcohol-acetone-H_2O-NH_4OH (50:50:30:0.04) as solvent. Examine visually and under UV light.

COTSIS, T. P., GAREY, J. C. Toilet Goods Assoc. *41*, 3-11 (1964). Determination of Lipstick Dyes by Thin-Layer Chromatography. Transfer about 1 g of lipstick to a flask containing 50 ml of benzene-acetone (3:1). Cover the flask with aluminum foil and reflux on a steam bath. Shake the lipstick suspension vigorously and then immediately apply 100 μl of it as a ¼ X 1½-in. band on a 2 in. X 8-in. glass plate coated with Adsorbosil-1 (Applied Science Laboratories, P. O. Box 140 , State College, Pa.). Most colorants present can be resolved using benzene-MeOH-NH_4OH (65:30:4). For those that can't, use benzene-n-amyl alcohol-HCl (65:30:5) or benzene-PrOH-NH_4OH (60:30:10).

DESHUSSES, J., DESBAUMES, P. Mitt Geb. Lebensm. Hyg. *57*, 373-376 (1966). Thin-Layer Chromatographic Identification of Lipstick Dyes. Extract 0.1-0.2 g of lipstick three times with petroleum ether centrifuging and decanting the supernatant liquid each time. Dissolve the sample residue in 96% EtOH, centrifuge, and chromatograph the supernatant liquid on silical gel G (0.2 mm, according to Stahl) using PrOH-NH_4OH (90:10) as eluant.

JORK, H., LEHMANN, G., RECKTENWALD, U. J. Chromatog. *107*, 173-

179 (1975). Quantitative Determination of Eosin in Cosmetics. Dissolve 0.1-0.4 g of lipstick in 10 ml of dimethylformamide and extract fats with 15 ml of light petroleum (40-60° boiling range). Dilute the remaining solution with 10 ml of water and adsorb the eosin on 10 g of a sand-polyamide (5:1) mixture packed in a 20 cm X 17-mm glass column. Elute the the eosin with 100 ml of methanol-25% aqueous NH_3 (20:1). Spot 1 μl of eluate onto a Kieselgel thin-layer plate and develop for about 1 hr with ethyl acetate-methanol-25% aqueous NH_3 (5:2:1). Detect spots under 366-nm radiation.

KALINOWSKI, D. Roczyn. panst. Zakl. Hig. 27, 403-409 (1976). Thin-Layer Chromatographic Separation and Identification of Triphenylmethane Dyes in Cosmetics. Standard mixtures of 18 dyes, including FD&C Blue No. 1, were isolated from cosmetics by column chromatography on alumina and then resolved by TLC. Procedures are given for removing surface-active constituents of the cosmetics and for separating alkaline and acids dyes.

LEGATOWA, B. Roczniki Panstwowego Zakladu Hig. 16, 453-459 (1965). Separation and Identification of Dyes from Cosmetics. Fluorescein dyes are separated by column chromatography using Celite as the column packing and EtOH-H_2O (1:1) as the eluant. The eluates are evaporated to dryness, made up in 1% NH_4OH, and resolved by two-dimensional paper chromatography using 1% aqueous NH_4OH saturated with isoamyl alcohol as the first solvent and BuOH-EtOH-H_2O-NH_4OH (100:20:44:1) as the second solvent.

LEHMANN, G., EINSCHUTZ, H., COLLET, P. Z. Lebensm. Forsch. 143, 187-191 (1970). The Concentration and Separation of Synthetic Dyes in Lipstick and Facepowder.

PERDIH, A. Z. analyt. Chem. 260, 278-283 (1972). Analysis of Cosmetic Dyes. III. Identification of Synthetic Organic Dyes in Lipsticks by Thin-Layer Chromatography. Schemes are presented for the separation of dyestuffs either directly by TLC using a variety of substrates and solvent mixtures, or by solvent extraction with dimethylformamide followed by TLC.

RUDT, U. Riechstoffe-Kosmetika—Seifen 71, 22 (1969). Fluorometric Determination of Xanthene Coloring Materials in Lipsticks. A method is described for TLC of lipstick dyes on silica with n-PrOH-NH_4OH (9:1).

SHANSKY, A., CARRUBBA, P. P. Am. Perfumer Cosmet. 78, 13-14 (1963). Qualitative Determination of Coal-Tar Dyes in Commercial Cosmetic Products. Solvent extraction and spectrophotometry are used to determine colorants in commercial cosmetic preparations.

SILK, R. S. JAOAC 48, 838-843 (1965). Separation of Synthetic Organic Colors in Lipsticks by Thin-Layer Chromatography for Quantitative Determination. Prepare the following reagents. For buffer solution, prepare a 0.1 M K_2HPO_4 solution and add a few drops of toluene as a preservative. Prepare a 0.1 M KH_2PO_4 solution and add a few drops of toluene. Mix 5.3 ml of the first solution with 94.7 ml of the second solution. Dilute to 200 ml with water.

Solvent A: Mix 20 ml of l-butanol, 4 ml of ethanol, and 3 ml of concentrated ammonium hydroxide.

Solvent B: Mix 15 ml of ethyl acetate, 3 ml of methanol, and 3 ml of 3:7 ammonium hydroxide: water. Prepare fresh.

Apply 0.2 ml of buffer solution in a 1/4-in. band 2 cm from the bottom of a 4-in. × 8-in. glass plate coated with a 375 μm-layer of silica gel G. Air dry for about 20 min. Remove the shiny surface from the rounded end of the lipstick sample with tissue and streak 5-8 mg of it just below the buffered zone of the warmed plate.

Line a No. 11 museum jar with paper and saturate it with dichloromethane. Allow the tank to equilibrate for a few minutes. Place the warm plate in the tank and develop it in the dark until the solvent reaches the top of the plate. Remove and dry the plate. Redevelop two to four more times to remove oils and waxes to the top of the plate. Unsulfonated pigment colors separate in zones in the following descending order: D&C Red No. 36, D&C Orange No. 17, and D&C Red No. 35. Scrape colored zones from the plate, leach with chloroform, and determine visible spectra against standards.

Place the same plate in a covered, unlined Desaga tank (No. 25-10-20) containing 189 ml of Solvent A. Develop to a height of 5 cm. Dry the plate with heat and air. Repeat development once or twice until a 1/4-in. zone appears above the D&C Red No. 7. The D&C Red No. 7 remains close to the baseline while other colors present move through the buffer. Scrape the zone containing Red No. 7 from the plate, leach with 30% acetic acid, and determine the visible spectrum vs a standard.

Line three sides of a Desaga tank with paper. Pour 315 ml of Solvent B over the lining and equilibrate for 10 min. Place the plate in the tank with the adsorbent layer facing the liner. Add glass beads to the tank until the solvent reaches the edge of the continuous coated portion of the plate. Develop to a height of 15-17 cm. If necessary, dry the plate and redevelop. Remaining colors separate in the following descending order: D&C Red No. 19, D&C Red No. 8 plus D&C Red No. 10, D&C Orange No. 4, D&C Red No. 27, D&C Red No. 3, D&C Red No. 21, tribromofluorescein, D&C Orange No. 5, monobromofluorescein, fluorescein, and FD&C Blue No. 1. Scrape colored zones from the plates and leach as follows: D&C Red No. 8 plus D&C Red No. 10 in ethanol, D&C Red No. 19 in 30% (v/v) acetic acid and halogenated fluoresceins in 1:9 ammonium hydroxide. Determine the visible spectra against known standards.

If the sample contains no D&C Red No. 7, it is not necessary to treat the plate with buffer or to develop it in Solvent A. To detect D&C Red No. 7, develop a 2 in. × 4-in. plate in a covered 500-ml tall-form beaker. Develop twice in dichloromethane and then once in 42 ml of Solvent B. The appearance of multiple bands that darken on drying to a dull, nonfluorescent, deep red color and overlap other colors suggest the presence of D&C Red No. 7. A dark red zone at the baseline also indicates its presence. This color is frequently found in dark red or purple lipsticks.

SILK, R. S. JAOAC 46, 1013-1017 (1963). Column Chromatographic Determination of Certifiable Colors in Lipstick. Line a chromatographic tank with filter paper and equilibrate with eluant for 1 hr.

Streak 8-10 mg of lipstick across a 20 cm × 22-cm sheet of Whatman No. 3MM chromatographic paper. Develop (ascending) 1½ hr using methyl-ethyl ketone-acetone-H_2O-NH_4OH (700:200:200:2). Remove the sheet and air dry it in semidarkness. Examine it in visible and UV light against standards similarly chromatographed. Colors that can be identified in this manner include:

D&C Orange No. 17 or D&C Red No. 36
FD&C Yellow No. 5
D&C Orange No. 5 or D&C Orange No. 10
D&C Red No. 21
FD&C Red No. 3
D&C Red No. 27
D&C Orange No. 4
D&C Red No. 8 or D&C Red No. 10
D&C Red No. 19

Subsidiary dyes of halogenated fluoresceins also separate. Based on information obtained above, analyze samples by column chromato-graphy using Procedure 1 when no D&C Red No. 7 is present or Pro-cedure 2 when D&C Red No. 7 is present.

Reagents:

(a) Immobile phase—EtOH-aqueous (1+9) NH_4OH (1:1).

(b) Dilute NH_4OH—NH_4OH-H_2O (1:19).

(c) 30% Acetic acid—Dilute 30 ml of glacial acetic acid to 100 ml with H_2O.

(d) Alkaline heptane-benzene—Dilute one volume of heptane with one volume of benzene. Saturate with 10 ml of (a) per 100 ml of mixture. Discard the lower phase.

(e) 1,1,1-Trichloroethane—Shake each 100 ml with 20 ml of (a). Dis-card the upper phase.

(f) Alkaline 1,2-dichloroethane—Saturate each 100 ml of 1,2-dichloro-ethane (DCE) with 20 ml of (b). Discard the upper phase.

(g) 30% n-Butanol in DCE—Dilute 70 ml of (f) with 30 ml of n-butanol. Shake with enough additional (b) to saturate the solution at room temp.

(h) 40% n-Butanol in DCE—Use appropriate volume. Prepared as described above.

(i) 50% n-Butanol in DCE—Use appropriate volume. Prepared as described above.

(j) 60% n-Butanol in DCE—Use appropriate volume. Prepared as described above.

(k) 80% n-Butanol in DCE—Use appropriate volume. Prepare as describ-ed above.

(l) Acid heptane-benzene—Dilute one volume of heptane with one volume of benzene. Saturate each 100 ml of mixture with 20 ml of (c).

(m) Acid DCE—Saturate each 100 ml of DCE with 20 ml of (c).

(n) 60% Acetic acid—Dilute 60 ml of glacial acetic acid to 100 ml with H_2O.

(o) Celite 545—Wash with chloroform and then with alcohol. Dry at 130°C and then air dry.

Procedure 1: Weigh 10 g of (o) into a 250-ml beaker. Mix thoroughly with 4 ml of (a) and pack into a 1.8 cm-id × 45-cm glass column using a plunger.

Remove the shiny surface from the tip of the lipstick with a tissue. Smear a known weight (0.025-0.03 g) over the inner surface of a 4-oz. mortar.

Thoroughly mix 3 g of (o) and 1.2 ml of (a). Transfer 1/3 of the mixture to the mortar; grind thoroughly. Then mix and grind in the remaining Celite, about 1 g at a time. Transfer the sample Celite mixture to the top of the column. Flush the mortar with an additional 1 g of (o) plus 0.4 ml of (a). Pack down the column to 15-15.5 cm high. Transfer any remaining sample to the column with a little of the first eluant. Elute under 4-5 lb pressure with the appropriate eluants. Use a flow rate of about 10 ml/min (see table that follows for eluants to use). The volume listed are approximate, the exact volume will depend on the concentration of the components to be eluted.

Eluants for Procedure 1	Comments
1. 100-150 ml of (d). Allow the column to wet before applying pressure	D&C Red No. 36 or D&C Orange No. 17 and D&C Red No. 19 elute together. Dilute the solution containing the mixture with an equal volume of heptane and extract the D&C Red No. 19 with 30% acetic acid
2. 150-200 ml of (e)	D&C Red No. 8 elutes with about 150 ml of (e); D&C Red No. 10 requires a larger volume
3. 50 ml of (f)	A small fraction, apparently esterified halogenated fluorescein colors, elutes
4. 100-150 ml of (g)	D&C Red No. 27 elutes; D&C Red No. 27 subsidiary dye does not elute
5. 100 ml of (h)	D&C Red No. 27 subsidiary dye elutes followed closely by D&C Red No. 21
6. 100 ml of (i)	Tribromofluorescein elutes
7. 100-200 ml of (j)[a]	D&C Orange No. 5 and/or D&C Orange No. 10 elute
8. 100 ml of (k)[b]	Monobromofluorescein plus fluorescein elute

[a]If FD&C Yellow No. 5 is present, elute the column through step No. 7 and then elute with 100 ml of (m). This procedure removes any remaining fluorescein colors. FD&C Yellow No. 5 can be removed with dilute NH_4OH.
[b]70% n-BuOH in DCE saturated with (b) elutes monobromofluorescein before fluorescein.

Procedure 2: Prepare a sample and column as in Procedure 1, except use 7 g of (o) and 2.8 ml of (a) for the column. Elute as follows.

Eluants for Procedure 2	Comments
1. 100-150 ml of (d)	Pigment color and D&C Red No. 19 elute. Separate by extraction as in Procedure 1
2. 100 ml of (l)	Solvents (l) and (m) elute fluorescein colors
3. 50 ml of (m)	See (2)
4. 50 ml of (l)	Removes (m) from the column
5. 50 ml of (n)	Elutes D&C Red No. 7

Evaporate fluorescein colors to dryness. Prepare a 10-g column as in Procedure 1. Mix 3 g of (o) with 1.2 ml of (a). Transfer the fluorescein colors to the column as in Procedure 1. Separate the colors as in Procedure 1 starting with step 3, using 50 ml of (f).

Determine the visible spectra of D&C Red No. 19 and D&C Red No. 7 in 30% acetic acid. Evaporate the other solution to dryness. Examine the water soluble dyes at a neutral pH. Dissolve the D&C Orange No. 17 and the D&C Red No. 36 in $CHCl_3$. Dissolve the D&C Red No. 8 and the D&C Red No. 10 in 95% EtOH. Dissolve halogenated fluoresceins in dilute NH_4OH.

TEWARI, S. N. Arch. Kriminol. *126*, 26-32 (1960). Paper-Chromatographic Investigation of Inks, Dyes and Lipsticks. Lipstick is dissolved in warm 40% AcOH and the mixture is filtered then extracted with petroleum ether. The ether extract is evaporated to dryness then taken up on 50% EtOH and chromatographed.

TONNET, N. Mitt. Geb Lebensm. Hyg. 66, 443-472 (1975). Extraction and Identification of Colours Used in Lipsticks. After a review of the bibliography, a scheme based on liquid-liquid extraction is described for the separation of colorants into chemically defined groups.

UNTERHALT, B. Z. Lebensm. Forsch. *144*, 109-112 (1970). Determination of Lipstick Dyes. Extract the sample three times with light petroleum, centrifuging each time. Extract the residue with EtOH. Chromatograph the EtOH extract on Kieselgel G or H (0.25-mm layer) with ethyl acetate-BuOH-NH_4OH (4:11:5) or PrOH-NH_4OH (1:1) as solvent.

DAIRY PRODUCTS

BENK, E., WOLFF, I. Alkohol Ind. *77*, 16-20 (1964). Detection in Egg Liquors of Carotenoids Foreign to Eggs. Carotenoids are separated on Al_2O_3 columns using mixtures of petroleum ether, benzene, and ether and are then identified by TLC using silica gel G plates as the substrate and petroleum ether-benzene-AcOH-Me_2CO (80:20:1:2) as eluant.

DALGAARD-MIKKELSEN, S., RASMUSSEN, F. Intern. Dairy Congress Proceedings 16th, Copenhagen, 1962, Section C, pp 465-473. Tracer Dyes for Rapid Detection of Antibiotics in Milk. As little as 0.03 ppm of some triphenylmethane dyes were detected in milk by the colored zone formed when 10 ml of sample were passed through a column packed with resin.

D'ALMEIDA, A. J. M. Rev. form. Bahia 2, 6-8 (1958). Micromethod for the Determination of Annatto in Cheese. Shake 10 g of grated cheese with 30 ml of EtOH, evaporate 10 ml of the extract to dryness, dissolve the residue in 10 ml of benzene, and centrifuge. Pass the solution through a microcolumn of Al_2O_3 and identify the adsorbed dye by the blue color produced with concentrated H_2SO_4.

DHAR, A. K., GUBA, K. C., ROY, B. R., MITRA, S. N. Ind. J. Dairy Sci. 24, 202-207 (1971). Detection of Added Colour in Milk and Milk Products. Two procedures are given for separating fat-soluble and acidic and basic water-soluble coal-tar dyes and natural dyes before identification by conventional methods.

Procedure 1: Repeatedly shake the milk with the same volume (or 3 volumes if formaldehyde is present) of ethanol-ethyl ether (1:1) until the lower phase is colorless, filter the organic extracts, and evaporate almost to dryness. Extract the residue with hot H_2O to test for water-soluble dyes, or with ether to test for fat-soluble dyes, annatto, and turmeric. Alternatively, make the milk alkaline with aqueous NH_3 and extract with ether before adding the ethanol-ether mixture; an extra volume of ethanol must then be added. This allows separate extraction of the basic coal-tar dyes, oil-soluble dyes, and annatto before extraction of the acidic dyes and turmeric.

Procedure 2: Acidify the warmed milk with acetic acid (1:3) and boil the mixture for a few minutes. Collect the casein in a fine cloth, wash it with hot H_2O, and leave it in ether overnight to extract fat-soluble dyes, annatto, turmeric, and basic dyes. For water-soluble acidic dyes, dry the extracted ppt, heat it with 80% ethanol containing 1% aqueous NH_3, filter, and evaporate the solution almost to dryness. Caramel is left on the casein ppt.

ESPOY, H. M., BARNETT, H. M. Food Technol., 357 (August 1955). The detection of annatto, β-carotene, Yellow AB, and Yellow OB in butter and margarine.

FEAGAN, J. T., GRIFFIN, A. T., BRAY, R. Aust. J. Dairy Technol. 20, 22-23 (1965). An Improved Test for the Detection of Marker Dyes in Milk. Dilute 100 ml of milk with 100 ml of hot deionized water. Vacuum filter through two 1.25-in sediment pads with 0.2 g of Dowex AC (Cl^- form) ion exchange resin between them. View the resin for color. As little as 0.0025 mg of FD&C Blue No. 1 per liter of milk can be detected using this procedure.

HARTMAN, C. P., PICHAMUTHU, S. J. Inst. Chem. Calcutta 42, 114-117 (1970). Paper Chromatographic Method for the Detection of Metanil Yellow and Vanaspati in Butter.

HORWITZ, W., (Ed.) *Official Methods of Analysis of the Association of Official Analytical Chemists*, 12 ed., pp. 268, 275, 276, 291. Color Additives in Ice Cream, Cream, Milk and Evaporated Milk. Ice Cream—Curdle

150-200 g of melted sample by adding an equal volume of water and 10-20 ml of HOAc. Heat to 70-80° with stirring and then allow to cool. Continue as below beginning with "Gather curd, when possible. . . ."

Milk, cream, and evaporated milk—Warm about 150 ml of milk in a casserole over a flame, add approximately 5 ml of HOAc (1 + 3), and continue to heat slowly nearly to boiling point while stirring. Gather curd, when possible, into one mass with stirring rod and pour off whey. If curd breaks up into small flecks, separate from whey by straining through sieve or colander. Press curd free from adhering liquid, transfer to small flask, macerate with about 50 ml of ether, keeping flask tightly corked and shaking at intervals, and let stand for several hours, preferably overnight. Decant ether extract into evaporating dish, remove ether by evaporation, and test fatty residue for annatto as follows.

Pour on moistened filter paper an alkaline solution of color obtained by shaking out oil or melted and filtered fat with warm 2% NaOH solution. If annatto is present, paper absorbs color, so that when washed with gentle stream of H_2O it remains dyed straw color. Dry paper, add drop of 40% $SnCl_2$ solution, and again dry carefully. If color turns purple, presence of annatto is confirmed.Curd of uncolored milk and milk colored with annatto is prefectly white after complete extn with ether. If extracted fat-free curd is distinctly orange or yellowish, synthetic dye is indicated. In many cases if lump of fat-free curd in test tube is treated with little HCl, color changes to pink, indicating presence of dye similar to aniline yellow or butter yellow or perhaps one of the acid azo yellows or oranges.

In some cases presence of synthetic dyes can be detected by directly treating about 100 ml of milk with equal volumes of HCl in porcelain casserole, giving dish slight rotary motion. In presence of some dyes separated curd becomes pink.

JAX, P., AUST, H. Milchwiss. Ber. 145-189 (1953). The Chromatography of Butter and Cheese Dyes and the Dyes of Other Dairy Products. Procedures are described for separating and identifying mixtures of fat- and water-soluble dyes alone or in butter or cheese.

KANEMATSU, H., NIIYA, I., IMAMURA, M., KAWAKITA, H. Bitamin 33, 52-56 (1966). The Quantitative Determination of β-Carotene and Vitamin A in Margarine. β-Carotene is eluted from a column of activated alumina using acetone-petroleum ether (1:49). Vitamin A and other pigments remain on the column.

LEHMANN, G., EINSCHUETZ, H., COLLET, P. Z. Lebensm. Forsch 143, 187-191 (1970). Analysis of Dyes. III. Enrichment and Isolation of Artifical Dyes in Cheese-Coating Materials and Lipsticks. Cheese-coating materials—The waxy coating material is dissolved in light petroleum and the solution is extracted with H_2O-98% formic acid (2:1). The aqueous extract is passed down a microcolumn of polyamide powder, and the column is washed with H_2O until the washings are neutral. The adsorbed dyes are eluted with 5 ml of eluant [0.1% NaOH solution in 70% methanol, or concentrated aqueous NH_3-methanol (1:19)] and then with 5 ml of methanol. The combined eluates are acidified with methanol-acetic acid (1:1) and evaporated under reduced pressure to $\simeq 1$ ml. The dyes are then identified by TLC on cellulose powder, with 2.5% aqueous Na citrate-concentrated aqueous NH_3-methanol (20:5:3) as solvent, by comparison of R_f values with those of standards.

Lipsticks—The sample is triturated with methanol-formic acid-acetone (3:2:1) and Celite in a porcelain mortar. After evaporation of the solvent the powder is transferred to a microtube. Lipophilic dyes are eluted from the column with light petroleum, the solvent is evaporated in a rotatory evaporator, the residue is dissolved in a little warm methanol, and the solution is filtered. Other dyes are then eluted from the column with methanol followed, if necessary, by methanol-formic acid (3:2). The combined solutions are purified on a microcolumn as in the procedure described above, and the dyes are identified by TLC on Kieselgel GF_{254} with ethyl acetate-methanol-concentrated aqueous NH_3 (5:2:1) as the eluant.

LEHMANN, G., COLLET, P. Z. Lebensm. Forsch *143*, 348-350 (1970). Contribution to the Analysis of Dyes. V. Detection of Synthetic Dyes in Liquid Eggs. The sample is treated with acetone, and the dyes adsorbed on the fat- and H_2O-free ppt are desorbed with concentrated aqueous NH_3-methanol (1:19). The alkaline extract is acidified with acetic acid to pH $\simeq 5$, and the dyes are subjected to purification on a microcolumn of polyamide powder and reextracted with aqueous NH_3-methanol; the extract is acidified and evaporated to a small volume, and the dyes are identified by paper or thin-layer chromatography. The acetone extract, containing fat, fat-soluble dyes, and, in part, the acid dyes, is diluted with H_2O, the acetone is removed by distillation under reduced pressure, and the fat-soluble dyes are extracted with light petroleum and identified by paper or thin-layer chromatography. The aqueous phase is acidified to pH $\simeq 6$ and purified on a microcolumn of polyamide powder. The adsorbed basic dyes are eluted with acetone and identified.

LEHMANN, G., COLLET, P. Z. Lebensm. Forsch *144*, 32-34 (1970). Detection of Synthetic Dyes in Milk Products. Procedures essentially the same as those used for liquid eggs (see) were applied to yogurt, cream, ice cream, and milk shakes.

LEONE, J. L. JAOAC 56, 535-537 (1973). Collaborative Study of the Quantitative Determination of Titanium Dioxide in Cheese. Weigh 10 g of sample into a 100-ml Petri dish and char under an IR lamp. Place in a cold furnace and ignite at $850°C$ to a white ash.

Cool, add about 1.5 g of anhydrous Na_2SO_4 and 10 ml of H_2SO_4, cover with a watch glass, and bring to a boil on a hot plate to dissolve. Turn heat off and let cool on the hot plate. Cautiously rinse cover, add 30 ml of H_2O, and mix with a stirring rod to disperse insoluble salts. Heat on a steam bath if insoluble material forms a cake on the bottom of the dish. Transfer quantitatively to a 100-ml volumetric flask using about 40 ml of H_2O. If the solution is cloudy heat on a steam bath or in a boiling H_2O bath. Cool; dilute to volume with H_2O. Pipette 3 ml of sample solution into a 5-ml volumetric flask and then dilute to volume with H_2SO_4 (1 + 9). Add 0.2 ml of 30% H_2O_2, mix well, then determine the sample's absorbance at the maximum near 408 nm. Compare against standards similarly prepared.

MARMION, D. M. JAOAC 54, 131-136 (1971). Analysis of Allura* Red AC Dye (A Potential New Color Additive) FD&C Red No. 40 in Ice Cream. Weigh 10 g of well-mixed melted ice cream into a 1 in. × 4.5-in. centrifuge tube. Add 35 ml of SD No. 30 alcohol and stir well. Centrifuge until clear and decant supernatant liquid into a 150-ml beaker. Repeat extraction and centrifuging, and combine supernatant liquids. Add 1 ml of galcial acetic acid to combined extracts and boil for 1 min. Let cool for 10 min and then place in ice bath for 30 min; stir occasionally. Filter through thick pad of alcohol-washed cotton into a 100-ml volumetric flask. Using chilled alcohol, wash all color from pad into flask (filtrate must be clear). Dilute to volume with SD No. 30 alcohol; mix. Similarly extract sample of ice cream containing no color. Using a suitable spectrophotometer, immediately determine absorbance of each solution in 5-cm cell (vs. SD No. 30 alcohol) at maximum near 505 nm and at 680 nm. Sample absorbance at maximum near 505 nm = A_1; sample absorbance at 680 nm = A_2; blank absorbance at 505 nm = A_3; blank absorbance at 680 nm = A_4.

Percent Allura* Red AC dye

$$= \frac{(A_1 - A_2 - A_3 + A_4) \times 100}{100 \times 5 \times 52.9} = A \times 0.00378$$

where 52.9 = absorptivity of Allura* Red AC dye at 505 nm in liters/g-cm; 100 = factor for conversion to percent; 100 = effective sample concentration in g/liter; and 5 = cell path length in cm.

PEREDIH, A., PRIHAVEC, D. Z. Lebensm. Forsch. 134, 239-242 (1967). Isolation of Water-Soluble Food Dyes. A method is designed for isolating food colors from protein materials, including eggs, meat, fish, and milk products. Strongly polar lipids are first defatted with $CHCl_3$-EtOH (2:1); viscous liquids or liquids containing greater than 50% alcohol are diluted with water; water-soluble samples are dissolved in water; and water-insoluble samples are ground with water to form an easily extractable suspension.

Transfer 10 g of sample into a centrifuge tube and then add 2-5 ml of H_2CO and 10 ml $CHCl_3$. Shake well and then centrifuge for 2 min at 2500 rpm. Repeat this extraction four times, and then treat the residue with pH 9-9.5 NaOH or NH_4OH and reextract with $CHCl_3$. Test extracts for fat soluble and basic colorants. Then add 0.5-2 ml of 10% alkyldimethylbenzylammonium chloride (or other similar quaternary ammonium salt) to the sample residue, mix well, add 10 ml $CHCl_3$, and shake and centrifuge as described above. Repeat the $CHCl_3$ extractions as needed. Wash the combined $CHCl_3$ extracts with water, concentrate, and then chromatograph on Na alkyl sulfate or Na alkylarenesulfonate-impregnated paper.

*Registered trademark of Buffalo Color Corporation.

RAMAMURTHY, M. K., BHALERO, V. R. Analyst 89, 740-744 (1964). A Thin Layer Chromatographic Method for Identifying Annatto and Other Food Colours. Extraction of color from butter: Dissolve 10 g of sample in 50 ml of diethyl ether. Pass the solution through a 7.5 cm X 1.5-cm-diameter glass column packed with aluminum oxide (E. Merck & Co., Inc.) prepared according to Brockmann. Annatto and curcumin are adsorbed on the column, whereas other fat-soluble dyes pass through. Concentrate the eluate by evaporating the ether and then saponify the residue with alcoholic potassium hydroxide. Extract the color with three portions of diethyl ether. Wash the combined extracts with water, dry over anhydrous sodium sulfate, evaporate to concentrate, and examine by TLC. Elute annatto and curcumin from the column with 25 ml of ethanol-ammonia (2:1). Acidify the eluant with 2N HCl, dilute it with water, and extract it three times with diethyl ether. Wash the extract with water, dry over anhydrous sodium sulfate, evaproate to concentrate, and examine by TLC.

SADINI, V. Intern. Dairy Congr. Proc. (16th, Copenhagen) 3, 474-486 (1962). Detection of Food Dyes in Dairy Products.

SCHWARZ, G., MUMM, H., WOERNER, F. Molkerei-u. Kaserei-Ztg. 9, 1430-1433 (1958). Coloring Cheeses with Annatto and Carotene Dyes and Their Detection. Extract 25-50 g of minced cheese for 20 hr with acetone. Evaporate the extract to dryness and then extract the residue with 15 ml of benzene. Dry the extract with Na_2SO_4 and transfer to a column packed with Al_2O_3. Elute carotene from the column with benzene and then elute annatto with chloroform.

USHER, C. D., FAVELL, D. J., LAVERY, H. Analyst 93, 107-110 (1968). A Method for the Determination of Vitamin A, α- and β-Carotene in Margarine, Including the Results of a Collaborative Test. Weight 10 g of margarine into a 250-ml flat-bottomed flask. Add 20 mg of quinol, 60 ml of ethanol, 10 ml of 60% w/v potassium hydroxide solution, and 10 ml of light petroleum. Boil under reflux for 30 min, protecting the flask from light. (Use flasks covered with a shield of aluminum foil.) Cool, and add 80 ml of distilled water. Transfer the solution into a 500-ml separatory funnel; rinse the flask into the funnel with an additional 80 ml of water. Extract the unsaponified material with 100 ml and three 50-ml portions of diethyl ether. Combine the ether extracts and wash with four 50-ml portions of distilled water; carry out the first washing by swirling and the following three by gentle shaking. Using a stream of inert gas, evaporate the unsaponifiable extract to dryness on a water bath at $50°C$. The last stages of the evaporation require full attention, because the residue in the flask must not be allowed to remain dry longer than is absolutely necessary. Immediately after all of the diethyl ether has been removed, add 2 ml of absolute ethanol and again evaporate to dryness in a current of inert gas; if the residue appears wet, repeat the addition of absolute ethanol and evaporation to dryness. Immediately dissolve the residue in 5 ml of light petroleum and again evaporate to dryness in a current of inert gas. Repeat the dissolution in light petroleum and evaporation to dryness twice more. Finally, dissolve the residue in 2-3 ml of light petroleum for chromatography.

Pretreat magnesia by heating magnesium oxide (heavy) at 100°C for 2 hr. Cool in a desiccator and set aside for 3-4 days in an airtight bottle.

Place a pledget of cotton wool in the tip of the chromatographic tube shown in Fig. 17. Add petroleum ether (boiling range, 40-60C°) to a level half-way up the center section and add 3 g of magnesia. Drain the ether just to the surface of the packing.

Using 2 ml of petroleum ether, transfer the sample solution to the column. Develop the chromatogram, under pressure if necessary, with light petroleum ether containing 4-12% ethyl ether. The exact amount of ethyl ether necessary varies with different batches of magnesia and must be determined by experience. α-Carotene elutes first as a pale yellow band. β-Carotene elutes next as a deeper orange-colored zone. After the α-carotene elutes, 1:1 ethyl ether-light petroleum may be used to speed up the elution of β-carotene. Determine both materials spectrophotometrically against standards.

VERMA, M. R., RAI, J., GANGOPADHYAYA, N. Ind. J. Technol. *1*, 358-360 (1963). Chromatographic Method for the Separation of Dyes from Butter and their Identification. The sample is dissolved in benzene, adsorbed on a column of alumina, and eluted with benzene than EtOH. Annatto that remains adsorbed on top of the column is removed with alcoholic ammonia.

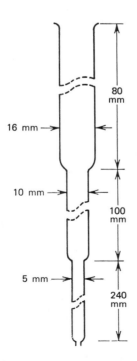

16 mm

10 mm

5 mm

80 mm

100 mm

240 mm

Figure 17   Chromatographic Tube

HAMMOND, E. G., CHANG, J., REINBOLD, G. W. J. Dairy Sci. *58*, 1365-1366 (1975). Colorimetric Method for Residual Annatto in Dry Whey. Mix 1 g of sample with 2 ml of 30% aqueous NH_3-H_2O (1:4) in a 15-ml stoppered tube for 1 min. Add 10 ml of anhydrous EtOH, shake the tube well, and then centrifuge for 3 min at 2500 rpm. Transfer the supernatant solution to a similar tube, add two drops of phosphate solution (17.1 g of $NaH_2PO_4 \cdot H_2O$ plus 10.8 g of Na_2HPO_4 in 100 ml of H_2O), shake the mixture, and then set it aside for 30 min. Centrifuge for 3 min and then measure the absorbance of the upper phase at 450 nm against a blank containing no whey.

DRUGS

ALARY, J., DUC, C. L., COEUR, A. Bull. Trav. Soc. Pharm. Lyon. *10*, 78-86 (1966). Identification of Synthetic Colorants in Drugs. Colorants in drugs are identified by ascending paper chromatography using BuOH-EtOH-NH_4OH-H_2O (50:25:10:25) or by TLC on Kieselgel using BuOH-MeOH-C_6H_6-H_2O. Both separations are performed in subdued light.

BALATRE, P., TRAISNEL, M. Bull. Soc. Pharm. Lille *1*, 41-47 (1965). Identification of Pharmaceutical Dyes by Thin-Layer Chromatography of their Complexes with a Quaternary Ammonium Compound. Colorants were extracted from the drugs, complexed with a quaternary ammonium derivative, and separated by TLC on Kieselgel G both with and without $NaCO_3$ binder, on alumina with $NaCO_3$ binder, or on cellulose MN300. The best separations were obtained with BuOH-EtOH-H_2O (2:1:1) and EtOAc-pyridine-H_2O (7:3:1) as eluants.

BALATRE, P., MULLEMAN-MARSY, D., TRAISNEL, M. Ann. pharm. fr., *25*, 649-653 (1967). Identification and Determination of Natural Dyes of Vegetable Origin in Drugs. Transfer 1 ml of aqueous sample solution to a separatory funnel. Add 2 ml of 10% aqueous Na_2CO_3. Shake well. Add 20 ml of surfactant solution [0.1 g of hexadecyl-(2-hydroxycyclohexyl) dimethylammonium bromide, 0.8 g of benzyl-lauryldimethylammonium bromide, and water to 100 ml] and 20 ml of $CHCl_3$. Shake the mixture for 10 min, centrifuge, and filter the $CHCl_3$ layer through absorbent cotton. Fat-soluble dyes, such as carotenes, xanthophylls, and chlorophyll, are not readily extracted. Indigotins and caramel are extracted from neutral solution, and the anthocyanins of bilberry, from acid solution containing $(NH_4)_2SO_4$. The dyes are identified and determined from the colors of the extracts under daylight and UV radiation, the absorption maximum, and the $E_{1cm}^{1\%}$. The $CHCl_3$ solution can be subjected to TLC on Kiesegel G (applied to the plates as a suspension in 1% aqueous Na_2CO_3) using butanol-ethanol-H_2O (2:1:1) or ethyl acetate-pyridine-H_2O (7:3:1) as the eluant.

JEKABSONS, E. JAOAC *52*, 110-112 (1969). Fluorometric Analysis of Sodium Fluorescein in Ophthalmic Solutions. Dilute the sample with water so that it contains about 1 μg of sodium fluorescein per milliliter. Transfer 3 ml of sample solution and 20 ml of borate buffer (0.05 *M* boric acid in 0.05 M KCl adjusted to pH = 9 with 0.2 M NaOH) to a 100-ml volumetric flask and dilute to volume with water. Mix well and then measure the sample's fluorescence at 515 nm with excitation at 460 nm and compare with that of standards.

JENTZSCH, K., SPIEGL, P., KAMITZ, R. Scientia pharm. *38*, 50-58 (1970). Qualitative and Quantitative Investigations on *Curcuma* (Turmeric) Colouring Matters in Zingiberaceae Drugs. II. Quantitative Investigation. Transfer 0.1 g of finely powdered sample to a Soxhlet apparatus and extract for 30 min with 10 ml of 96% ethanol. Evaporate the extract to 2 ml and then dilute to 5 ml with 96% ethanol. Chromatograph 220 μl of solution on Kieselgel H (0.25 mm) using CHCl$_3$-benzene-ethanol (1:8:1). Dry the chromatogram for 30 min at 90-100°C. Examine under UV light, remove the appropriate areas from the plate, extract the colorants with 96% ethanol, and determine spectrophotometrically.

LEHMANN, G., COLLET, P. Arch. Pharm. Berl. *303*, 855-860 (1970). Analysis of Dyes. VIII. Identification of Synthetic Dyes in Drugs. Dyes are adsorbed on Polyamide MN SC6 powder. The powder is transferred to a 150 mm × 15 mm microchromatographic column, eluted, and then identified by TLC on cellulose layers using 2.5% aqueous ammonium citrate-aqueous NH$_3$-methanol (20:5:3) as eluant.

MARES, V., STEJSKAL, Z. Cslka. Farm. *16*, 474-479 (1967). Identification of Dyes Used for Coloring Drugs. Dyes are extracted from drugs with quinoline and then separated on Whatman No. 1 paper (descending) using 2.5% aqueous Na citrate-25% aqueous NH$_3$ (4:1) plus 3% triethanolamine, or BuOH-acetic acid-H$_2$O (1:1:1).

MERKUS, F. W. H. M., SAGEL, J. Pharm. Weekblad 99, 1098-1116 (1964). The Use and Analysis of Synthetic Dyes in Pharmaceutical Products. Colorants are extracted with quinoline, amyl alcohol, or BuOH and the extracts are chromatographed on paper using BuOH-EtOH-H$_2$O (1:1:1) or 2% Na citrate in 5% NH$_4$OH.

PELLERIN, F., GAUTIER, J. A., CONRARD, A. M. Ann. Pharm. Franc. *22*, 621-627 (1964). Identification of Authorized Synthetic Organic Dyes in Pharmaceuticals. The sample is extracted with 10 ml of water and filtered. Then 1 g of Na$_2$SO$_4$ and 1.5 ml of 1:5 H$_2$SO$_4$ is added to the filtrate (except for alizarin-erythrosine, which is extracted at neutral pH), a 1-cm-wide piece of polyfiber (Colcombet) is added, and the solution is heated for 30 min in boiling water. The ribbon is washed thoroughly with H$_2$O at 40-45°C and then dried below 50°C. Colorant is stripped from the ribbon with 2-5 ml of 10% aqueous NH$_4$OH, the extract is evaporated to dryness on a water bath, and the residue is dissolved in 0.5 ml of water and then resolved by paper or thin-layer chromatography.

PELLERIN, F., KIGER, J. L., CAPORAL-GAUTIER, J. Synthetic Organic Colours in Plastic Packaging Materials for Pharmaceutical Use. II. Identification in Plastics and Detection of Their Release into Drugs. Ann. pharm. fr. *32*, 427-431 (1974). The plastic, cut into fine slivers, is dissolved in 10 ml of benzene, toluene, or acetone for polyalkenes, in 1,4-dioxan or tetrahydrofuran for poly (vinyl chloride), in formic acid or cyclohexane for polyamides, and in dichloroethane or acetone for cellulose acetate. The dyestuffs are identified in the solution by: (1) TLC on silica gel, with $CHCl_3$-xylene (3:1), benzene-$CHCl_3$ (4:1) or $CHCl_3$-acetic acid (200:1) as solvent, and development for 12 to 15 cm; (2) spectrophotometric measurements; (3) ppt of the plastic by adding another solvent (e.g., $CHCl_3$, ethanol, or H_2O) and then TLC of the filtrate; and (4) ppt of the plastic with H_2O, after dissolution in acetone or H_2SO_4, with the dyestuff examined for ion-pair formation in $CHCl_3$ by reaction with dodecyl sulfate or cetylpyridinium salt. The method is also applied to storage tests on semisynthetic glycerides in the presence of dyed plastic packaging materials.

PLA DELFINA, J. M., MACIAN, R. S. Galenica Acta 9, 243-286 (1956). Chromatography of Synthetic Colors in Pharmaceutical Preparations Used Internally. Samples (10 g) were extracted with water at $60°C$. Any insoluble residue was reextracted with 5% tartaric acid at $60-80°C$, adsorbed on wool, eluted with 0.02N NH_3, concentrated, and then chromatographed on Schleicher & Schull 2043A paper using water-saturated BuOH or $(ClCH_2)_2CHOH$. The water extract was split in two. One portion was acidified with 5% tartaric acid, adsorbed on and stripped from wool as described above, and then extracted with AmOH. Both layers were examined for colorants. The second portion of the water extract was treated with 5% NaCl and the colors were adsorbed on wool, eluted with normal tartaric acid, and extracted with $CHCl_3$. The $CHCl_3$ extract was examined for colorant.

SERINI, G. Chimica 34, 95-96, 144-145, 197-200 (1958). Separation and Identification by Paper Chromatography of Dyes Added to Aliments.

SITZIUS, F., RENTSCH, H. Pharm. Ind. Berl. 35, 148-150 (1973). Detection of Colouring Matter in Capsules and Sugar-Coated Tablets. A suitable number of empty gelatin capsules is dissolved in 5 ml of 10% acetic acid and the mixture is passed through a 1-cm column containing 1.5 g of alumina (Brockmann). The gelatin is removed by passing 10 ml of H_2O thru the column using gentle suction. Colorant is eluted with 0.1% aqueous NH_3, the eluant is evaporated to dryness on a steam bath, and the residue is dissolved in a few drops of methanol and examined by TLC on G1440 cellulose plates.

STORCK, J. Ann. Pharm. Franc, 23, 113-115 (1965). Detection of Dyes in Pharmaceutical Gelatin Capsules. Five gelatin capsules are dissolved in 25% HOAc and placed on an alumina column. The colorant is eluted with NH_4OH-H_2O (1:100), concentrated to 5-10 ml, and then chromatographed (descending) on Whatman No. 1 paper using tri-Na citrate dihydrate-NH_4OH-H_2O (2 g, 20 ml, dilute to 100 ml), or on 0.25 mm of Kieselgel G using Et_2NH-MeOH-EtOH (10:35:55).

UNTERHALT, B., KREUTZIG, L. Dt. Apoth Ztg. 112, 449-450 (1972). Detection of Dyestuffs in Cough Linctuses. Dilute 10 ml of sample with 40 ml of H_2O and acidify with $KHSO_4$ or HOAc. Adsorb colorant onto wool fibers or onto a column of polyamide powder (0.5 g). Elute colorant with methanolic NH_3, evaporate eluate to dryness, dissolve in two drops of H_2O, and chromatograph on a layer of Cellulose MN300 using aqueous NH_3-2.5% aqueous monosodium citrate (1:4), propanol-ethyl acetate-H_2O (6:1:3), or ethyl acetate-pyridine-H_2O (3:1:1).

WOJCIK, Z. Farmacja pol. *25*, 419-425 (1969). Chromatographic Identification of Synthetic Dyes in Pharmaceutical Preparations. Scrape the colored coating from 5-10 tablets, dissolve the scrapings in 10-20 ml of H_2O, add 1-2 ml of 10% HCl, and mix in 2 g of alumina. Filter the mixture on a sintered-glass filter, wash with 100 ml of H_2O, and then extract colorant with 5 ml of 0.5% aqueous NH_3. Evaporate the extract to dryness, dissolve the residue in a few drops of water, and chromatograph on Whatman No. 1 paper using 2.5% aqueous Na citrate-25% aqueous NH_3-triethylamine (80:20:3) and then BuOH-HOAc-H_2O (1:1:1). Examine under daylight and under UV light.

WOJCIK, Z. Farmacja pol. *26*, 723-729 (1970). Thin-Layer Identification of Azo Dyes Permitted in Poland for Use in Pharmaceutical Preparations. Colorants are extracted as described in the preceding paragraph, applied to a plate of MN300 cellulose powder, activated at 100°C for 1 hr, and then developed with 2.5% aqueous Na citrate-aqueous NH_3 (7:3), PrOH-ethyl acetate-H_2O (5:2:3), or BuOH-HOAc-H_2O (25:5:12).

FATS AND OILS

BOSE, P. K., ROY, B. R., MITRA, S. N. J. Food Sci. Technol. *7*, 112-113 (1970). Analysis of Oil-Soluble Dyes from Foods Using Clean-Up by Adsorption. Oil containing natural or synthetic dyes is diluted with light petroleum (boiling range 60-66°C) and then sufficient chromatographic-grade silica gel is added to adsorb the colorants. The solvent is removed by decantation and then the silica is washed with light petroleum. The adsorbed dyes are extracted from the silica gel with methanol and identified by reverse-phase chromatography.

DAVIDEK, J., JANICEK, G. Qual. Plant. Mater. Veg. *16*, 253-257 (1968). Thin-Layer Chromatographic Separation of Fat-Soluble and Water-Soluble Food Dyes. The colored fat is saponified and the dyes are extracted with light petroleum ether and then separated by chromatography.

JONES, F. B. JAOAC *49*, 674-678 (1966). Synthetic Organic Colors in Oils. Prepare the following columns using 20 mm-ID X 300-mm glass tubes.

(a) Florisil column—Activate 60-100-mesh Florisil (Floridin Co., Englewood Cliffs, N. J.) at 650°C. Store at 130°C. For use, add 1.5 ml of H_2O to 100 g in a stoppered bottle, shake to break up lumps, mix thoroughly, and let stand overnight. Pack the column 4 in. high and wash with petroleum ether.

(b) Alumina column—Heat 80-200-mesh alumina for 1 hr at 400°C. Add 50 ml of petroleum ether to a closed chromatographic tube. Add 18 g of alumina, mix, and drain the ether to the top of the column.

(c) Magnesia column—Mix equal weights of Sea Sorb-43 and Celite 545.

Pack 9 g of the mixture as described above for the alumina column. Compress with slight air pressure.

(d) Silicic acid column—Add 4 in. of a mixture of equal weights of 100-mesh silicic acid and Celite 545. Wash with n-hexane, using pressure.

Dilute 10 ml of sample with 10 ml of petroleum ether and place on the Florisil column. Elute with petroleum ether, collecting the colored zone (eluate No. 1). Elute with ether, collecting the colored zone (eluate No. 2). Elute with 1:3 ethanol-ethyl ether, collecting each resolved color separately; usually the natural base-oil color, which is discarded, is followed by D&C Violet No. 2, and then D&C Yellow No. 11. Elute with acetonitrile and collect any D&C Red No. 35.

Evaporate the individual ethanol-ethyl ether and acetonitrile eluates to dryness, dissolve the residues in chloroform, and determine their visible spectra. Evaporate eluate No. 2 to dryness. Add eluate No. 1 to the residue and evaporate to about 15 ml. Transfer the solution to the alumina column and wash with 50 ml of petroleum ether; discard the eluate. Add two 10-ml portions of chloroform. If the chloroform eluate is green or blue, add it to the following alcohol-CHCl$_3$ eluate; if the chloroform eluate is colorless, discard it. Elute the column with 1:3 ethanol-chloroform until the eluate is colorless. Evaporate the eluate; dissolve any residue in petroleum ether.

Add this solution to the magnesia column. Wash with 25 ml of petroleum ether, discarding the eluate. Elute with chloroform collecting the individual colored zones. The first fraction contains D&C Green No. 6 and Ext. D&C Blue No. 5. The second fraction contains Ext. D&C Orange No. 4 and Ext. D&C Red No. 14. Elute with 1:3 ethanol-chloroform, collecting the individual colored zones. They are D&C Red No. 18, Ext. D&C Yellow Nos. 9 and 10, and D&C Red No. 17. Evaporate the individual fractions to dryness and compare the visible spectra of the residues in chloroform with standards.

If the spectrum of the blue-green portion does not conform to a known color, evaporate the chloroform and dissolve the residue in n-hexane. Transfer it to the silicic acid column and elute with 1:1 n-hexane-benzene. Collect the eluate until it is colorless. Elute with benzene and collect the eluate until it is colorless. Evaporate the individual fractions to dryness and compare the visible spectra of the residues in chloroform with standards. If the spectrum for D&C Red No. 17 has a minimum at 385 nm, Ext. D&C Yellow Nos. 9 and 10 may be present. To resolve, evaporate the chloroform, dissolve the residue in a minimum volume of petroleum ether, and transfer it to the magnesia column. Elute with 1:3 ethanol-chloroform and collect the individual colored fractions. Evaporate each fraction to dryness and compare the visible spectra of the residues in chloroform with standards.

HORWITZ, W., Ed. *Official Methods of Analysis of the Association of Official Analytical Chemists*, 12th ed. 1975, p. 279. Color Additives in Fat. Pour about 2 g of filtered fat, dissolved in ether, into each of two test tubes. To one tube add 1-2 ml of HCl (1 + 1) and to other about the same volume of 10% NaOH solution. Shake the tubes well and let stand. In the presence of some azo dyes the acid solution turns pink to wine-red, whereas the alkaline solution in the other tube shows no color. However, if annatto or some other vegetable color is present, the alkaline solution is yellow, whereas no color is apparent in the acid solution. (Red changing to yellow, especially on warming, in alkaline solution may be due to presence of gallate antioxidants.)

LINDBERG, W. Z. Lebensm. Forsch. *103*, 1-14 (1956). Detection and Identification of Fat-Soluble Coal-Tar Dyes in Food Products. The fat or oil is dissolved in petroleum ether and the colorant is extracted with acid solution (20 ml of HCl, 10 ml of H_2O made to 100 ml with HOAc, or 40 ml of H_2SO_4, 10 ml of H_2O, and 90 ml of HOAc). The acid extract is extracted with ether, the ether is evaporated, and the residue is saponified. The unsaponified material is isolated with EtOAc and the colorant therein identified by chromatography.

MARK, E., MC KEOWN, G. G. JAOAC *41*, 817-818 (1958). Isolation of Oil-Soluble Coal-Tar Colors from Foods. Dissolve 10 g of sample in 50 ml of petroleum ether. Filter, if necessary, into a separatory funnel. Extract with three 20-ml portions of *N,N*-dimethylformamide (DMF); discard the ether layer. Combine the DMF solutions and extract with four 25-ml portions of petroleum ether, back extracting each time with 5 ml of DMF. Discard the ether extracts. Dilute the combined DMF solutions with an equal volume of water and extract with 30 ml, and then 10 ml of chloroform. Discard the aqueous DMF layer. Combine the chloroform extracts and wash them with water to remove any dissolved DMF. Evaporate the chloroform solution to dryness under vacuum at room temperature. Dissolve the residue in 25 ml of DMF and transfer it to a separatory funnel. Add 25 ml of water and extract the solution with three to five 25-ml portions of petroleum ether. Discard the aqueous DMF layer and wash the combined ether layers wih water. Evaporate the ether solution under vacuum at room temperature. Examine the residue chromatographically or spectrophotometrically.

FRUITS

ANONYMOUS. Identifying Artificial Color on Oranges. Chemistry *43*, 29-30 (1970). Rinse the colorant from the surface of the orange with 25 ml of $CHCl_3$, evaporate the solution to dryness on a steam bath, and then dissolve the residue in 3 ml of $CHCl_3$. Chromatograph the solution for 1 hr on paper impregnated with a solution of 5 g of mineral oil in 95 ml of Et_2O using 65% Me_2CO.

DRAPER, R. E. JAOAC *56*, 703-705 (1973). Separation and Determination of FD&C Red No. 4 and FD&C Red No. 40 in Maraschino Cherries by Column Chromatography.

Reagents:

(a) Solvents A and B—Add 500 ml of 5% Amberlite LA-2 resin (Rohm & Haas Co., Philadelphia, Pa.) in *n*-butanol, 200 ml of water containing 7.5 ml of acetic acid, and 12.5 ml of saturated $(NH_4)_2SO_4$ solution to a separatory funnel. Shake vigorously for 1 min and let phases separate. Lower layer is solvent A; upper layer is solvent B.

(b) Hydrochloric acid— 0.75% (1 + 49).

(c) Buffer solution— pH 1.5. Mix 50 ml 0.2 M KCl (14.911 g of KCl/liter of water) and 41.4 ml of 0.2N HCl in a 200-ml volumetric flask and dilute to volume with water. Check to ensure a pH of 1.5 ± 0.02.

(d) Solvents C and D— Add 400 ml of n-butanol-CCl$_4$ (1:1) and 200 ml of buffer solution (c) to a separatory funnel. Shake vigorously for 2 min and let phases separate. Lower layer is solvent C; upper layer is solvent D. Prepare fresh daily.

(e) Resin-n-hexane— Add 500 ml of 5% Amberlite LA-2 resin in n-hexane and 100 ml of HCl (1 + 49) to a separatory funnel. Shake for 1 min. Discard lower phase.

(f) Adsorbent— Celite 545, acid-washed, rinsed to neutrality, and dried.

Procedure: Drain packing liquid as completely as possible from cherries and chop cherries for 15 min in a Hobart 84141 food cutter or the equivalent. Mix thoroughly while chopping. Transfer to Mason jar with tight-fitting lid.

Weigh 5-g sample into 8-oz glass mortar, add 3 ml of solvent A, and carefully grind with pestle for 2 min. Add 15 g of adsorbent and carefully grind for an additional 2 min. Scrape off pestle and thoroughly mix sample with spatula. Transfer mixture to a 300 mm × 23 mm-ID glass-chromatographic column containing small plug of glass wool (silanized, Applied Science Laboratories, State College, Pa.) and firmly pack with tamping rod. Wipe off mortar, pestle, and spatula with piece of glass wool and add wipe to column. Rinse mortar with 10 ml of solvent B and add rinse to column. After rinse has entered column, elute column with 90 ml of solvent B, collecting eluate in a 125-ml separatory funnel containing 1 ml of water. Add 30 ml of hexane, shake, and let separate. Discard lower layer. Add 10 ml water (carefully rinsing around stopper and neck of separatory funnel) and 2 ml of NH$_4$OH. Extract color by shaking for 2 min. Allow to separate and drain lower layer into second 125-ml separatory funnel, rinsing stem with a small portion of water. Completely extract color from first separatory funnel with an additional 10 ml of water and 1 ml of NH$_4$OH and add lower layer to second separatory funnel. Rinse first funnel with 5 ml of water and add rinse to second separatory funnel. Wash combined aqueous extracts with two 25-ml portions of CHCl$_3$, discarding CHCl$_3$ completely each time. Render acidic with 2 ml of HOAc and extract color with 50 ml of n-butanol. Continue extraction with 10-ml portions of n-butanol until color is visually completely extracted (3-6 extractions are usually sufficient). Combine extracts in a 150-ml beaker, rinsing each separator with 2 ml of butanol. Add 15-25 ml of ethanol, mix with stirring rod, and evaporate just to dryness on steam bath under current of air.

Mix 5 g of adsorbent and 3 ml of solvent D in a 100-ml beaker and transfer to chromatographic column containing small plug of glass wool. Pack with tamping rod. Dissolve color residue in 1 ml of HCl (1 + 49), being sure to dissolve color on sides of beaker. Add 2 g of adsorbent, thoroughly mix, and transfer to prepared column. Pack with tamping rod. Dry wash beaker with 0.5 g of adsorbent and add wash to column. Wipe beaker with a piece of glass wool and add wipe to column. Rinse beaker with three 5-ml portions of solvent C and add rinses to column, allowing each to enter column before next one is added. Completely elute FD&C Red No. 4 with an additional 180-235 ml of solvent C, depending on amount of color in sample, but not exceeding 250 ml total. Collect in either a 200-ml or 250-ml volumetric flask. Dilute to volume with solvent C and determine spectrophotometrically at the maximum near 502 nm. After complete elution of FD&C Red No. 4, pass 20 ml of n-hexane through column and discard. Elute FD&C Red No. 40 with 50 ml of resin n-hexane, collecting eluate in a 100-ml volumetric flask. Dilute to volume with resin-n-hexane, filter through glass wool if cloudy, and determine spectrophotometrically at the maximum near 500 nm.

PRZYBYLSKI, W., SMITH, R. B., MC KEOWN, G. G. JAOAC 43, 274-278 (1960). Determination of Coal-Tar Colors on Oranges. Using 250 ml of chloroform, wash the color from 10 oranges. (Surface waxes, oils, and some natural pigments also wash off.) Combine the washings and dilute to 250 ml. Evaporate a 50-ml aliquot on a steam bath and dissolve the residue in about 25 ml of petroleum ether.

Fill a 2.5 cm × 10-cm glass column with petroleum ether. Sift adsorbent alumina (Fisher A-540) into the column to a height of about 4 cm.

Pass the petroleum ether solution of the sample into the column. Wash with 50 ml of petroleum ether followed by 200 ml of carbon tetrachloride. Discard the washings. Elute the coloring matter with ethanol. Evaporate the eluate to dryness on a steam bath.

Dip a 7 in. × 22½-in. strip of Whatman 3MM paper into 5% (w/v) light mineral oil in ethyl ether. Air dry. Dissolve the eluted sample in a few drops of chloroform and spot as a 6-in. band 2½ in. from the bottom of the paper. Develop for 3 hr by descending chromatography using 6:4 acetone-water as an eluant. Examine the chromatogram for coal-tar colors.

Dry the chromatogram and extract the individual colors from the paper with ethanol. Examine each spectrophotometrically against standards.

ADAMS, J. B., BUTLER, R. Analyst 101, 140-142 (1976). A Rapid Method for Detecting Erythrosine in Canned Red Fruits. Weigh 20 g of macerated sample. Add 5% aqueous sodium heptahydrate to decolorize any anthocyanins present and then increase the pH of the sample to 4-6 to ensure the solubilization of the FD&C Red No. 3 (Erythrosine). Shake the mixture vigorously with 5 ml of 3-methylbutan-1-ol and centrifuge. Determine the visible spectrum of the upper (alcohol) layer from 700 nm to 300 nm. A sharp peak at 545 nm indicates the presence of Erythrosine. Identification can be confirmed by noticing the almost complete loss of absorbance at 545 nm after one or two drops of HCl are added to the sample in the absorption cell.

GRAIN AND GRAIN PRODUCTS

ANDRZEJEWSKI, H. Pr. Zakresu Towarozn. Chem., Wyzsza. Szk. Ekon. Poznaniu, Zesz. Nauk Ser. I. No. 25, 5-39 (1966). Determination of Riboflavin in Cereal Products. The powdered sample is heated and stirred for 1 hr at 160°C in 50% aqueous LiCl. The solution is placed on a column of K-28 cation-exchange resin (H^+ form), the column is washed with an eluant containing Li salts, and then the riboflavin is eluted with Me_2CO-water (1:1) and determined fluorometrically as lumiflavine.

CIRILLI, G., SANDRI, M. Tec. Molitoria 22, 42-48 (1971). Chromatographic and Colorimetric Method for the Determination of β-Carotene. Mix 2 g of powdered corn or lucerne with 20-30 ml of benzene (or hexane)-acetone (7:3) and 0.5 ml of water. Allow the sample to stand in the dark for 15-16 hr. Then, protecting the sample as much as possible from light, dilute it to 100 ml with benzene and mix well.

Chromatograph 10 ml of the supernatant liquid on a column packed from bottom to top with 5 cm of alumina, 8 cm of Celite-magnesium oxide (1:1), and 8 cm of Na_2SO_4. Elute the β-carotene with benzene (or hexane)-acetone (9:1), dilute the eluate to 100 ml with eluant, and determine β-carotene spectrophotometrically at 450 nm.

HORWITZ, W., Ed. *Official Methods of Analysis of the Association of Official Analytical Chemists*, 12 ed., 1975, p. 242. Extraction, Separation, and Identification of Coloring Matter in Macaroni Products. Transfer 0.5 g of coarsely ground sample to a 1-liter Erlenmeyer flask, add 700 ml of 80% alcohol, and shake at intervals for 24 hr or until no more color is extracted. Place the sample in a refrigerator overnight to permit dissolved protein to precipitate, filter, and then evaporate the filtrate to 100 ml. Add 25 ml of 25% NaCl solution and a slight excess of NH_4OH to the filtrate, cool, and transfer the sample to a separatory funnel. Extract the sample with equal volumes of petroleum ether (boiling point, <60°C) until no more color is extracted. If colored, reserve the lower layer for further treatment.

Combine the ether extracts and wash them with several small portions of NH_4OH (1 + 50). The ether solution contains fats and oil-soluble dyes that may be identified as in (a), below. If colored, immediately acidify the aqueous alkaline solution with acetic acid and extract it with ether. Any color remaining in the ether solution may be turmeric, annatto or saffron. These may be identified as in (b).

If the original aqueous solution, freed from ether-soluble colors, is still colored and water-soluble dyes are suspected, extract the aqueous solution with 50-ml portions of isoamyl alcohol to remove any residual saffron as well as various orange dyes and Martius yellow; to separate these, proceed as in (c). Drain the lower aqueous layer, which, if colored, may contain naphthol yellow S, FD&C Yellow No. 5, and FD&C Yellow No. 6. Extract these dyes with isoamyl alcohol after acidifying the solution with HCl to about 1N. Remove the FD&C Yellow No. 5 from the solvent with 0.25N HCl. FD&C Yellow No. 6 is removed with slightly lower acid concentrations. Naphthol Yellow S is removed from nearly neutral solution.

(a) Extract the original petroleum ether extracts with two or three 10-ml portions of HCl-HOAc (1:5).

If yellow AB or yellow OB are present, the solution will be pink or red. A few drops of 40% $SnCl_2$ added to a small portion of the acid extract should cause either decolorization or a decided fading of such colors. These colorants can be removed from the acid extract by diluting it with water, rendering it slightly alkaline, and extracting it with petroleum ether. Any remaining colors in the petroleum ether extract may be due to natural coloring matter of wheat or eggs. The coloring principle of egg yolk, lutein, when heated with alcoholic $FeCl_3$, produces a green solution. This test is not specific, however, since carotene and xanthophyll produce similar reactions.

(b) Wash the ether extracts with 5-ml portions of water to remove excess acid. To remove annatto and traces of saffron, wash successively with 20-ml portions of 5% $NaHCO_3$ soltuion. Divide the alkaline solution into two portions. Heat one to 60°C on a steam bath, dye the color on unmordanted cotton, and compare spot tests with a standard. Acidify the remaining portion of the alkaline annatto solution with HOAc and reextract with ether. Divide the ether extract into two small casseroles and evaporate to dryness. Dissolve the contents of one casserole in 10 ml of NH_4OH (1 + 9) and impregnate a strip of cotton or filter paper with the solution. An orange-yellow to orange-red stain is obtained, depending on the amount of dye present. Dry the filter paper or cotton, add a drop of 40% $SnCl_2$ solution, and dry again. If annatto is present, a purple stain is produced. Spot the contents of the other casserole with H_2SO_4 and HNO_3, which yield blue and greenish-blue colors, respectively.

Transfer two 10-ml portions of the original ether extract (from which annatto has been removed) into separate test tubes. Treat one with an equal volume of 10% NaOH and the other with an equal volume of HCl (1 + 1). In the presence of turmeric (*Curcuma*), the alkaline solution is reddish brown; the acid solution is red.

Turmeric can be further confirmed by its behavior with H_3BO_3. Test by shaking a portion of the original ether extract with an equal volume of 70% alcohol; add 1/10 the volume of HCl, mix, and divide equally into two test tubes. Then to one tube add a few crystals of H_3BO_3 and shake. Use the other tube as a control. In the presence of turmeric the solution turns red after a short time.

(c) To separate and identify saffron and the orange synthetic dyes, dilute the isoamyl alcohol extract with two volumes of petroleum ether and extract the mixed dyes with several 10-ml portions of water. To a small portion of this aqueous extract add 1/10 its volume of HOAc and a few milligrams of dry sodium hyposulfite to reduce the azo dyes. Extract the saffron with isoamy alcohol, wash the extract with several small portions of water, evaporate the alcohol to dryness, and confirm the presence of saffron with spot tests.

HORWITZ, W. Ibid., p. 243. FD&C Yellow No. 5 in Macaroni Products. Place 800 ml of cold water and 5 ml of NH_4OH in a 1-liter Erlenmeyer flask and add 200 g of unground sample. Stopper the flask and shake at intervals over a 3-4-hr period. Use a glass rod to dislodge material caking on the botton. Centrifuge and decant the clear supernatant liquid into a 1-liter flask. Add a solution of 50 g of $MgSO_4 \cdot 7H_2O$ dissolved in 100 ml of water, 10 ml of 12% silicotungstic acid solution, and 10 ml of HCl. Shake well and let stand for 1 hr to allow protein to precipitate. Then centrifuge the solution and examine the supernatant liquid spectrophotometrically.

HORWITZ, W. Ibid., p. 243. Total Carotenoids and Carotene in Flour, Semolina, Macaroni, Egg Noodles and Egg Yolk. Grind macaroni and noodles to as near the fineness of flour as possible.

Weigh 20 g of flour, semolina, or macaroni, or 10 g of egg noodles, or 2 g of egg yolk into a 125-ml Erlenmeyer flask. Add 50 ml of 10% (w/v) alcoholic KOH and boil on a steam bath for 30 min under a reflux condenser. Occasionally rotate the flask (as carefully as possilbe) to keep the sample from collecting on the sides of the flask. Remove the flask and cool to room temperature. Filter through a Buchner medium-porosity fritted glass filter into a 250-ml suction flask, using suction, transferring most of the material with a few milliliters of alcohol from a wash bottle. Turn off the suction, rinse the flask with 25 ml of ether, pour the rinsing onto the glass filter, and stir the material with a rod to allow the ether to contact the entire sample. Filter and then repeat this operation twice.

Transfer the filtrate to a 250-ml separatory funnel and rinse with about 25 ml of ether, disregarding any soapy material in the flask. Add 175 ml of water and carefully invert and rotate the flask several times. When the layers separate remove the lower aqueous-alcoholic layer and extract this layer again with 25 ml of ether. Discard the lower layer and add the ether layer to the original ether solution. Wash the ether layer by pouring 50 ml of water through it. After the layers separate, withdraw and discard the aqueous layer. Add 50 ml of petroleum ether to the ether solution and wash with five 50-ml portions of water, carefully inverting and rotating the separator. Discard all the aqueous layers (slight emulsions usually clear in a few minutes but may be discarded, especially if there is no significant yellow tinge).

Transfer the ether-petroleum ether mixture to a 250-ml distillation flask, rinsing the separator with petroleum ether; place the flask in a beaker of water at 45-50°C. Stopper the flask, connect the side arm with vacuum, and concentrate to about 5 ml to remove ether. Filter through an Allihn-type adsorption tube with a coarse fritted glass plate containing about a 3-mm layer of anhydrous powdered Na_2SO_4, or through a 5.5-7-cm filter paper half filled with Na_2SO_4 (use a small, long-stemmed funnel reaching through the neck of the flask) into a 25-ml volumetric flask. Dilute to volume with petroleum ether that has been used to rinse the distillation flask and then passed portionwise through the filter containing Na_2SO_4. Mix the sample well and determine the carotenoid spectrophotometrically against a standard.

MITRA, S. N., ROY, S. C. Current Sci. (India) *26*, 89 (1957). Detection of Metanil Yellow in Pulses Dal. Treat a small amount of whole pulse with a little concentrated hydrochloric acid. If metanil yellow is present, the acid will turn purple.

To 20 g of broken (not powdered) sample add 150 ml of water and a few drops of NH₄OH and boil the mixture for a few minutes. Decant the colored solution from the pulse, render the solution just acid by the dropwise addition of 3*N* HCl, add a few strands of white wool, and heat the mixture on a boiling water bath for 30-40 min. Stir occasionally. Wash the wool well with tap water then boil for a few minutes in 100 ml of water containing two drops of 3*N* HCl. Wash the wool again under tap water and then strip the color from it using weak, hot ammonia.

Acidify a portion of this solution with 3*N* HCl, add fresh strands of wool to it, and boil to adsorb the dye. If metanil yellow is present, the wool will turn violet when treated with concentrated hydrochloric or sulfuric acid.

Concentrate a second portion of the above-described ammoniacal solution on a water bath and chromatograph it for 18 hr against a standard on Whatman No. 1 paper using iso-butyl alcohol-ethanol-water (4:1:4) as the eluant. Dry the paper and test for metanil yellow using hydrochloric acid.

MITRA, S. N., ROY, B. R. Sci. Culture *25*, 539-554 (1960). Further Studies on the Detection of Metanil Yellow in Pulses Dal. To eliminate interference from large amounts of starch, the extraction described in the previous paragraph is done using several portions of 80% alcohol instead of aqueous NH₄OH. A new chromatographic procedure using phenol-water (80:20) as the eluant is also described.

MUTONI, F., TASSI-MICCO, C. Rend. Inst. Super. Sanita *25*, 567-573 (1962). Chromatographic Identification of Dyes in Macaroni. II. Mix 10 g of finely ground sample for 10 min with 25 ml of 50% ethanol. Centrifuge the mixture for 10 min at 5000 rpm. Acidify the clear solution with 8-10 drops of 2% tartaric acid. Pour onto a column 1 cm in diameter, containing a 1-cm layer of dry, ground gluten. Add 10 drops of 1.5% (v/v) ammonium hydroxide and elute. Transfer the eluate to chromatographic paper and elute with 2:1:1 butanol-ethanol-water.

OSADCA, M., ARAUJO, M., DE RITTER, E. JAOAC *55*, 110-113 (1972). Determination of Canthaxanthin in Concentrates and Feeds. Weigh 45 g of feed into a 250-ml Erlenmeyer flask, add 100 ml of warm 7% ammonium hydroxide solution containing 0.5% propyl gallate, and mix well with a glass rod. Place the flask for 15 min. in a 65°C water bath. Using 150 ml of ethanol, rinse the contents of the flask into a 1-liter blender jar, cap tightly, and blend for 5 min at a speed adjusted to keep the mixture well below the cap. Add 450 ml of extracting solution (30-60°C petroleum ether-peroxide-free diethylether (2:1) containing 1 g each of butylated hydroxyanisole (BHA, United Oil Products) and butylated hydroxytoluene (BHT, Shell)/liter of mixture) to the blender, cap tightly, and blend for about 5 min with stops of about 10 sec after 1.5 min and 3 min. Vent blender occasionally. Stop the blending, allow the blender contents to settle, decant the supernatant liquid into a 1-liter separatory funnel, allow the phases to separate, and drain and discard the lower aqueous phase.

For samples containing 1 g of canthaxanthin per ton of feed, transfer 100 ml of clear upper phase (50 ml for 2 g/ton feed, 25 ml for 4 g/ton feed) into a 125-ml amber, round-bottomed flask. If the upper phase is not clear, filter rapidly through a funnel containing a glass-wool plug. Evaporate in a water bath at 45°C under a stream of N_2 until no odor of ether or petroleum ether is detectable; about 2-5 ml of liquid will remain. Add 10 ml of petroleum ether, 10 ml of 50% KOH, and 0.5 g of propyl gallate to the flask and swirl carefully. Keep 15 min at room temperature; swirl occasionally.

Quantitatively transfer the contents of the flask to a 125-ml separatory funnel, rinsing with two 5-ml portions of water and two 5-ml portions of alcohol. Add the rinsings to the funnel. Finally, rinse the flask with two 10-ml portions of petroleum ether and add these rinsings to the separatory funnel. Add 25 ml of water to the funnel (do not shake) and let the phases separate. If necessary, use small amounts of alcohol to break any emulsion. Discard the aqueous phase and retain the entire ether phase in the funnel. Wash the petroleum ether extract three times with 50-ml portions of water, swirling gently each time and discarding the aqueous phase as completely as possible without losing any ether phase. Add about 3 g of anhydrous granular Na_2SO_4 to the washed extract and mix carefully. Filter the extract through glass wool into a 100-ml Erlenmeyer flask. Rinse the Na_2SO_4 into the separatory funnel with three successive 10-ml portions of petroleum ether; filter each extract through the glass wool into the flask.

Pack an 18 mm × 200-mm glass chromatographic column with 8 cm of 100-200-mesh Florisil (Fisher Scientific Co.) and then top the column with a 1-cm layer of Na_2SO_4. Prepare the column immediately prior to use. Wash the column with 10 ml of petroleum ether and then add the ether extracts and washings to the column. Rinse the flask with two 10-ml portions of ether and add the washings to the column. Elute the column with diethyl ether (40-50 ml) until a broad yellow band appears. The brown-red band of canthaxanthin should remain close to the top of the column. Elute the canthaxanthin with 30% acetone in petroleum ether and collect the colored fraction in a 40-ml or 50-ml conical centrifuge tube. Evaporate the eluate containing the canthaxanthin (and other smaller components) almost to dryness on a 45°C water bath under a stream of N_2. Dissolve the residue in 0.5 ml of benzene.

Coat a 20 cm × 20 cm glass thin-layer plate with a 0.75-mm layer of silica gel G. (To prepare five plates, blend 120 ml of water with 60 g of silica gel G for 2 min. Air dry at room temperature for at least 4 hr and then oven dry for at least 4 hr more. The plates should be used within minutes after removal from the oven.) Streak the above benzene extract across the TLC plate as a band of 0.5 cm or less wide. Dry under a stream of nitrogen. Rinse the beaker with several 0.2-ml portions of benzene adding the washings to the TLC plate. Chromatograph in a 27 cm × 7 cm × 27 cm covered chromatographic tank using 191 ml of benzene-diethyl ether-methanol-pyridine (160:20:10:1) containing 0.25 g each of butylated hydroxyanisol (BHA, United Oil Products) and butylated hydroxytoluene (BHT, Shell). Let the plate develop (ascending) until the canthaxanthin appears separated by 0.25-0.5 cm from any interfering bands when viewed for a few seconds under white light. Scrape the canthaxanthin band from the plate through a small funnel into a 50-ml glass-stoppered centrifuge tube containing 5 ml of alcohol. Swirl for 10 sec and then pipette 20 ml of benzene through the funnel into the tube. Cap the centrifuge tube and shake for 7 min on a mechanical shaker. Add 1 g of Celite filter aid to the tube, shake for 2-3 min, and then centrifuge for 3 min at 2000 rpm. Remove the tube from the centrifuge, swirl to wash any particles adhering to the walls, and then centrifuge for an additional 10 min at 2000 rpm. Decant this solution into a second glass-stoppered tube.

Pipette 20 ml of the solution into a 25-ml amber volumetric flask. Add 0.1 ml of freshly prepared 0.3% w/v methanolic iodine solution and mix. Stopper the flask loosely and immerse the tube for 15 minutes in a 65°C water bath. Cool to room temperature and mix well.

Determine the sample's absorbance in a 5-cm absorption cell at the absorption maximum near 480 nm against a reagent blank of alcohol-benzene (1:4).

$$\text{Grams of canthaxanthin/ton} = \frac{A \times D \times 9070}{1840 \times 5 \times 0.85}$$

where D is dilution factor $[(450 \times 25)/(45 \times V) = 250/V]$, V is ml of original extraction solution taken for evaporation, 9070 is factor for converting result to g/ton, and 0.85 is recovery factor.

JAMS AND JELLIES

DEL BIANCO, F. M., TRABACCHI, G. Chem. e ind. (Milan) *41*, 896-898 (1959). Extraction and Identification of Synthetic Colors in Sweet Foods. Extract the dye with a neutral or highly acid aqueous medium. Condense the extract and chromatograph it on paper using Na citrate-NH_4OH-phenol (10:10:80), or BuOH-EtOH-H_2O (2:2:1).

DAVIDEK, J., DAVIDKOVA, E. Z. Lebensm. Forsch. *131*, 99-101 (1966). Application of a Polyamide in the Investigation of Water-Soluble Food Dyes. II. Isolation of Dyes From Food by Paper Chromatography. The sample solution is acidified with 10% tartaric acid of 10% $KHSO_4$, polyamide powder is added to adsorb the dye, and the sample is filtered. The colorant is stripped from the powder with NH_4OH-MeOH(5:95), concentrated on a steam bath, and resolved chromatographically.

GILHOOLEY, R. A., HOODLESS, R. A., PITMAN, K. G., THOMSON, J. J. Chromatog. *72*, 325-331 (1972). Separation and Identification of Food Colours. Weigh about 5 g of sample into a beaker, add 50 ml of water, and warm into solution on a water bath. Acidify the mixture with acetic acid. Plug a 15 mm × 250-mm glass-chromatograhic column with poly-amide staple fiber (Nylon 66, 3.3 g per 10,000 m of fiber) and then pour enough water suspension of polyamide powder (MN CC6, Macherey, Nagel and Co.) into the column to obtain a settled height of about 20 mm. Rinse the column walls with a small amount of acetone, and then cap the column with about a 6-mm layer of acid-washed sand.

Pour the hot sample solution through the column and then wash the column with six 10-ml portions of hot water and three 5-ml portions of acetone. Elute the colors from the column with a minimum volume of fresh acetone-ammonia-water (40:9:1), rejecting the eluate until the colors elute. Remove the ammonia by blowing a stream of air over the surface of the liquid and then reduce the volume by about one-half on a steam bath. Add an equal volume of water then adjust the pH to 5-6 with hydrochloric acid.

Pour the solution through a column of polyamide powder packed in a 10 mm × 100-mm chromatographic tube packed as described above and then wash the column with five 5-ml portions of hot water. Elute the dyes with a minimum of acetone-ammonia solution. Remove the ammonia as before and evaporate the solution to near dryness on a steam bath. Dissolve the residue in a few drops of $0.1N$ HCl and use this solution for TLC. (IF FD&C Red No. 3 is present, dissolve the residue in water.)

MEAT

GILHOLLEY, R. A., HOODLESS, R. A., PITMAN, K. G., THOMPSON, J. J. Chromatog. 72, 325-331 (1972). Separation and Identification of Food Colours. Chop about 25 g of sample on a glass plate, add 5 g of acid-washed sand, and grind the mixture to a paste. Add 10 g of Celite 545 and mix with a palette knife into a homogeneous mixture.

Transfer the mixture to a Soxhlet thimble and extract it with chloroform for 2 hr. Remove the sample from the thimble and place it in an evaporating dish to allow residual chloroform to evaporate.

Place a plug of polyamide staple fiber (Nylon 66, 3.3 g per 10,000 m of fiber) in the end of a 22 mm × 300-mm glass-chromatographic tube and add the powder sample to the tube, tapping the column gently to aid in packing. Pass methanol-ammonia-water solution (90:5:5) through the column until all the dyes are eluted.

Add 5 ml of 1% aqueous polyoxyethylene sorbitan monooleate solution to the eluate and evaporate the solution on a steam bath, blowing a stream of air over the surface of the liquid until all the ammonia and methanol are removed. Add an equal volume of water and adjust the pH of the solution to 6 with hydrochloric acid.

Place a plug of polyamide staple fiber in the end of a 15 mm × 250-mm chromatographic tube and add a suspension of polyamide powder (MN CC6, Macherey, Nagel and Co.) in water to the tube to give a height of about 20 mm. Rinse the walls of the tube with a small volume of acetone to aid the settling of the polyamide and then add a 6-mm layer of sand on top of the polyamide.

Pour the solution of dyes through the column and then wash the column with three 10-ml protions of water, two 5-ml volumes of acetone, two 5-ml portions of chloroform-absolute ethanol-water-formic acid (100:90:10:1), and two 5-ml portions of acetone. Elute the dyes from the column with a minimum of acetone-ammonia-water (40:1:9), rejecting the eluate until the dyes are eluted. Remove the ammonia by blowing a current of air over the surface of the liquid and then reduce the volume by about half on a steam bath. Add an equal volume of water and adjust the pH to approximately 6 with hydrochloric acid. Pour the solution through a column of polyamide in a 10 mm × 200-mm chromatographic column prepared and washed as previously described. Elute the dyes with a minimum volume of acetone-ammonia solution. Remove the ammonia by blowing a current of air over the surface of the liquid and then evaporate the solution to near dryness on a steam bath. Dissolve the residue in a few drops of 0.1N hydrochloric acid and use this solution for TLC. (If FD&C Red No. 3 is present, dissolve the residue in water.)

LEHMANN, G., COLLET, P. Z. Lebensm. Forsch. *144*, 107-109 (1970). Detection of Synthetic Dyes in Meat and Meat Products. Grind meat paste or homogenized minced meat, sausage, or salami in a mortar with sand, Celite, and acetone. Remove the acetone by filtration and repeat the extraction until no more color is removed. Grind the residue, dry it to remove solvent, and transfer it to a small chromatographic column packed with polyamide powder. Elute the column with NH_3-methanol (1:19). Acidify the eluate, evaporate it to a small volume, and separate the dyes present by paper or thin-layer chromatography. Dilute the acetone extract with water, remove the acetone by distillation under reduced pressure, extract fat-soluble dyes with light petroleum, concentrate the extract, and identify the dyes present by paper or thin-layer chromatography. Acidify the aqueous phase to pH = 6 and purify it on a microcolumn of polyamide powder. Elute adsorbed basic dyes with acetone and identify.

MARMION, D. M. JAOAC 54,131-136 (1971). Analysis of Allura* Red AC (A Potential New Color Additive). Determination is wieners. Slice a 10-g length from the middle of a dyed wiener. Blend well in a small Waring blender with 100 ml of $CHCl_3$. Filter; discard the $CHCl_3$ extract.

Return the cake to the blender and blend well with 100 ml of warm mixed solvent (SD No. 30 alcohol-water-NH_4OH, 80:20:1); filter. Return the cake to blender and extract again with 100 ml of fresh mixed solvent. Wash the blender and filter cake with two 50-ml portions of warm mixed solvent. Combine mixed solvent extracts and washings in a 500-ml volumetric flask, add 5 ml of acetic acid, and heat solution to incipient boil. Dilute to volume with SD No. 30 alcohol, mix, and let stand overnight.

Adjust the volume with SD No. 30 alcohol, mix, and filter by gravity

*Registered trademark of Buffalo Color Corporation.

through Whatman No. 42 paper (filtrate must be clear). Similarly extract sample containing no color.

Using a suitable spectrophotometer, determine absorbance of each solution in a 5-cm cell (vs. SD No. 30 alcohol) at the maximum near 505 nm and at 680 nm. Sample absorbance at maximum near 505 nm = A_1, sample absorbance at 680 nm = A_2, blank absorbance at 505 nm = A_3, blank absorbance at 680 nm = A_4.

Parts per million Allura* Red AC dye

$$= \frac{(A_1 - A_2 - A_3 + A_4) \times 1,000,000}{20 \times 5 \times 52.9} = A \times 189.0$$

where 52.9 = approximate absorptivity of Allura* Red AC dye at 505 nm (in liters/g-cm), 1,000,000 = factor for conversion to ppm; 20 = effective sample concentration (in g/L), and 5 = cell path length (in cm).

Mc NEAL, J. JAOAC 59, 570-577 (1976). Qualitative Tests for Added Coloring Matter in Meat Products. Slurry the meat with a minimum amount of warm water or 80% ethanol, let the mixture stand for 5 min, and then filter. Divide the filtrate into three equal portions and evaporate each just to dryness on a steam bath; do not boil. Dissolve the residue from one

R_f Values of Selected Colorants

Dye	Color Index No.	In Inorganic Solvent[a]	In Organic Solvent[b]
FD&C Red No. 1[c]	16155	0.15	0.32
FD&C Red No. 2[c]	16185	0.55	0.20
FD&C Red No. 3	45430	0.05	0.70
FD&C Red No. 4[d]	14700	0.42	0.50
FD&C Red No. 40	16035	0.35	0.45
FD&C Yellow No. 1[c]	10316	0.70	0.50
FD&C Yellow No. 4[c]	11390	0.20	0.96
FD&C Yellow No. 5	19140	0.85	0.21
FD&C Yellow No. 6	15985	0.77	0.35
FD&C Blue No. 1	42090	0.95	0.46
FD&C Blue No. 2	73015	0.18	0.21
FD&C Green No. 2[c]	42095	1.00	0.39
FD&C Green No. 3	42053	1.00	0.46
FD&C Violet No. 1[c]	42640	0.80	0.65
Methyl violet	42535	0.03	1.00
Orange B	19235	0.57	0.45
Orange No. 1	14600	0.36	0.61
Orange No. 2	15510	0.36	0.64
Rhodamine B	45170	—	—

[a] NH_4OH-2.5% sodium citrate-water (45 + 10 + 45).
[b] n-Propanol-ethyl acetate-water (6 + 1 + 3).
[c] These colors are no longer permitted for use in foods, drugs, and cosmetics.
[d] Permitted in externally applied drugs and cosmetics only.

portion in water and dissolve the second in $0.2N$ HCl, and the third in $0.2N$ NaOH. Filter if necessary and determine the spectra of the solutions against those of knowns. Alternately, the extract from above can be filtered, concentrated, and chromatographed on Whatman No. 1 paper using inorganic or organic solvent systems.

Test for natural coloring agents as follows. Run against appropriate standards and blanks.

Cochineal (carminic acid, carmine red)—Weigh about 25 g of meat into a beaker. Add 100 ml of hot (80°C) 5% aqueous borax solution, mix on a steam bath for 30 min, and filter. A purple filtrate indicates the presence of cochineal; yellow is negative. The addition of borax will give a positive test if >0.1% cochineal is present. Beet powder—Slurry the sample with $1N$ H_2SO_4. A purple color indicates the presence of beet powder. To confirm this, filter the slurry and divide the filtrate into three portions. Adjust these to pH = 2, 5, and 9, respectively, with dilute H_2SO_4 and NaOH and determine the spectrum of each solution from 700 nm to 400 nm. Peak maxima should be at 535 nm, 537 nm, and 544 nm at pH = 2, 5, and 9, respectively. Annatto and saffron—Mix 25-50 g of sample with 200 ml of ethyl ether and 2 ml of concentrated HCl and filter the slurry through anhydrous Na_2SO_4 in a funnel with a glass-wool pledget. Extract 10 ml of the dried ether extract with about 3 ml of 2% NaOH. Absorb any color present on a strip of filter paper and air dry. Dip the dried paper in concentrated H_2SO_4. A blue color indicates the presence of annatto or saffron. To differentiate between the two, add 40% $SnCl_2$ to another strip on which color has been absorbed, and let air dry. If annatto is present, the paper will turn pink to purple. If annatto is absent, or if the previous test was positive due to saffron only, there will be no change in color.

Paprika and turmeric—Pack a 10-mm-id glass chromatographic column with 10 cm of Florisil (Fisher Scientific Co., No. 100) topped with about 2 cm of anhydrous Na_2SO_4. Prewet the column with ethyl ether. Mix 50 g of sample with 200 ml of ethyl ether and 2 ml of concentrated HCl. Let the mixture stand for 5 min and then filter the extract onto the chromatographic column through a funnel containing 10 g of anhydrous Na_2SO_4. Allow the extract to percolate through the column at 3 ml/min and then wash the column at the same rate with 50 ml of petroleum ether. If paprika is present, a red band will appear at the interface of Na_2SO_4 and Florisil. This will turn yellow and elute from the column with ethyl ether. If turmeric is present, a yellow band will appear at the interface. Elute this band with 150 ml of acetone, mix the acetone eluate with 300 ml of water, and add three or four drops of concentrated HCl and a few crystals of boric acid. A red color confirms the presence of turmeric. If both coloring agents are suspected, prepare two columns and run each of the above procedures separately.

Alkanet-Extract 25-50 g of sample with 100 ml of ethanol and filter. Add 10 ml of 10% NaOH solution. A blue color indicates the presence of alkanet.

Carotene—Blend 30 g of sample for 4 min with 40 ml of water, 40 ml of methanol, and 80 ml of $CHCl_3$. Let the blend stand for 5 min and then filter through glass wool. Dilute 5 ml of the lower ($CHCl_3$) layer to 100 ml with $CHCl_3$ and compare spectrophotometrically against knowns.

SPELL, E. Fleischwirtschaft 52, 75-77 (1972). Detection of the Beetroot Pigment Betanin in Jellied Meats Containing Red Wine. Suspend the sample in H_2O at 30° C, strain and centrifuge the suspension, and then place it in a refigerator and allow the gelatin to set. Remove the fat layer and then separate the betanin and the red-wine color in the gelatin by ion-exchange chromatography. Resolve the isolated colorants by thin-layer electrophoresis using cellulose as the support and pH = 4.5 citrate buffer as the electrolyte.

VENTURINI, A., NOVI, M. Boll. Lab. Chim. Provinciali 16, 175-180 (1965). Identification of Synthetic Water-Soluble Coloring Compounds from Cochineal in Meat and Sausages. Place 20 g of finely ground meat in a mortar, add 30 g of quartz sand and 30 ml of Cl_3CCOOH, and grind well for 7-8 min. Filter through a layer of 3-4 mm of asbestos (use suction), and collect 20-30 ml of filtrate. If the filtrate is clear or slightly yellow, cochineal, enocianin, and acid azoic dyes are absent. If the filtrate is red, place 3-4 ml of it in a test tube and add NH_4OH. A greenish color indicates the presence of enocianin. Cochineal gives a purple color, whereas acid azoic dyes give no color change at all. The presence of cochineal can be confirmed by the green color formed with 5% uranium acetate. The acid azoic dyes can be separated on Al_2O_3 and identified by paper or thin-layer chromatography. If all three types of colorants are present, the azoic dyes must first be separated on Al_2O_3.

SPICES AND CONDIMENTS

BENK, F., PHILIPP, W. R. Gordian 69, 537-540 (1969). Detection of Permitted Natural Coloring Matter in Mayonnaise. Extract 20 g of sample with a mixture of 100 ml of petroleum ether and 100 ml of MeOH, saponify the extract, remove any lipids, and chromatograph the extract on a column of highly activated Al_2O_3. Chromatograph the fractions on a thin-layer plate coated with Kieselgel G using light petroleum-benzene-acetone-acetic acid (80:20:2:1) or benzene-ethyl acetate-methanol-H_2O (2:5:2:1) as eluant. Extract the colorants from the plate and examine spectrophotometrically.

LEHMANN, G., GERHARDT, U., COLLET, P., GUTER, J. Fleischwirtschaft 50, 946-948 (1970). Detection of Foreign Pigments in Spice Extracts Used in the Manufacture of Meat Products. Suspend the sample in water and extract fat-soluble synthetic and natural dyes with light petroleum ether and identify the isolated colorants by TLC. Isolate the water-soluble colorants by adsorption on polyamide power, DEAE-cellulose, or carboxymethylcellulose and, after desorption, identify them by TLC.

LEHMANN, G., GERHARDT, U., COLLET, P. Z. Lebensm. Forsch 144, 345-348 (1971). Analysis of Dyes. XII. Detection of Synthetic and Curcuma Dyes in Mustard. Mix the sample with sand and extract it with acetone to remove fat and water- and acetone-soluble dyes. Extract the residue with methanol-NH_4OH (19:1), adjust the extract to pH = 5-6 with acetic acid, and transfer it to a microcolumn packed with polyamide powder. Elute the acid dyes with hot H_2O and identify the eluted colors by paper or thin-layer chromatography. Concentrate the acetone filtrate and chromatograph it by TLC on Kieselgel using $CHCl_3$-methanol (18:1) as the eluant. Examine the TLC plate for basic fat-soluble and *Curcuma* dyes.

MITRA, S. N., ROY, S. C., CHATTERJI, R. K. J. Ind. Chem. Soc., Ind. & News Ed. *19*, 155-158 (1956). Detection of Coal Tar Dyes in Turmeric. Synthetic and natural coloring matters in turmeric are distinguished by an acid-wash technique and subsequent paper chromatography. Strip 3-5 g of powdered sample by boiling with 100 ml of dilute NH_4OH, filter, acidify the filtrate with HCl, and boil the filtrate with four or five strands of pure white wool. Wash the wool with water and boil with very dilute HCl. Boil the wool for 15 min with dilute NH_4OH and divide the solution. Acidify one portion and use it to dye fresh strands of wool and for spot tests with HCl, H_2SO_4, 10% NaOH, and 12% NH_4OH; characteristic colors are produced with Orange AG, Sunset Yellow, Naphthol Yellow, Tartrazine, and Metanil Yellow. Chromatograph a portion of the concentrated extract against knowns on Whatman No. 1 paper using the organic phase from a mixture of iso-$BuOH$-H_2O-$EtOH$ (4:4:1).

MITRA, S. N., ROY, S. C. J. Proc. Inst. Chemists *29*, 155-157 (1957). Detection of the Presence of Small Amounts of Turmeric in Other Spices. Triturate 20 g of sample several times with petroleum ether to remove as much oil as possible. Mix the residue with EtOH and allow the mixture to stand for 15 minutes; swirl occasionally. Filter the ether extract and concentrate the filtrate to near dryness. Spot a few drops of the filtrate on filter paper, allow it to dry, treat it with aqueous boric acid solution, and heat in an air oven for 10 min. A characteristic rose-red color indicates the presence of turmeric. This may be confirmed by the greenish blue color formed when a drop of ammonia is added to the red spot.

To further confirm the presence of turmeric, condense a portion of the ether extract, chromatograph it on Whatman No. 1 paper using the organic phase prepared by mixing iso-$BuOH$-$EtOH$-H_2O (4:1:2), and spot the resolved bands with boric acid and NH_4OH as described above.

MITRA, S. N., SEN GUPTA, P. N., ROY, B. R. J. Proc. Inst. Chemists *33*, 69-73 (1961). The Detection of Oil-Soluble Coal-Tar Dyes in Chilli (Capsicum). Separate portions of powdered sample are shaken with Et_2O, petroleum ether, and 90% alcohol, the extracts are treated with various concentrations of HCl and H_2SO_4, and the resultant color reactions are observed. As a confirmatory test, fresh extracts are filtered and concentrated and chromatographed on Whatman No. 1 paper that has been soaked in 5% liquid paraffin in 60-80° petroleum ether, air dried, and then dried at 100°C for 30 min. The eluant is 80% alcohol. With uncolored chili, only dull brown spots are resolved.

SACCHETTA, R. A. Rev. Asoc. Bioquim. Agric. *25*, 187-194 (1960). Paper Chromatography of Red Paprika Powders. Powdered samples are extracted with EtO_2 and the extracts are concentrated and chromatographed on Whatman No. 1 paper using EtOH as the eluant.

STELZER, H. Nutr. Bromatol. Toxicol. *2*, 177-179 (1963). Identification of Synthetic Coloring in Paprika. Extracts of paprika are chromatographed on thin-layer plates using EtOH-AcOH (95:5) as the eluant. The plates are prepared by coating glass with a suspension of talc-wheat starch-water (7:0.04:30) and then drying the plates for 24 hr at ambient temperature.

MISCELLANEOUS

DEL BIANCO, F. M., TRABACCHI, G. Rass. Chim. *13* (2), 17-19 (1961). Method for Extraction of Colorants from Food Products. The procedure used by the authors for analyzing jams and jellies (see) was modified by using powdered leather treated with HCHO to adsorb colorant from weak acid solutions (pH = 5).

DEVON, B., LAUR, J. Ann. fals. fraudes, *52*, 155-161 (1959). Determination of Coloring Matter in Foods With Quaternary Ammonium Compounds. Basic colorants may be extracted directly with $CHCl_3$. For acid colorants, a sample containing 5-10 μg of colorant is adjusted to pH = 9 with Na_2CO_3 and shaken for 10 min with 10 ml of $CHCl_3$. The sample is centrifuged, and the $CHCl_3$ layer containing any basic colorants is removed. The aqueous layer including any solids that have formed at the interface of the liquid layers is mixed with a large excess of 0.1% cetyl-cyclohexyldimethylammonium bromide, shaken with $CHCl_3$, and centrifuged. The $CHCl_3$ solution is drawn off and evaporated at low temperature. The residue is dissolved in 0.5 ml of $CHCl_3$, and a known amount is chromatographed for 24 hr by descending chromatography using 95% $EtOH-H_2O-NH_4OH$ (50:25:25).

GRAICHEN, C., Molitor, J. C. JAOAC 46, 1022-1029 (1963). Determination of Certified FD&C Color Additives in Foods and Drugs.

REAGENTS

Resin-Hexane—Dissolve 50 ml of Rohm and Haas Amberlite LA-2 resin in 950 ml of *n*-hexane. Shake the solution with 200 ml of 1:4 acetic acid. Discard the lower phase.

Resin-Butanol—Dissolve 100 ml of Amberlite LA-2 resin in 900 ml of butanol. Shake the solution with 400 ml of 1:4 acetic acid and 15 ml of water saturated with ammonium sulfate. Discard the lower phase.

pH = 7.5 Buffer—Mix 75 ml of 0.1 M citric acid with 925 ml of 0.2 M Na_2HPO_4.

Resin-Butanol, pH = 7.5—Dissolve 50 ml of Amberlite LA-2 resin in 950 ml of butanol and 3 ml of glacial acetic acid. Shake the solution with 400 ml of the pH = 7.5 buffer. The pH of the lower phase should be 7.3-7.7. If it is not, repeat the preparation adjusting the amount of acetic acid. Discard the lower phase.

pH = 3 Buffer—Mix 101.5 ml of 0.2 N hydrochloric acid and 250 ml of 0.2 M potassium acid phthalate solution. Dilute to 1 liter.

Resin-Butanol, pH = 3—Mix 50 ml of Amberlite LA-2 resin, 950 ml of butanol, and 8 ml of concentrated hydrochloric acid. Shake the mixture with three successive 200-ml portions of the pH = 3 buffer. The pH of the aqueous phase should be 2.8-3.2. If it is not, repeat the preparation, adjusting the amount of hydrochloric acid used.

SAMPLE PREPARATION

Weigh 5 g of sample into a tissue blender. The sample should contain at least 0.2 mg of each color but no more than 5 mg of total color. Add 25 ml of 1:4 acetic acid and blend into a fine mixture. Transfer the suspension to a mortar and grind in about 5 g of Celite. Add more Celite as necessary to give the proper texture for the particular sample. The mixture should be wet enough to pack under pressure but dry enough to crumble when disturbed. With many samples the entire sample preparation can be done in a mortar.

CHROMATOGRAPHIC SEPARATION

Mix 20 g of Celite and 8 ml of 1:4 acetic acid. Pack about 15 g of the mixture into a 5.2 cm- ID $\times$ 20-cm glass chromatographic column. Transfer the sample onto the column and pack using the weight only of a 1300 g aluminum plunger (see Fig. 18). Flush the mortar with the remaining Celite-acid mixture, transfer it to the column, and pack as described above. Cover the surface with a porous disc. Elute fats and chloroform-soluble colors from the column with 100 ml of chloroform followed by 50 ml of hexane. Elute FD&C Red No. 2[a], FD&C Red No. 4[b], FD&C Yellow No. 5, and FD&C Yellow No. 6 from the column with 200 ml of the resin-hexane solution. Next, elute FD&C Blue No. 1, FD&C Green No. 2[a], FD&C Green No. 3, and FD&C Violet No. 1[a] from the column with the first resin-butanol solution.

Alternate Method A: This method is preferred when FD&C Blue No. 2 or FD&C Red No. 3 is present. All colors elute. Grind the sample and pack the column as described above except use pH = 7.5 buffer in place of 1:4 acetic acid. Elute fats and chloroform soluble colors with chloroform. Elute FD&C Blue No. 2 and FD&C Red No. 3 with pH = 7.5 resin-butanol.

Alternate Method B: This method is best when aluminum lakes of colors are present. Allowing the column to stand for several hours or overnight in contact with the pH = 3 resin-butanol eluant improves the extraction of lakes. Grind the sample and pack the column as described above except use pH = 3 buffer in place of 1:4 acetic acid. Elute fats and chloroform-soluble colors with chloroform. Eliminate the chloroform wash if FD&C Red No. 3 is to be determined. Elute colors with the pH = 3.0 resin-butanol solution.

[a]These colorants are no longer permitted in foods, drugs or cosmetics in the U.S.
[b]Permitted in externally applied drugs and cosmetics, only.

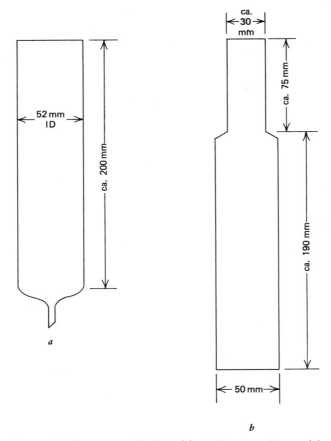

Figure 18   Chromatographic Tube (a) and Aluminum Plunger (b)

ISOLATION OF COLORS FROM THE RESIN SOLUTIONS

From Resin-Hexane—Wash 200 ml of extract with three 100-ml portions of water. Add 30 ml of water. Add concentrated ammonium hydroxide dropwise until the sample is alkaline, as indicated by the extraction of color into the aqueous phase. Extract all the color with 10-ml portions of dilute ammonium hydroxide. Quickly wash the combined aqueous extracts with 20 ml of chloroform, discard the chloroform, and acidify the aqueous layer with acetic acid.

From Resin-Butanol and Resin-Butanol at pH = 3—Dilute 100 ml of extract with 200 ml of hexane. Discard the aqueous layer which separates. Then extract as described above.

From Resin-Butanol at pH = 7.5. Wash the organic layer with several portions of dilute ammonium hydroxide. Acidify the aqueous layer and extract with ethyl ether. Extract the ether solution with dilute ammonium hydroxide to isolate FD&C Red No. 3 from other FD&C colors.

KARASZ, A. B., DE COCCO, F., BOKUS, L. JAOAC 56, 626-628 (1973). Detection of Turmeric in Foods by Rapid Fluorometric Method and Improved Spot Test. Mix 2 g of salad dressing or mashed pickle in a beaker with 3 g of Hyflo Super-Cel filter aid to a uniform mix. Add 50 ml of water-saturated n-butanol and stir thoroughly. Let stand for 15 minutes with occasional stirring and then filter through Whatman No. 42 paper. If the sample is a bread, pulverize 10 g and transfer it to a flask containing 50 ml of water-saturated n-butanol. Stopper the flask, shake well, and let stand for 15 min. Shake again and filter as described above.

Transfer 20 ml of filtrate to a separatory funnel, add 10 ml of NaOH solution (150 g of NaCl + 4 g of NaOH/liter), and shake vigorously for 1 min. Draw the aqueous layer and any red droplets at the interface into a second separatory funnel. Add 1 ml of glacial acetic acid and 200 ml of $Na_2S_2O_4$ and swirl to dissolve the salt. Add 20 ml of water-saturated n-butanol and shake vigorously for 1 min. Filter the butanol extract and determine its spectrum in a spectrophotofluorometer within 15 min as follows.

Set the fluorometer excitation scale at 435 nm and the emission scale at 520 nm. Fill the cuvette with reference solution prepared by diluting 5 ml of 0.03% curcumin in ethanol to 500 ml with water saturated n-butanol and then adjust slits, meter multiplier, and sensitivity to obtain 100% full-scale deflection on the recorder. Replace the reference solution with sample extract and, keeping the excitation scale at 435 nm, record its emission spectrum. The emission maximum for turmeric appears at 520 nm.

To confirm the presence of turmeric, evaporate a portion of the butanol extract to dryness, dissolve the residue in a minimum of ethanol, and spot a sufficient amount on Whatman No. 1 paper to produce a distinct yellow spot. Dry the paper in an oven at 100°C for 2 min and then add 3-4-μl portions of boric acid reagent to the yellow area. A red color that develops within 2 min at room temperature indicates the presence of turmeric.

To prepare the boric acid reagent, dissolve 1 g of H_3BO_3 and 5 ml of HCl in 95 ml of ethanol. Dry over anhydrous Na_2SO_4 and filter.

LEHMANN, G., COLLET, P., HAHN, H.-G., ASHWORTH, M. R. F. JAOAC 53, 1182-1189 (1970). Rapid Method for Detection and Identification of Synthetic Water-Soluble Coloring Matters in Foods and Drugs. Acid dyes are leached from foods with ammoniacal alcohol, acidified, and adsorbed onto polyamide powder. Protein-containing foods are treated with acetone to remove fat and water and to coagulate soluble protein. The residue is packed into a special chromatographic tube (see Fig. 19), and the colorants are eluted with ammonical alcohol, whereas the protein remains on the column. Water-soluble forms of natural colorants such as

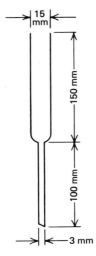

Figure 19   Microchromatographic Tube

chlorophyll, carmine, annatto, alkanna red, betanin, and grape-juice red pigment can also be adsorbed on polyamide powder.

Basic dyes are adsorbed on carboxymethyl cellulose.

LEHMANN, G., HAHN, H.-G., COLLET, P., SEIFFERT-EISTERT, B., MORAN, M. Z. Lebensm. Forsch. *143*, 256-263 (1970). Analysis of Dyes. II. Rapid Determination of Water-Soluble Dyes in Foods. Dyes are Extracted from samples by methods that depend on whether the sample is soluble in or miscible with water or contains natural coloring matter, starch, pectin, or protein. The extracts are purified on microcolumns of polyamide powder, ion-exchange resin, carboxymethylcellulose, bentonite, and Fullers's earth.

LEHMANN, G., MORAN, M., NEUMANN, B. Z. Lebensm. Forsch. *155*, 85-87 (1974). Analysis of Dyes, XV. Detection of Beetroot Dye (Betanin) in Foods. Betanin is isolated from H_2O-soluble samples by chromatography on a microcolumn of polyamide powder using formic acid-methanol (2:3) as eluant. Protein-containing samples are treated with Celite and sand in the presence of acetone to precipitate protein and extract fat, water, and lactoflavine. The dried solids are then transferred to a polyamide column and the betanin is eluted with methanol-concentrated aqueous NH_3 (19:1). The eluate is neutralized and treated with DEAE-cellulose and the adsorbed betanin is eluted with formic acid-methanol (1:4). The concentrated eluate is chromatographed on Kieselgel using propanol-acetic acid-H_2O (3:1:1) as the eluant.

LEHMANN, G., HAHN, H.-G. Z. Analyt. Chem. *238*, 445-456 (1968). Detection and Determination of Water-Soluble Synthetic Food Dyes with Polyamide Powder. Polyamide powder is used to quantitatively adsorb dyes from aqueous or aqueous-alcoholic solutions. The dyes are washed from the polyamide with a solution of 0.5 g of NaOH in 1 liter of 70% MeOH and identified spectrophotometrically.

MATHEW, T. V., BANERJEE, S. K., MUKHERJEE, A. K., MITRA, S. N. Res. Indust. (New Delhi) *14*, 140-142 (1969). Isolation and Estimation of Synthetic Foods Colours by Alumina Adsorption and Paper Chromatography. The sample is diluted with water and stirred with neutral alumina powder. The alumina is removed by filtration and then extracted with isoamyl alcohol-95% ethanol-5% aqueous NH_3-H_2O (4:4:1:2) or, if indigo carmine is present, with butanol-anhydrous acetic acid-H_2O (20:5:12).

ONRUST, H., HOEKE, F. Chem. Weekblad *54*, 465-470 (1958). Identification of Synthetic, Water-Soluble Food Colors. The following procedure is recommended for the analysis of foods with high sugar contents, alcoholic beverages, and milk products.

Mix one part of solid food with 4 parts of NaOAc-HOAc buffer (pH = 3), or mix 10 g of liquid sample with 20 ml of buffer. Extract the mixture with 10 ml of quinoline and cetrifuge to remove the water layer. Wash the quinoline layer twice with water and then shake with 30 ml of ether, 1 ml of H_2O, and 2 ml of 10% aqueous NH_3. Centrifuge to remove the quinoline-ether layer and then wash the colored aqueous layer with ether and analyze it chromatographically.

OSADCA, M., DERITTER, E., BUNNELL, R. H. JAOAC 49, 1078-1083 (1966). Assay of Apocarotenal and Canthaxanthin in Foods. Carotenoids are extracted by blending or shaking the sample with an appropriate solvent and then separated from naturally occurring pigments and other added coloring agents by selective solvent extraction and/or column chromatography.

SINGH, M., GRAICHEN, C. JAOAC 56, 1458-1459 (1973). Determination of FD&C Red No. 3 in Rat-Blood Serum. The Dye is extracted from the acidified sample with acetone-ethyl ether, extracted into dilute aqueous NH_3, and then measured spectrophotometrically.

SOHAR, J. Z. Lebensm. Forsch. *132*, 359-362 (1967). Extraction of Dyes from Food with Quaternary Ammonium Compounds. Colorants are complexed with cetyltrimethylammonium bromide and then extracted with organic solvent. The complexes are decomposed with cupferron and the free dyes are determined by the usual procedures.

INDEX

344 INDEX